Towards Responsible Plant Data Linkage:
Data Challenges for Agricultural Research
and Development

Hugh F. Williamson • Sabina Leonelli
Editors

Towards Responsible Plant Data Linkage: Data Challenges for Agricultural Research and Development

Springer

Editors
Hugh F. Williamson
Exeter Centre for the Study of the Life
Sciences (Egenis)
University of Exeter
Exeter, UK

Sabina Leonelli
Exeter Centre for the Study of the Life
Sciences (Egenis)
University of Exeter
Exeter, UK

University of Exeter

ISBN 978-3-031-13278-0   ISBN 978-3-031-13276-6   (eBook)
https://doi.org/10.1007/978-3-031-13276-6

This Springer imprint is published by the registered company Springer Nature Switzerland AG
The registered company address is: Gewerbestrasse 11, 6330 Cham, Switzerland

Acknowledgments

The seeds for this book were planted over 15 years ago, when Leonelli started to investigate the infrastructures, semantic systems, organizational strategies, and political economy underpinning efforts to link biological databases with each other. Among the hundreds of researchers she has interacted with since then, she wishes to thank in particular her colleagues in the steering committee of GARNet, an organization that for two decades played a crucial role in coordinating plant data resources within the UK and beyond. Much was learnt from these exchanges (e.g., Bastow & Leonelli, 2010; Leonelli et al., 2012, 2013; Perry et al., 2020), not least the value of working together and across disciplines to think about thorny socio-technical issues, which is the main presupposition of this volume. During those years, our connections with volume contributors were also forged through several workshops, conferences, and reciprocal visits. Interacting with these researchers and their respective organizations has been a privilege and cemented our conviction that data linkage strategies and tools, too often underestimated as mere "service" and "infrastructure," are actually a fundamental and consequential component of research on food security and environmental sustainability in the face of climate change. We thank all the contributors to this volume for years of fruitful discussions and for their commitment to producing a book that documents the significance of plant data linkage to data-intensive agriculture, and charts its future directions and challenges. We also wish to acknowledge our colleagues at the Exeter Centre for the Study of the Life Sciences (Egenis), particularly Özlem Yılmaz, John Dupré, Nick Smirnoff, David Studholme, and Britta Kuempers for convivial "plant life" discussions and Chee Wong for consistently brilliant administrative help, and at the Institute for Data Science and Artificial Intelligence for their generous support and lively collaborative exchanges – Richard Everson deserves a special shout-out for his open-minded, generative, and long-sighted leadership. Springer warmly welcomed and supported this project from the start and provided an excellent platform for its dissemination as an Open Access volume, for which we are truly thankful. Michel Durinx carved out time under difficult circumstances to design a cover image that brilliantly captures the book's theme, and deserves our major gratitude. Last but certainly not least, this volume would not exist without generous funding from the Alan Turing Institute, which supported this effort and the cost of Open Access publishing as a core component of the project "From Field Data to Global Indicators" (under EPSRC grant EP/N510129/1); the European Research Council, who funded Leonelli's work on plant databases between 2014 and 2019 through grant award 335925 (DATA_SCIENCE) and supported the final phase of writing and revisions to this volume in 2021–22 through grant award 101001145 "A Philosophy of Open Science for Diverse Research Environments" (PHIL_OS); and the Wissenschaftskolleg zu Berlin, which provided Leonelli with the ideal space and time to complete the volume (and most crucially, our introduction and own chapter within it). We dedicate our book to all those who work every day to improve spaces of collaboration and dialogue, without which there is no hope for development, justice or understanding (scientific and otherwise).

References

Bastow, R., & Leonelli, S. (2010). Sustainable digital infrastructure. *EMBO Reports, 11*(10): 730–735.

Leonelli, S., Charnley, B., Webb, A., & Bastow, R. (2012). Under one leaf. A historical perspective on the UK plant science federation. *New Phytologist, 195*(1): 10–13.

Leonelli, S., Smirnoff, N., Moore, J., Cook, C., & Bastow, R. (2013). Making open data work in plant science. *Journal for Experimental Botany, 64*(14): 4109–4117.

Perry, G., Yoselin, B. A., Gibbs, D., Grant, M., Harper, A., Harrison, J., Kaiserli, E., Leonelli, S., May, S., McKim, S., Spoel, S., Turnbull, C., van der Hoorn, R., & Murray, J. (2020). How to build an effective research network: Lessons from twenty years of the GARNet plant science community. *Journal of Experimental Botany,* eraa307.

Contents

The Research Data Alliance Interest Group on Agricultural Data: Supporting a Global Community of Practice 289
Patrícia Rocha Bello Bertin, Cynthia Parr, Debora Pignatari Drucker, and Imma Subirats

Cultivating Responsible Plant Breeding Strategies: Conceptual and Normative Commitments in Data-Intensive Agriculture 301
Hugh F. Williamson and Sabina Leonelli

Abbreviations

ABS	Access and benefit sharing
ADC	Ag Data Coalition
AHDB	Agriculture and Horticulture Development Board (UK)
AHTEG	Ad-Hoc Technical Expert Group
AI	Artificial intelligence
APHA	Animal and Plant Health Agency (England and Wales)
API	Application programming interface
ATP	Agricultural technology provider
BBSRC	Biotechnology and Biological Sciences Research Council (UK)
CARE	Collective benefit, Authority to control, Responsibility, Ethics
CBD	Convention on Biological Diversity
CGIAR	Consultative Group for International Agricultural Research
CIAT	International Center for Tropical Agriculture
CIP	International Potato Center
CIRAD	Agricultural Research Centre for International Development (France)
COMECON	Council for Mutual Economic Assistance
CoP	Community of practice
COPO	Collaborative Open Plant Omics
COPP	Community of practice partnership
CRISPR	Clustered regularly interspaced short palindromic repeats
CSA	Community supported agriculture
CTA	Technical Centre for Agricultural and Rural Cooperation ACP-EU
DBMS	Database management system
DCT	Demand creation trials
DCTF	Data Coordination Task Force
DFW	Designing Future Wheat
DGO	Digital Genetic Object
DLT	Distributed ledger technology

DOI	Digital Object Identifier
DSI	Digital sequence information
DUS	Distinctness, Uniformity and Stability
ECP/GR	European Cooperative Programme on the Conservation and Exchange of Crop Genetic Resources
ELIXIR	European Infrastructure for Life Science Data
EMPHASIS	European Infrastructure for Plant Phenotyping
EMR	Electronic medical records
ENA	European Nucleotide Archive
EOSC	European Open Science Cloud
EPPN	European Plant Phenotyping Network
e-RA	Electronic Rothamsted Archive
EUCARPIA	European Association for Research on Plant Breeding
EVA	European Variation Archive
FAIR	Findable, Accessible, Interoperable, Reusable
FAO	United Nations Food and Agriculture Organization
GDPR	General Data Protection Regulation
GEF	Global Environment Facility
GFAR	Global Forum on Agricultural Research
GiSC	Grower Information Services Cooperative
GLIS	Global Information System of the ITPGRFA
GLTEN	Global Long-term Agricultural Experiments Network
GODAN	Global Open Data for Agriculture and Nutrition
GRU	Genetic Resource Unit
GS	Genomic Selection
GURT	Genetic use restriction technologies
GWAS	Genome-wide association study
GxE	Genotype by environment interactions
HTP	High throughput
IBPGR	International Board of Plant Genetic Resources
ICARDA	International Center for Agricultural Research in the Dry Areas
ICBG	International Cooperative Biodiversity Groups
ICO	Initial coin offering
IDF	Israel Defense Forces
IG	Interest group
IGAD	Interest Group on Agricultural Data
IITA	International Institute of Tropical Agriculture
IMF	International Monetary Fund
INRAE	National Research Institute for Agriculture, Food and Environment (France)
INSDC	International Nucleotide Sequence Database Collaboration
IoT	Internet of Things
IP	Intellectual property
IPR	Intellectual property rights
ISA	Investigation, Study, Assay

ISTA	International Seed Testing Association
ITPGRFA	International Treaty on Plant Genetic Resources for Food and Agriculture
KTBL	Küratorium für Technik und Bauwesen in der Landwirtschaft
LGU	Land Grant University
LIMS	Laboratory information management system
LTE	Long-term experiment
LTP	Low throughput
MAGIC	Multi-parent Advanced Generation InterCross
MCPD	Multi-Crop Passport Descriptors
MENA	Middle East and North Africa
MIAPPE	Minimum Information About a Plant Phenotyping Experiment
MoA	Ministry of Agriculture (Palestinian Authority)
NAFTA	North American Free Trade Agreement
NARC	National Agricultural Research Center (Palestine)
NARS	National agricultural research systems
NCBI	National Center for Biotechnology Information
NIAB	National Institute of Agricultural Botany (UK)
NIRS	Near-infrared spectroscopy
NL	National List
oPt	Occupied Palestinian territory
OSTS	Official Seed Testing Station
PA	Palestinian Authority
PECO	Plant Experimental Conditions Ontology
PGR	Plant genetic resource
PGRFA	Plant Genetic Resource for Food and Agriculture
PID	Persistent identifier
PII	Personal Identifiable Information
PRDP	Palestinian Reform and Development Plan
PVS	Participatory varietal selection
QC	Quality control
QTL	Quantitative trait loci
RDA	Research Data Alliance
RDM	Research data management
RI	Research infrastructure
RL	Recommended list
RTB	Roots, tubers, and banana
SES	Socio-ecological systems
SMTA	Standard material transfer agreement
SNP	Single nucleotide polymorphism
SOP	Standard operating procedures
TDWG	Taxonomic Databases Working Group
TK	Traditional knowledge
TEK	Traditional ecological knowledge

TRUST	Transparency, Responsibility, User focus, Sustainability, Technology
UAV	Unmanned aerial vehicle
UAWC	Union of Agricultural Work Communities (Palestine)
UK-COSMOS	UK Cosmic-ray Soil Moisture Monitoring Network
UKRI	UK Research and Innovation
UNCTAD	United Nations Conference on Trade and Development
UNDP	United Nations Development Programme
UPOV	Union for the Protection of Varieties
VCU	Value for cultivatable use
WDI	Wheat Data Interoperability working group
WG	Working group
WISP	Wheat Institute Strategic Programme
WTO	World Trade Organization
YITEDEV	Youths in Technology and Development Uganda

List of Figures

List of Tables

Introduction: Towards Responsible Plant Data Linkage

Sabina Leonelli and Hugh F. Williamson

Abstract This chapter provides a framing for this volume by reviewing the significance and the organisational, technical and social opportunities and challenges related to plant data linkage. We review what "responsible practice" means in relation to the plant environments being documented, the infrastructures used to circulate data, the institutions involved in data governance and the communities involved in plant data work. We show how, across these domains, responsible plant data linkage involves consideration of technical, legal, ethical and conceptual dimensions, thereby: (1) creating and maintaining digital infrastructures, technical standards and discussion venues focused on critical data reuse; (2) developing adequate legal and institutional frameworks that work transnationally; (3) identifying and implementing guidelines for what constitutes acceptable data use, together with systems to monitor and allocate responsibility for breaches and mistakes; and (4) considering the variety of views on what constitutes agricultural development in the first place and how plant research can sustainably, reliably and responsibly contribute to achieving food security. The production of sustainable, responsible and reliable agricultural solutions in the face of climatic and political change depends on the flourishing of transnational, interdisciplinary collaborations such as those represented in this volume.

1 Introduction: Why Care About Plant Data Linkage

Global challenges such as climate change and the needs of a rapidly growing population have led to the emergence of new priorities in plant science and agricultural research. There is increasing interest in crops from the Global South that have been relatively neglected in previous agricultural development schemes, especially

S. Leonelli (✉) · H. F. Williamson
Exeter Centre for the Study of the Life Sciences (Egenis), University of Exeter, Exeter, UK
e-mail: s.leonelli@exeter.ac.uk; h.williamson@exeter.ac.uk

© The Author(s) 2023
H. F. Williamson, S. Leonelli (eds.), *Towards Responsible Plant Data Linkage: Data Challenges for Agricultural Research and Development*,
https://doi.org/10.1007/978-3-031-13276-6_1

1

those perceived to have less commercial value yet remain of great importance to smallholders. Improving research and understanding on heritage and orphan crops, as well as the wider set of crop varieties, are now recognised as important goals (Ribaut & Ragot, 2019). Given rapidly changing environmental and climatic conditions, deepening our understanding of genotype by environment interactions (GxE) also constitutes a key goal, especially the impact of environmental stressors on phenotypic traits.

The acquisition, curation and interpretation of data about plants, their environments and their human consumers play a central role in these efforts. Research in the plant sciences is marked by a high volume and heterogeneous range of data formats and sources, including quantitative, observational as well as imaging data generated by field trials, breeders, agricultural machinery, agribusinesses and seed distribution companies, publicly funded scientists, and national/regional institutions. These data are certainly "big", and yet they are neither easy to access nor easy to use. Making these data *accessible* to those who may wish to analyse them is proving an intricate challenge, with large efforts around the world devoted to expanding data access for research purposes and complications emerging from the privatisation and commercialisation of such data. An even greater challenge is to foster fruitful data analysis and interpretation given the countless forms of expertise, goals and perspectives involved: in other words, to make those data *usable* despite their heterogeneous provenance and even more heterogeneous re-purposing, and ensure the reliability and effectiveness of the resulting knowledge, technologies and interventions.

This is why *data linkage*, understood as the ability to connect and jointly analyse diverse datasets, has emerged as a key global challenge for agricultural research and development in the twenty-first century. Agriculture has long depended on the exchange of biological materials and knowledge, but the opportunities for data collection, dissemination and analysis opened up by computational technologies have dramatically expanded the potential of data-intensive research in this domain. Linking heterogeneous data helps to conduct analyses that address the multiple scales that impact plant growth and traits, from the molecular through the physiological to the social and environmental. This in turn facilitates understanding of the complex, scale-spanning phenomena underpinning sustainable food production and environmental management under rapidly shifting climatic and socio-political conditions.

The roots of the multiple challenges involved in linking data of relevance to agriculture are cultural. The landscape of plant data production, circulation and use is marked by the encounter between different *cultures of data exchange*, which in turn creates substantive technical, legal and social challenges to data linkage. At the scientific end of this spectrum, plant science has long sat at the intersection of the laboratory and the field, with a growing emphasis on integrating agronomic research with fundamental plant science and -omics data in order to understand the molecular mechanisms that underpin key crop traits, variation and performance, as well as to make use of molecular technologies for breeding and other applications (Harfouche et al., 2019; Sperschneider, 2019; Dobrescu et al., 2020; Wang et al., 2020).

Moreover, the last two decades have seen the creation of hybrid research spaces, such as smart glasshouses and digital farm platforms, that utilise new sensing and imaging technologies to capture features of the environment with unprecedented precision and scale (Coppens et al., 2017; Tardieu et al., 2017; Giuffrida et al., 2018). Each of these research spaces hosts different constellations of interdisciplinary work, whose diverse methods and outputs can be challenging to consider as a single body of evidence. Add to these scientific concerns the legal challenges presented by frequent (and frequently unresolved) clashes between different intellectual property regimes. More egregiously, there is a tension between publicly and privately funded research efforts. Much plant research takes place under the auspices of the agrotech industry, whose tendency to keep data in-house, due to its commercial sensitivity, differs substantially from the Open Science ethos characterising much publicly-funded plant science, where large-scale research around model organisms like *Arabidopsis thaliana* resulted in an extensive set of standards, conventions and platforms devoted to effective data sharing and the idea of data as "knowledge commons" (Leonelli, 2016a; Henkhaus et al., 2020). Tensions between competing claims to national sovereignty over biological materials and related data, as well as the jurisdiction of different types of licenses, patenting systems and copyright agreements, further complicate this landscape. Last but not least, at the social end of the spectrum plant-related work involves many contributors beyond professional research circles, including farmers and their communities, breeders, food producers, and policy-makers involved in agricultural policies at the regional and national levels and trade agreements at the international level. These diverse participants tend not to communicate effectively with each other. Differences in skills and goals, social divides, persisting power asymmetries and the sheer quantity of relevant stakeholders make it particularly hard for farmers to provide input and feedback to researchers and policy-makers, and thus to contribute to discussions around what counts as scientific findings, what those may signify for agricultural development within local territories, and what role digital technologies can and should play in land management and food production.

This volume starts from the recognition that *scientific, legal, political and socio-economic challenges such as these are inextricable from each other and have a decisive impact on which plant data get to circulate, to whom and for which purposes*. An immediate implication of this premise is that confronting these challenges requires an awareness of the complex landscape in which they emerge, including some understanding of their historical roots. This volume is intended as a multidisciplinary, transnational entry point to that landscape. It assembles a wide range of practitioners from data science, ethics and the law, history and social studies of science and agronomy, which together represent some of the key initiatives in plant data linkage and curation in the world. The volume thus examines the opportunities and challenges of plant data linkage and re-use as experienced by contributing authors who have spent decades working in this domain. Our goal is to chart and support data exchanges that are not only scientifically and agronomically productive, but also responsive to the social circumstances in which data and plants are collected and used – and in that sense, are both effective and responsible.

In this introduction, we provide essential background to this work. In the next section, we examine the different meanings that the idea of responsible practice can take at the four key sites of plant data governance: the plant environments being documented (the *field*), the infrastructures used to circulate information (the *data*), the entities involved in data governance (the *institutions*) and the variety of expertise and interests involved in plant data work (the *communities*). In Sect. 3, we then outline what we regard as four crucial steps towards achieving responsible data linkage: (1) the building of infrastructures to foster critical data reuse; (2) the development and implementation of transnational legal and institutional frameworks; (3) the formulation of effective ethical guidance and related monitoring systems; and (4) the creation of mechanisms to identify and regularly evaluate assumptions made about agricultural development and the contribution of agricultural science to society, and to consider alternative frameworks. In conclusion, we emphasise the importance of giving equal consideration to these four steps not just in developing but most importantly in maintaining responsible and fruitful practices of plant data linkage in the long term – a crucial factor in making such practices trustworthy and dependable.

2   Dimensions of Responsible Plant Data Governance

There is increasing awareness of the enormous resources and labour required to develop data infrastructures through which data and knowledge about plants can be garnered and harnessed appropriately. These include tools that can foster harmonious data exchange and mining, such as semantic systems, formatting standards, metadata categories and tailored databases. Developing such tools is a technical challenge that has kept thousands of computer, plant and data scientists busy for decades. As many contributions to this volume illustrate (Bertin et al.; Devare et al.; Rawlings and Davey; Pommier et al.; Ostler et al.), such efforts have yielded impressive progress, with substantive innovations emerging to help curate and organise plant data for future re-use. Nevertheless, we remain far from the seamless global systems for data collection and access that were envisaged already at the turn of the last century, when organisations like the League of Nations started to promote systematic efforts to garner and integrate scientifically relevant information from across the world (Hewson, 1999; Edwards, 2010). The vision of all-encompassing automated data analysis linked to the rise of computing in the 1960s and 1970s has not yet materialised, despite the resources devoted to building digitised data infrastructures and the hype surrounding the mining of big data (Williamson et al., 2021).

A key reason for this gap between expectations and reality is that assembling reliable data systems is not only a technical issue, and making plant data amenable to reuse is more than a technical challenge. The creation and curation of interoperable data involves a range of conceptual and social challenges that are inseparable from the technical aspects. For example, in order to make given plant traits amenable to large-scale computational analysis, it is necessary to have suitable labels for the data clusters relevant to investigating such traits. This requires the development of reliable and standardised trait descriptors, which in turn involves consultations across breeders, farmers, researchers and consumers concerning which traits are most significant for investigation and which labels are most appropriate in defining them – a fraught set of questions to ask within a cross-cultural, multilingual environment plagued by power differentials and inequity between the parties involved (Arnaud et al., 2020; Leonelli, 2022; Curry & Leonelli, 2022). Additionally, analysing data on phenomena ranging from ecological stressors to host-pathogen interactions requires having sufficient metadata about the conditions of origin and the legitimate range of possible uses of such data (Shaw et al., 2020); and linking data from many different sources (whether genomic and experimental data from public or corporate research, knowledge of plant strains and environments held by farmers and breeders, or data related to stored germplasm collections) requires sharing, access and reuse agreements among stakeholders as well as venues in which such agreements can be forged. These are very complex requirements given the diverse regimes of intellectual property, commercial sensitivity, research incentives, cultural ownership and trade to which data are subjected, and the existing tensions around the goals, motivations, and implications of data disclosure and re-use.

All this makes the idea of 'responsible practice' in plant data management difficult to understand and operationalise. What does responsibility mean here, given how distributed and diverse plant data stakeholders, contributors, infrastructures and users are?[1] Our starting point in answering this question is to acknowledge that responsibility means different things depending on the setting and goals it needs to serve. Thus, rather than trying to settle on a unique and common definition for this notion, we review four key dimensions of data linkage, and examine what responsible practice may signify within each. These dimensions of data practice also provide the main structure for this volume, which is divided into four parts accordingly.

2.1 The Field: Documenting Variability in Plants and Their Environments

A recurring concern in the management and curation of plant data is the extensive variability encountered both in the plant specimens and in the environments in which they grow, including the intersections of such environments with human

---

[1] In approaching the topic of responsibility in scientific and technical domains, we follow in the path of extensive work undertaken under the aegis of 'responsible research and innovation' (e.g. Owen et al., 2012). As presented in this introduction, our framing of issues of responsibility is tailored to the specific issues in the fields of plant and agricultural science. For a related summary of 'responsible data' issues in agriculture, see Ferris and Rahman (2016); for considerations of responsible data governance within highly distributed technical systems, see Edwards et al. (2011), Lagoze (2014) and Leonelli (2016b).

communities. It is critical for plant data systems to capture accurate information about which species and varieties are being documented and which seeds are collected, as well as which environmental features are most relevant to plant development and yield. And there is broad agreement on the prominent role that genetic information has come to play in supporting this effort, and therefore on the significance of sharing digital sequence information as a gateway to understanding agrodiversity (Morgera et al., 2020). Nevertheless, the variability in the characteristics of plants and their environments (including, crucially, the soil) is extensive and highly dynamic, particularly under conditions of climate change. Moreover, such environmental variability is flanked by variability in the methods and procedures used to generate data and curate relevant materials (such as germplasm), as well as social variability in the preferences, assumptions and conceptual commitments held by data producers and stewards. Settling on data practices and standards to capture such information is a priority and a serious challenge, with important repercussions on the systems used to evaluate performance, productivity and success of agricultural strategies.

When considering the processes involved in extracting data from local fields, crops, seed systems and their environments, the central concern around responsible data practice thus relates to decisions around which kinds of variability need to be reported into data systems. Responsible data linkage involves explicitly asking how different kinds of variability feature in data systems and the ways in which the success of such systems is assessed, and ensuring that the decisions taken in response to this question are regularly scrutinized and reviewed across stakeholders. As the chapters in the first part of this volume make clear, this is hard to implement in practice for a variety of reasons. One is scientific disagreements on how data may be interpreted, which Radick's chapter discusses under the heading of "Theory-Ladenness as a Problem for Plant Data Linkage" and elegantly exemplifies with reference to the history of Mendelian plant genetics. Another is the cost and technical intricacy of harmonizing various types of environmental data with data about crops, especially considering the evolution of seed trade, intellectual property and public-private relations underpinning modern breeding – as beautifully illustrated, through a narrative spanning the whole of the twentieth century, by Harrison and Caccamo's chapter "Managing Data in Breeding, Selection and in Practice: A Hundred Year Problem That Requires a Rapid Solution". A third consists of the diverse political conditions under which specific taxonomies of seeds may come to be defined and valued as objects of analysis, which in turn determines the characteristics of related data collections. This is poignantly exemplified by Fullilove's and Alimari's analysis of wheat breeding and preservation projects on the West Bank in their chapter "Baladi Seeds in the oPt: Populations as Objects of Preservation and Units of Analysis". And last but not least, there are concerns around how the apparatus devised to extract and manage data intersect with breeding practices on the ground, which call for the establishment of effective and mutually respectful dialogue between data linkage experts and those who run field trials and provide key materials and observations. Efforts in this direction are exemplified and discussed by Agbona and colleages in relation to Root, Tubers and Banana crop breeding programmes in

Africa in the chapter "Data Management in Multi-Disciplinary African RTB Crop Breeding Programs".

These challenges are not only about the technical assemblage of data sources, though this is certainly a crucial problem in this domain. Among the broader issues raised by these studies, we find a systematic questioning of the extent to which data management methods focused on digital sequence information can fruitfully serve broader phenotype and environmental datasets; of how breeding strategies are identified and chosen, and with what implications (for example when privileging ex situ breeding over in situ efforts, as long done by many research institutes around the world; see also Curry, 2017 and Curry's chapter in this volume); and whether and how data systems can and should pay more attention to marginal environments where uniform crop varieties do not perform consistently, rather than prioritizing data collection on selected high yield varieties.

2.2 The Data: Developing Scalable and Interoperable Infrastructures

These issues become even more pronounced when shifting attention from the field and circumstances of data collection to the nature of the data themselves, and how the characteristics of data affect efforts to develop and link data infrastructures. A key concern in this respect is how to bring data together in the first place. The idea of *integration* is often used to refer to the ability to aggregate and analyze different datasets as if they were a single body of evidence. Yet integration conceptualized in this way is very demanding: it requires making specific choices as to what the best ways to format and visualize data may be, which may be well-suited to the question at hand but not to other forms of data re-purposing; and yet it may be difficult to disaggregate the data once they have been fully integrated. These concerns are the reason why *interoperability* has taken the place of integration as a crucial and potentially more responsible form of data linkage. Interoperable databases are those that enable their users to ask common queries, thereby supporting links between datasets without reducing users' ability to ask different questions and re-purpose the data accordingly. Interoperability can foster the accountability of data practices, by making it easier to track who has selected and co-analysed which data, from where and how – and thereby being more responsive to the diverse interests and goals of data users. Effective interoperability requires, in turn, at least some level of standardization in both datasets and data infrastructures, which can facilitate common searches and make data comparable in the desired respects.

This is where the desire for interoperable data meets the problem of *scalability*. It is hard enough to set up a data infrastructure able to capture, store and disseminate data obtained from different field trials carried out by a specific institution on a particular crop – such as the UK-based work on wheat documented by Rawlings and Davey's chapter "From Farm to FAIR: The Trials of Linking and Sharing Wheat

Research Data". But as Rawlings and Davey show so effectively in their discussion of the Designing Future Wheat research programme, the thorniest issues emerge when trying to link such data to data obtained on field trials carried out elsewhere or on other crops, or even to other types of data on crops (such as phenotyping or experimental data). The success and scalability of data practices depend on the effectiveness of Field-to-Lab-to-Field cycles, in Rawlings and Davey's words, where those involved in generating, standardizing and interpreting the data have means to interact regularly and give each other feedback.

Of course, data work can be scaled up further by going beyond the field and lab environments to include environmental research and statutory data produced through agronomic governance, as discussed at length in Harrisons' and Caccamo's chapter. Yet another way to scale up data practices is to take a longitudinal view of agricultural research, and link data produced in the present with data generated by the many decades of experimentation which preceded the digital era, while also paying attention to how data collected from very long-running experiments should be managed to enhance their usability now and in the future. The chapter "Linking Legacies: Realising the Potential of the Rothamsted Long-Term Agricultural Experiments" by Ostler and collaborators from Rothamsted Research, one of the longest-running agricultural research stations in the world, closely examines means of facilitating data scalability and interoperability in time as well as in space, and challenges emerging when considering legacy data. These are crucial concerns at a moment where many data infrastructures are set up to serve specific projects through short cycles of funding, leaving the future maintenance of those databases in limbo, and agricultural institutions around the world host precious, non-digital data collections stretching back several decades, whose potential value to plant research is limited by their inaccessibility. Among the solutions developed at Rothamsted to these challenges, including the design of the electronic Rothamsted Archive (e-RA) database, the emphasis placed on *skilled data curation* is particularly notable. As evident in almost all contributions to this volume, data curators play a key role in mediating between the archive and would-be users, bringing expert knowledge of datasets and experimental narratives (i.e. the history and purposes of each experiment) to bear where standardisation alone is insufficient to ensure effective and responsible reuse.

A fruitful way to conceptualize and explore standards for data linkage, and support the development of interoperable systems, is to consider whole *data lifecycles*. This involves rejecting a strict compartmentalization of different types of data practices, such as for instance data production, cleaning, formatting and modelling, and instead understanding such data practices in terms of how they relate to each other within and beyond the world of research (Borgman, 2019; Leonelli, 2019). In their chapter "Plant Science Data Integration, From Building Community Standards to Defining a Consistent Data Lifecycle", Pommier and colleagues reflect on the ongoing attempt to develop data standards that are meaningful and useful to specific communities of plant researchers, while also taking account of how such standards may support subsequent stages of the data lifecycle in a consistent manner. As they note, data standards are only effective in promoting interoperability as long as they are successfully implemented, and the conditions of successful implementations prominently include the degree to which standards are tractable, trusted and perceived to be useful among users.

2.3   The Institutions: Overseeing the Dissemination and Use of Plant Data and Materials

Concerns around which kinds of expertise, venues and social arrangements are most appropriate to facilitating data linkage have already repeatedly come to the fore in our discussion, and it is therefore no surprise that the third domain we wish to highlight is that of the institutions responsible for devising and implementing data governance strategies. Responsible practice here includes not only the design of rules and regulations that may support – rather than hinder – data work, but also regular monitoring of the extent to which these systems are being implemented, and most importantly, of their impact on plant research as well as agricultural and food systems. Ultimately, responsibility in this domain means taking ownership of both the positive and negative social consequences of specific data practices, and taking action whenever a given governance method fails to support agricultural and social development. This in turn requires ongoing consideration of what constitutes desirable development, and for whom.

What organizational and governance structures are fit to address such a challenge? Devare and colleagues consider this question in their chapter "Governing Agricultural Data: Challenges and Recommendations" through a discussion of the forms of leadership, strategy and management required to support data linkage within the CGIAR, a large international organization comprised of 15 agricultural research institutes around the world. The history and current structure of CGIAR effectively exemplifies the opportunities and obstacles created by the requirement to link highly diverse data, coming from culturally, geographically and socio-economically distant communities, in ways that inform agricultural development on a global scale. The central coordination efforts within CGIAR depend on a plethora of other institutions, ranging from the individual CGIAR institutes themselves (each of which has its own governance structure, which is in turn responsive to the specific territory and political situation in their host countries) to the various private and public funders involved in sponsoring projects carried out by CGIAR institutes, the many collaborative networks and consortia set up in relation to specific initiatives and crops, and the international regulations under which this quintessentially transnational work takes place.

Beyond such fragmentation and multiplicity, a central governance challenge for international institutions such as CGIAR is the large *inequity* that characterizes agricultural research across different locations, with many parts of the developing world (and particularly ex-colonies) routinely serving as providers for biological materials and related data and botanical knowledge, and yet not playing an active role in using the data to produce agricultural innovation (cf. Kloppenburg, 2004; Hayden, 2003; Soto Laveaga, 2009). Unless exploitative practices are appropriately identified and challenged in the course of data work, there is a substantive risk that data linkage strategies may help to further entrench existing systems of unfair data collection and predatory data re-use (Miles, 2019). Fullilove and Alimari's chapter highlights such issues in relation to contemporary seed systems and agricultural development in the occupied Palestinian territories (oPt), thus underscoring how countering in-built inequity and the dominance of the Global North over the agricultural landscape is a priority when seeking to develop responsible systems for data linkage. The chapter "Digital Sequence Information and Plant Genetic Resources: Global Policy Meets Interoperability" by Manzella and colleagues presents some of the progress made in developing more equitable data systems in tandem with existing policy frameworks for the international governance of plant and agricultural science. These include the systems of access and benefit sharing (ABS) that form a key pillar of the International Treaty on Plant Genetic Resources for Food and Agriculture (ITPGRFA) and the Nagoya Protocol of the UN Convention on Biological Diversity (CBD). These policy systems and their underpinning legal structure have been challenged in recent years by the high availability of digital sequence information, which has the potential to undermine existing systems of ABS focused on access to germplasm and other biological materials (Morgera et al., 2020). Manzella and colleagues survey the current status of discussions regarding sequence data and ABS policy, focusing on the urgent need to enhance the interoperability of relevant data systems such that the origins and use of sequence data and the status of corresponding biological materials under the ITPGRFA can be easily identified.

Such technical solutions are born of careful consideration of the large political and ethical issues relating to the circulation of plant materials as well as data. The ability to link plant data transnationally is crucial to enhancing biological understanding of crop usage and food systems worldwide, and yet the imaginary of plant genetic resources and related data as 'common goods' remains in tension with national systems of governance for agricultural resources (Bonneuil, 2019). The very idea of (national) *sovereignty* associated to plant materials and data is itself a double-edged sword: it is important to acknowledge and respect, especially given postcolonial legacies of exploitation of specific countries, but it also supports highly restrictive understandings of who may own and use crop data. Responsible data linkage involves tackling these issues through the co-creation of governance and technical systems capable of mediating legal, ethical and social considerations. Kochupillai and Köninger's chapter, "Creating a Digital Marketplace for Agrobiodiversity and Plant Genetic Sequence Data: Legal and Ethical Considerations of an AI and Blockchain Based Solution", presents an ambitious proposal for how new digital systems could be put in place to overcome some of the current obstacles of supply and demand of agrobiodiversity for research and breeding. Central to their proposals is the need for cutting-edge technical solutions that not only facilitate in situ innovations for farmers as well as researchers, but also respond to current inequities in legal and regulatory regimes. This requires a wholesale rethinking of key components of contemporary regulation, such as the current dependence of benefit sharing mechanisms on downstream intellectual property rights. The economic implications of such a transformation are vast, both in their consequences for the seed and food markets and in their demands on current investment in data-intensive technologies and related practices.

The relevance of *economic strictures*, and the clash between the need for transformation and the ever more limited resources available for the development and maintenance of reliable data systems, is aptly illustrated by Curry's chapter "Data, Duplication, and Decentralisation: Gene Bank Management in the 1980s and 1990s". Her analysis of the 'rationalisation' of gene bank collections illustrates a recurring tension between idealised efforts of conservation and reuse of plant-related resources (such as the attempt to assemble comprehensive collections of viable seeds from all over the world) and the lack of the financial and organisational resources required to maintain such plant resources over time. This example shows how policy and organisational solutions implemented on the ground are rarely straightforward responses to data challenges, but are entwined with the need to respond to many competing imperatives, including expectations around what constitutes a profitable investment and the timescale of economic returns. The implications of the political economy of collecting, and the costs involved in long-term maintenance whose economic impact is hard to quantify, must always be borne in mind when evaluating and designing institutional and governance strategies for plant data management and linkage (Strasser, 2019). Whether practitioners explicitly acknowledge it or not, developing long-term data linkage strategies typically involves challenging short-term arguments for predefined and easily quantified sources of economic return, and emphasising instead the diffuse – and even more impactful – ways in which data governance systems may support economic growth and sustainable agricultural development (Leonelli, 2022).

2.4   The Communities: Perspectives from and Accountability to Farmers and Consumers

Perhaps most fundamental and challenging of all is the recognition that plant data – like all other forms of data – have multiple *values* depending on who handles them and for which purposes. Beyond their obvious scientific and commercial value, they may hold affective value, cultural value (if they document knowledge by local communities, for example) and political value (e.g. in disputes over ownership of biological resources). The constellation of relevant values will vary between stakeholders, and there are often tensions between different values held even by single individuals – let alone distributed networks of stakeholders. Recognition and debate around such values is crucial to responsible data practices, which play an essential role in connecting different stakeholders and facilitating communication across communities (Leonelli, 2016b).

When thinking about the governance, circulation and use of plant data, it is crucial to broaden the conversation from the more technical discussion of standards and curation strategies characterising data and plant science circles, and to bring in perspectives from farmers as well as consumers of crops (whether as food, medicine, fuel, fabric or other), and other stakeholders in seed and food systems. It seems trivial to assert that the rights, needs and goals of these communities need to be foregrounded and included in the processes through which infrastructures, governance regimes and policy directions are shaped; and yet, farmers are rarely consulted on and included into the design and governance of data exchange systems. The chapter from Zampati, "Ethical and Legal Considerations in Smart Farming: A Farmer's Perspective", examines the proliferation of ways to extract and monetise data from farmers' everyday activities, often resulting in exploitative technologies that may benefit the national economy but do not necessarily benefit individual farmers and their communities, and in fact remain unintelligible to farmers and far removed from their sphere of intervention. The chapter presents some models to increase farmer engagement, focusing especially on the adoption of codes of conduct that encourage a dialogue between farmers and the data experts and companies involved in smart farming.

Looking instead at efforts to meaningfully link data from diverse territories and crop varieties with each other, in their chapter "Communities of Practice in Crop Diversity Management: From Data to Collaborative Governance" Louafi and collaborators provide an example of what they call 'collaborative governance', whereby a *community of practice* is constituted to help address both technical and social challenges involved in data linkage. This is a case where the heterogeneity of stakeholders is transformed from a problem into an asset: regular consultation among different experts, including farmers as well as breeders, consumers and data experts from a variety of different territories, becomes a crucial way to understand and manage crop diversity, and thereby build plant data infrastructures that successfully incorporate wide-ranging knowledge sources of relevance to agricultural development. Another great example of a community of practice at work is provided by Rocha Bello Bertin and collaborators, whose chapter discusses the efforts to build such communities by "The Research Data Alliance Interest Group on Agricultural Data: Supporting a Global Community of Practice". This volunteer group has spent over a decade on efforts to identify and assemble communities of practice that can support long-term discussions and decisions around the standards and semantics to be used when linking crop data from around the world. Notably, this group has long been open to participation from any relevant stakeholder around the world, and yet – as they observe in the chapter – found several obstacles in integrating wide-ranging expertise and new voices into their work. Being included in data governance efforts often requires some expertise in, and understanding of, existing data systems, as well as the time and resources to find and engage with the right international groups. This is an additional burden on the shoulders of farming communities already under pressure to produce high yield under increasingly competitive and adverse conditions. Engaging in communities of practice can also be slow and sometimes tedious work, replete with discussions over what are sometimes minute aspects of data curation and standardisation – issues that may matter very little to some of the stakeholders, but crucially affect others.

Questions around the role and incentive structures for communities of data practice parallel long-held debates over the relation between the germplasm acquired from farmers and breeders and the digital data produced by researchers, industry, and governmental institutions. There are sometimes many degrees of separation between the biological materials produced by farmers and the various types of data (molecular as well as administrative and socio-economic) generated by those tasked with analysing and regulating food production and distribution. Given the diverse types of labour and contributions to innovation in such a complex system, it is important to ensure that benefits are equally distributed across the "data chain", including to farmers and other data providers, rather than being captured by certain end users or those who hold intellectual or other property rights. Equally important is problematising the question of what constitutes a benefit to different stakeholders in the first place, and under which circumstances. The question of adequate and appropriate benefits is one that is hard to address through purely quantitative analysis, and often requires the kind of context-sensitive inquiry that the qualitative social sciences and Science and Technology Studies (STS) are well-placed to carry out. Social scientists are also well-placed to collaborate with both data scientists and farming communities, and thereby help broker conversations and exchanges between different groups.

This is not only an exercise in inclusion for inclusion's sake. As argued by Radick's chapter as well as Williamson's and Leonelli's, the development of any data system unavoidably involves making strong *conceptual assumptions*, which affect and shape social relations, research goals and even the types of expertise which are regarded as relevant. These assumptions become entrenched into those technical systems and thus increasingly difficult to challenge. At the same time, however, the purpose and reach of those systems continues to change and expand, raising questions as to whether the initial assumptions made when creating those data infrastructures continue to be valid and fruitful. For instance, in their concluding chapter on "Cultivating Responsible Plant Breeding Strategies: Conceptual and Normative Commitments in Data-Intensive Agriculture", Williamson and Leonelli discuss how even apparently value-neutral, scientific concepts such as the notion of genetic gain in plant breeding – which is increasingly used as a measure for the productivity of specific crops – can embody a restrictive normative vision for what agricultural development means, how it can be measured and incentivised, and who it is supposed to benefit.

Which criteria are used to single out a desirable plant trait? Are farmers and breeders consulted on which plant trait is most valued by consumers in local markets? Is soil health factored into data systems meant to document field trials, or are the data focusing exclusively on genetic markers for the plant varieties themselves? Asking such questions is a way to critically question received views on the relationship between crops and their biological and social ecosystems, which may be implicitly embedded into data system and linkage tools. Data infrastructures are most often born of the need to compile and circulate an existing dataset, and are thereby often conceptualised as a neutral container – a black box whose only function is to preserve and spew out data whenever required, and whose functioning should not affect the data and the ways in which they are repurposed. As the chapters in this volume demonstrate, however, there is no such neutrality: rather, data infrastructures are unavoidably value-laden and replete with normative assumptions about what counts as sustainable ways to care for the environment, cultivate crops and produce food. Responsible data practice involves regularly opening and re-ordering the black boxes, checking that their components – including their conceptual apparatus – are fit for their constantly shifting purposes. Ultimately, data linkage systems are systems of relations: taking time to define and regularly re-evaluate what count as relevant *relata*, depending on one's goals, is therefore paramount.

3   Steps Towards Responsible Plant Data Linkage

Who is then responsible and accountable for decisions around data management and the re-use of data, and mistakes or problems associated with such decisions? For example, regarding the allocations of rewards and rights, we might ask who is responsible for "data production" in a given experiment. Is data production the result of growing the plant specimens, selecting strains, designing field trials, adopting novel measurement tools or designing data storage? The answer to this question will determine who is viewed as the legitimate owner of data and who has control over their use. Yet all of these activities have a legitimate claim to being part of data production. Indeed, the chapters of this volume demonstrate the diversity and pervasiveness of responsible practice across the main domains of plant data linkage, which raises urgent questions around the meaning of accountability in such fragmented and distributed systems of knowledge production. All those who partic-ipate in plant data analysis – and related benefits and profits – are arguably account-able for their work in some way: their contributions should be evaluated with an eye to their role and consequences within the whole system, and there should be mechanisms to reward good practice and discourage problematic or wrong deci-sions. However, evaluation of what may constitute responsible practice lags behind. It remains hard to determine what such distributed accountability means in practice; who may be held responsible – and with which implications – when things go wrong; and how to differentiate between human error, system bias and deliberate misuse. In this section, we point to four essential steps towards fostering responsible behaviour while also helping to identify and address problematic data practices.

3.1   Focusing on Critical Data Reuse

A starting point for responsible data linkage is the acknowledgment that the prob-lems of accessing and using data cannot be separated. In other words, Open Data can and should not be a goal in and of itself. Focusing solely or primarily on "putting data online", without worrying about who may access such materials, how and for which purposes, is a recipe for disaster. Notably, data linkage makes concerns around data access and re-use inextricable from each other – for while there is no opportunity for linkage without some level of data access, linkage methods unavoidably serve specific expectations of data may be re-purposed. Openness thus needs to be intelligent (Boulton et al., 2012); data infrastructures and tools for data analysis should be developed with at least some awareness of the ways in which data may – or not – be employed in the future, and the types of users who may be involved. Of course, the future of data is never certain or fully predictable, especially in the era of data-driven analysis (Leonelli, 2016a). This does not mean, however, that thoughtful consideration should not be given to the priorities and assumptions built into data linkage systems; in fact, the unpredictability of data use is a key reason to pay close attention to the design, maintenance and broad impact of data linkage systems.

An important step towards refocusing data practices on data reuse is exemplified by the FAIR principles for data management. These principles, now widely recognised worldwide, define effective data sharing as making data Findable, Accessible, Interoperable and Reusable (Wilkinson et al., 2016). This comes with the acknowledgment that Open Data are not always required and never sufficient to guarantee data re-use. Being able to access data that have been badly curated and annotated is often as bad as not having access at all, since data that are badly curated are near-impossible to re-use meaningfully. Furthermore, as is well-acknowledged in the biomedical domain, data can be made available for re-analysis and re-purposing even without direct access: for example, through data mining techniques such as DataShield which facilitate pooled data analysis without sharing individual-level data (Murtagh et al., 2012). The FAIR principles thus took attention away from sheer data access and re-focused instead on *the conditions for* "best data practice", which in turn involve critically investigating what data exist, whether or not they can or should be accessible, what mechanisms should be used to grant access and how such mechanisms will inform re-use. As part of such efforts, data history (including data provenance as well as the locations, methods and interests of those involved in data processing) is increasingly recognised as essential meta-data that needs to be adequately tracked and documented (Leonelli, 2020). Indeed, within the FAIR framework metadata are arguably more important than data themselves – without appropriate meta-data, data re-use is compromised and the opportunities to re-purpose data are radically restricted, if not altogether eliminated.

As repeatedly noted by our contributors, the FAIR Principles are widely recognised in plant science (Pommier et al., 2019; Reiser et al., 2018) and increasingly built into data collection at source, for example through the creation and use of digital fieldbooks that facilitate the standardisation and semantic interoperability of field data collection (e.g. Rife & Poland, 2014). They are also recognised at the infrastructural level, exemplified by the incorporation of FAIR data metrics into the CGIAR Big Data Platform's GARDIAN search tool, and there are now dedicated tools to assist in the management and deposition of FAIR data and metadata, notably the Collaborative Open Plant Omics (COPO) platform (Shaw et al., 2020). The extensive implementation of FAIR is a big step forward in the development of plant data infrastructures that facilitate extensive and responsible linkage, not least for recognising that data access should be carefully monitored and regulated (as often stressed by FAIR data proponents: "data should be as open as possible, as closed as necessary"). This is crucial to enable a more critical and nuanced understanding of the multiple social contexts of data sharing processes, and the potential implications of granting data access in the case of sensitive data. However, as we will see in the next section, this framework does not go far enough, and indeed does not directly include attention to ethical aspects such as equity and fairness in the provenance, ownership and distribution of data resources.

3.2   Encouraging Multiple Forms of Transnational Data Governance

The regulatory framework for plant data work is as yet vague and unclear, with few (if any) existing international agreements concerning the goals, rewards, responsibilities and rights pertaining to the generation, circulation and use of digital plant data. This situation contrasts stridently with the biomedical field, where such agreements have been at the centre of developments in genomics (Maxson-Jones et al., 2018) and the set-up of structures and regimes of data governance for health-related data (Hilgartner, 2017). This arguably owes much to the distinctive risks associated with the category of "personal data" about individual patients, a category which, with the exception of data documenting the socio-economic status of individual farmers, is of less relevance to the plant sciences. And yet, the dissemination and linkage of plant data bears its own social and ethical risks. First, as we discussed above, data sharing across countries remains an underregulated and yet sensitive matter, where data produced in the Global South is systematically harnessed and profitably re-used in the Global North and yet such appropriation often happens without proper attribution and compensation.[2] Second, large agrotech corporations dominate plant data production and re-use (including through remote sensing technologies incorporated within agricultural machinery) in ways that are rarely transparent and well-aligned with equivalent efforts in the public domain (Shiva, 2016; Fullilove, 2017; Miles, 2019). This makes dialogue around regulation, technical standards and socio-economic implications of data re-use even harder, as there is no overarching sense of the amount, variety and nature of existing data of relevance. Third, the commercial value and cultural capital associated with plant data – and particularly data about indigenous crops – is well-recognised by most countries/ governments as a national resource, and yet there is little clarity around whether the deployment of such resource does (or should) reflect national interests, and how this

---

[2]This phenomenon can be construed as the digital equivalent to bioprospecting: a "digital feudalism" building on centuries-long exploitation and discrimination built into the food production system (Scott, 1998; Hayden, 2003; Mazzucato, 2019).

sits vis-à-vis the conception of plant knowledge as a global common good (Kloppenburg, 2004; Krige, 2022).

All this indicates that technical means to enable plant data linkage need to be accompanied by an effective system of *transnational data governance*, comprising both the norms and the infrastructure needed to share and re-use data adequately and responsibly. Most contributions to this volume can be read as working towards this goal, whether by developing sharing standards, legal frameworks, governance venues, ethical norms or physical tools. The diversity of such work shows how benefits to be distributed across the data chain include economic gain (as in the proposal of a blockchain solution in Kochupillai and Köninger's chapter) as well as opportunities for shaping data work (through the communities of practices discussed by Rocha Bello Bertin, Louafi and their colleagues in their respective chapters) and be appropriately rewarded for that effort through proper acknowledgment, as fostered by current FAO efforts discussed by Manzella's chapter, and the tracking of data provenance promoted by GODAN, as exemplified in Zampati's chapter. What such suggestions will involve in terms of legal frameworks both nationally and internationally is a crucial problem whose resolution goes well beyond the scope of this volume, but which we hope these contributions may help to inform – particularly by highlighting the diverse levels of governance involved in making data linkage work for users (see also Welch et al., 2021), and fostering a better integration of socio-political concerns into technical efforts to develop plant data infrastructures.

3.3 Developing Guidance in Tandem with Incentives and Monitoring Systems

How to achieve such integration? Alongside the infrastructural work to facilitate critical data reuse and the regulatory work to ensure the legality of data exchanges especially at the international level, there have been increasing efforts to ensure that ethical considerations are built into the design and use of data infrastructures. One mechanism for this has been the creation of additional guiding principles, complementary to FAIR, that are focused on ethical issues.

One such set of principles are the TRUST principles proposed by the Research Data Alliance, which stand for: Transparency, that is the need to make data operations as easy as possible to understand and scrutinize; Responsibility; User Focus, which involves prioritising the needs, skills and concerns of users over the wishes of infrastructure developers; Sustainability, which implies attention to the long-term prospects and environmental impact of the infrastructure; and Technology, that is the importance of keeping an infrastructure up-to-date with evolving software and hardware requirements (Lin et al., 2020). Informed by such principles, there is an emerging trend in broader data science towards public and collective use of knowledge and infrastructures, with a number of data initiatives built with these values at their core (including for instance the Ada Lovelace Institute in London, the Centre for Technomoral Futures at the University of Edinburgh, the research line on Digital Infrastructures for the Public Interest at Stanford PACS, the PublicSpace coalition in Amsterdam and the Institute for Digital Public Infrastructure at UMass Amherst, to mention just a few). This trend is also visible in programmes that prioritise a responsible approach to research and innovation or to human centric and trustworthy data technology, prominently fostered by the European Commission. It is high time that such approaches are explicitly extended to the plant and agricultural domain, as our Exeter Centre for the Study of Life Sciences at the University of Exeter is attempting to do.

Another important development are the CARE Principles for Indigenous Data Governance, which were produced by the Global Indigenous Data Alliance in consultation with a very wide range of data subjects, producers and users. The CARE Principles draw attention to the implications of open data sharing for indigenous and other communities from whom data may be extracted,[3] by focusing on four key issues: (1) equitable distribution of Collective Benefits, that is of evaluating the impact of a given data intervention on groups and communities and ensuring that this impact is positive; (2) the recognition of communities' own Authority to Control, which points to the necessity to distribute power and control over the data across the stakeholders involved, rather than placing all control in the hands of one party (especially if this party consists of digital platforms or specific data users); (3) the Responsibility of researchers to communities, which involves the need to clearly acknowledge who is being held responsible when data work goes wrong; and (4) the foregrounding of Ethics at all stages of the data life cycle, which is a broad invitation to monitor the social and moral implication of any kind of data work.

Both TRUST and CARE principles are part of multiple efforts to introduce reflection on wider obligations and responsibilities into the workings of a given data infrastructure. All too often however the nature of such reflection and any changes resulting from it are left open to actors' own judgement and rely on voluntary adherence. This reflects concerns in the field of AI about "ethics washing" through the creation of sets of principles or guidelines that co-opt the flag of ethics but potentially do little to actually change how tech companies use data, as well as subsequent counter-critiques of "ethics bashing" (Bietti, 2020). As Kind (2020) has noted, moving beyond ethics washing and bashing requires treating the implementation of ethical principles not just as a narrow technical matter but as a socio-technical one that involves addressing local practice and organisation. Hence what we wish to highlight here is not only the significance of such ethical frameworks for future data work, but also the critical role of systems of incentives and monitoring in making it possible to concretely implement these frameworks. For instance, CARE principles need to be complemented by data labels and validation systems that help certify and monitor adherence to such principles.

A great example is provided by the Traditional Knowledge and Biocultural Labels Initiatives, which "allow communities to express local and specific conditions for

---

[3] https://www.gida-global.org/care

sharing and engaging in future research and relationships in ways that are consistent with already existing community rules, governance and protocols for using, sharing and circulating knowledge and data" (Liggins et al., 2021). The Biocultural Labels focus specifically on the handling of plant genetic resources derived from crops samples associated to traditional knowledge. These labelling systems have been devised by a consortium of researchers working closely with traditional communities in New Zealand as well as representative bodies for Indigenous Communities around the world, such as the Indigenous Data Sovereignty movement (Hudson et al., 2020). Having been successfully trialled within individual projects and specific collections, they are now being considered for adoption by several large data infrastructures around the world. Such an initiative is very important to data linkage initiatives relating to food and agricultural research, especially given the lack of international agreement on whether and how to govern data sharing through the framework of the International Treaty on Plant Genetic Resources for Food and Agriculture – whose Access and Benefit Sharing mechanisms do not include clear instructions on the status of digital sequence data (Aubry, 2019).

The case of TK and Biocultural Labels shows not only the significance of governance mechanisms to concretely implement principles such as CARE and TRUST, but also – going back to the previous section – the multiplicity of forms of governance required for such implementation. These include large-scale efforts from national governments, prominent research funders, corporations and international organisations such as the Food and Agricultural Organisation and the Convention for Biological Diversity, all of whom can consider mandating the use of these kinds of labels within the data infrastructures and policies that they support; as well as small-scale efforts such as individual projects, research centres and universities, whose reach may be limited but which are much closer to the data practices of interest on the ground.

3.4 Considering Alternatives

The final point we want to highlight is that developing responsible and effective data linkage systems requires bringing infrastructural and ethical strategies in line with the conceptual and normative dimensions of scientific and agricultural practice, including the ways in which both agricultural development and data-intensive research are framed. Major global challenges such as climate change, which require significant rethinking of large-scale systems, cannot be tackled without addressing the conceptual underpinnings of those systems and their implications. This in turn involves identifying the imaginaries of agricultural development and related data usage that are instantiated within existing systems, and asking what alternative ways of constructing and understanding the world could look like, what difference they would make to the principles and values supported by contemporary data linkage infrastructure, and what would be the technical implications and strategies involved in implementing such alternative frameworks.

The example of accelerating genetic gain in plant breeding, discussed at length in our own chapter at the end of the volume, is a case in point. The use of genetic gain as a key indicator for agricultural development needs to be situated in relation to the legacy of the Green Revolution, including the tendency to prioritize increased selection efficiency and breeding outputs over the extent to which diverse preferences, practices and contributions can be built into data practices that inform plant breeding. This generated a trend in twenty-first century science-led agriculture towards conceptualising the growth of molecular breeding and climate-smart agriculture as unrelated – or even antithetical – to farmer engagement and participatory methodologies. As we already discussed, this need not be the case; and yet such a conceptual commitment has severe implications for what responsible data practice is taken to be, and by whom. Responding to a wider set of gendered and other agricultural needs in diverse environments, for instance, will require data mining germplasm collections to spotlight non-elite materials that contain traits of potentially greater relevance to these needs, and then dedicating significant pre-breeding efforts to adapt this material such that they too can benefit from more intensive population improvement (cf. Fadda et al., 2020). When taking such an approach, concepts such as genetic gain may well retain an important role, but may not necessarily feature as a central priority around which all other activity is organised. There remains substantial scope for data-intensive breeding in the service of agricultural development and gender equality, without necessarily structuring major breeding decisions around an algorithmic rationality that conceptualises decision-making as a comparison of metric values.

Similarly, efforts such as the TK and Biocultural Labels are associated with a reconceptualization of the very workflow underpinning data-intensive methods, which challenge the idea of data as raw materials from which knowledge can be extracted through a linear process of analysis and interpretation, and instead support a cyclical understanding of how data are generated and used, with multiple feedback loops between data subjects, data collectors, data stewards and data users. As Devare and colleagues also point out in their chapter, considering a variety of perspectives on how data are used in research, and which workflows can best support the production of reliable knowledge, is a fundamental part of data linkage efforts. Conceptual commitments made in data science, plant breeding and agriculture typically structure – and constrain – the uses of plant data, their paths of travel, and the choice of participants (and related types of expertise) in data collection, circulation and use. When addressing responsible data linkage in the plant and agricultural sciences, it is therefore necessary to consider how different visions of data use may be amenable to achieving different goals, be they economic development, equality of participation or justice in food production systems.

4   Conclusion: Training for the Future

We have reviewed how transnational plant data circulation and re-use is subject to countless constraints and strictures from a variety of perspectives and levels of governance and monitoring. Far from being discouraging, acknowledging these constraints should foster an imagination of what may constitute socially responsive, sustainable data linkage systems, and an alertness to the variety of conceptual underpinnings such systems could have (think about the visionary quality of the European Open Science Cloud, whose attempt to federate existing research data infrastructures across Europe constitutes an unprecedented feat of data linkage within a highly disruptive and at times openly hostile political and economic environment). At a moment of enormous technological, social and geo-political transformation, it is particularly important to challenge long-held assessments of the impact of structural constraints on available courses of action. This is especially important since, at a practical level, the space to consider such alternative conceptualisations has radically shrunk in agricultural research within the last few decades, as seen in the relative decline of participatory methodologies.

A key tool to push this forward is education. We argued that the current data-intensive model of agricultural research and development is predicated on a distinctive set of conceptual and normative visions for agriculture, and that multiple forms of governance need to be implemented in order to enact responsible data linkage practices. By and large, however, neither scientists nor other stakeholders are trained to identify and evaluate such assumptions or to consider their implementation across different technical, social and political contexts. And yet the importance of training tools and programmes for data scientists, farmers, breeders, researchers working in this space – as well as policy-makers and businesses – was already evident during the Green Revolution, where training programmes such as those devised by the CGIAR centres were very effective in furthering a specific understanding of agricultural development and its implementation on the ground. What would it take to operate at the same scale in the realm of data? Who would be responsible for such training, to guarantee that responsible data practice sits at its core? Should industrial and corporate efforts incorporate these forms of education, and how? And can this be achieved without an acritical commitment to exclusionary approaches to genetic conservation and agricultural development? This volume does not provide exhaustive answers to these questions, but it is our hope that readers will be convinced of the significance of querying what constitutes responsible data linkage in the first place, and take inspiration from the multiple efforts described by our contributors in devising ever more data infrastructures and data sharing solutions to foster sustainable agriculture and a healthier planet.

References

Arnaud, E., Laporte, M. A., Kim, S., et al. (2020). The ontologies community of practice: An initiative by the CGIAR platform for big data in agriculture. *Patterns, 1*, 100105. https://www.cell.com/patterns/pdf/S2666-3899(20)30139-2.pdf

Aubry, S. (2019). The future of digital sequence information for plant genetic resources for food and agriculture. *Frontiers in Plant Science, 10*, 1046. https://doi.org/10.3389/fpls.2019.01046

Bietti, E. (2020). From ethics washing to ethics bashing: A view on tech ethics from within moral philosophy. *Proceedings to ACM FAT* Conference (FAT* 2020)*. https://papers.ssrn.com/sol3/papers.cfm?abstract_id=3513182

Bonneuil, C. (2019). Seeing nature as a 'universal store of genes': How biological diversity became 'genetic resources', 1890-1940. *Studies in History and Philosophy of Biology and Biomedical Science, 75*, 1–14. https://doi.org/10.1016/j.shpsc.2018.12.002

Borgman, C. L. (2019). The lives and after lives of data. *Harvard Data Science Review, 1*(1). https://doi.org/10.1162/99608f92.9a36bdb6

Boulton, G., Campbell, P., Collins, N., Hall, W., Elias, P., Laurie, G., O'Neill, O., Rawlins, M., Thornton, J., Vallance, P., & Walport, M. (2012). *Science as an open enterprise* Royal Society Science Policy Centre report 02/12. The Royal Society.

Coppens, F., Wuyts, N., Inzé, D., & Dhondt, S. (2017). Unlocking the potential of plant phenotyping data through integration and data-driven approaches. *Current Opinion in Systems Biology, 4*, 58–63. https://doi.org/10.1016/j.coisb.2017.07.002

Curry, H. (2017). From working collections to the World Germplasm Project: Agricultural modernization and genetic conservation a the Rockefeller Foundation. *History and Philosophy of the Life Sciences, 39*, 5. https://doi.org/10.1007/s40656-017-0131-8

Curry, H., & Leonelli, S. (2022). Describing crops in the CGIAR era. In H. Curry & T. Lorek (Eds.), *Research as Development: Historical Perspectives on Agricultural Science and International Aid in the CGIAR Era.* Cambridge University Press.

Dobrescu, A., Giuffrida, M. V., & Tsaftaris, S. A. (2020). Doing more with less: A multitask deep learning approach in plant phenotyping. *Frontiers in Plant Science.* https://doi.org/10.3389/fpls.2020.00141

Edwards, P. N. (2010). *A Vast Machine: Computer Models, Climate Data, and The Politics of Global Warming.* MIT Press.

Edwards, P. N., Mayernik, M. S., Batcheller, A. L., Bowker, G. C., & Borgman, C. L. (2011). Science friction: Data, metadata, and collaboration. *Social Studies of Science, 41*(5), 667–690.

Fadda, C., Mengistu, D. K., Kidane, Y. G., Dell'Acqua, M., Pe, M. E., & van Etten, J. (2020). Integrating conventional and participatory crop improvement for smallholder agriculture using the seeds for needs approach: A review. *Frontiers in Plant Science, 11*, 559515. https://doi.org/10.3389/fpls.2020.559515

Ferris, L., & Rahman, Z. (2016). *Responsible data in agriculture.* Global Open Data in Agriculture & CABI.

Fullilove, C. (2017). *The profit of the earth: The global seeds of American agriculture.* University of Chicago Press.

Giuffrida, M. V., Chen, F., Scharr, H., & Tsaftaris, S. A. (2018). Citizen crowds and experts: Observer variability in image-based plant phenotyping. *Plant Methods, 14*, 12. https://doi.org/10.1186/s13007-018-0278-7

Harfouche, A. L., Jacobson, D. A., Kainer, D., et al. (2019). Accelerating climate resilient plant breeding by applying next-generation artificial intelligence. *Trends in Biotechnology, 37*(11), 1217–1235. https://doi.org/10.1016/j.tibtech.2019.05.007

Hayden, C. (2003). *When Nature Goes Public: The Making and Unmaking of Bioprospecting in Mexico.* Princeton University Press.

Henkhaus, N., Bartlett, M., Gang, D., et al. (2020). Plant science decadal vision 2020-2030: Reimagining the potential of plants for a healthy and sustainable future. *Plant Direct, 4*(8), e00252. https://doi.org/10.1002/pld3.252

Hewson, M. (1999). Did global governance create informational globalism? In M. Hewson & T. J. Sinclair (Eds.), *Approaches to Global Governance Theory.* State University of New York Press.

Hilgartner, S. (2017). *Reordering life: Knowledge and control in the genomics revolution.* MIT Press.

Hudson, M. N. A., Garrison, R. S., et al. (2020). Rights, interests and expectations: Indigenous perspectives on unrestricted access to genomic data. *Nature Reviews. Genetics, 21*, 377–384. https://doi.org/10.1038/s41576-020-0228-x

Kind, C. (2020, August 23). *The term 'ethical AI' is finally starting to mean something.* Venture Beat. https://venturebeat.com/2020/08/23/the-term-ethical-ai-is-finally-starting-to-mean-something/

Kloppenburg, J. R. (2004). *First the seed: The political economy of plant biotechnology,* 1492–2000 (2nd ed.). University of Wisconsin Press.

Krige, J. (Ed.). (2022). *Knowledge flows in a global age: A transnational approach.* University of Chicago Press.

Lagoze, C. (2014). Big Data, data integrity, and the fracturing of the control zone. *Big Data & Society, 1*(2), 2053951714558281.

Leonelli, S. (2016a). *Data-centric biology: A philosophical study.* Chicago University Press.

Leonelli, S. (2016b). Locating ethics in data science: Responsibility and accountability in global and distributed knowledge production. *Philosophical Transactions of the Royal Society: Part A, 374,* 20160122. https://doi.org/10.1098/rsta.2016.0122

Leonelli, S. (2019). Data governance is key to interpretation: Reconceptualising data in data science. *Harvard Data Science Review,* 1.1. https://doi.org/10.1162/99608f92.17405bb6

Leonelli, S. (2020). Learning from data journeys. In S. Leonelli & Tempini (Eds.), *Data journeys in the sciences.* Springer. Open Access. https://www.springer.com/gp/book/9783030371760

Leonelli, S. (2022). How data cross borders: Globalising plant knowledge through transnational data management and its epistemic economy. In J. Krige (Ed.), *Knowledge flows in a global age: A transnational approach.* University of Chicago Press.

Liggins, L., Hudson, M., & Anderson, J. (2021). Creating space for Indigenous perspectives on access and benefit-sharing: Encouraging researcher use of the local contexts notices. *Molecular Ecology, 30*(11), 2477–2482. https://doi.org/10.1111/mec.15918

Lin, D., Crabtree, J., Dillo, I., et al. (2020). The TRUST principles for digital repositories. *Nature Scientific Data, 7,* 144. https://doi.org/10.1038/s41597-020-0486-7

Maxson Jones, K., Ankeny, R. A., & Cook-Deegan, R. (2018). The Bermuda Triangle: The pragmatics, policies, and principles for data sharing in the history of the human genome project. *Journal of the History of Biology, 51,* 693–805. https://doi.org/10.1007/s10739-018-9538-7

Mazzucato, M. (2019, October 2). *Preventing digital feudalism.* Project Syndicate. Available at https://www.project-syndicate.org/commentary/platform-economy-digital-feudalism-by-mariana-mazzucato-2019-10

Miles, C. (2019). The combine will tell the truth: On precision agriculture and algorithmic rationality. *Big Data & Society, 6*(1). https://doi.org/10.1177/2053951719849444

Morgera, E., Switzer, S., & Geelhoed, M. (2020). *Possible ways to address digital sequence information – Legal and policy aspects.* Consultancy study for the European Commission. Strathclyde Centre for Environmental Law and Governance.

Murtagh, M. J., Demir, I., Jenkings, K. N., et al. (2012). Securing the data economy: Translating privacy and enacting security in the development of DataSHIELD. *Public Health Genomics, 15,* 243–253. https://doi.org/10.1159/000336673

Owen, R., Macnaghten, P., & Stilgoe, J. (2012). Responsible research and innovation: From science in society to science for society, with society. *Science and Public Policy, 39,* 751–760.

Pommier, C., Michotey, C., Cornut, G., et al. (2019). Applying FAIR principles to plant phenotypic data management in GnpIS. *Plant Phenomics, 2019,* 1671403. https://doi.org/10.34133/2019/1671403

Reiser, L., Harper, L., Freeling, M., Han, B., & Luan, S. (2018). FAIR: A call to make published data more findable, accessible, interoperable and reusable. *Molecular Plant, 11,* 1105–1108. https://doi.org/10.1016/j.molp.2018.07.005

Ribaut, J.-M., & Ragot, M. (2019). Modernising breeding for orphan crops: tools, methodologies, and beyond. *Planta, 250,* 971–977. https://doi.org/10.1007/s00425-019-03200-8

Rife, T. W., & Poland, J. A. (2014). Field book: An open-source application for field data collection on android. *Crop Science, 54,* 1624–1627. https://doi.org/10.2135/cropsci2013.08.0579

Scott, J. (1998). *Seeing like a state: How certain schemes to improve the human condition have failed.* Yale University Press.

Shaw, F., Etuk, A., Minotto, A., et al. (2020). COPO: A metadata platform for brokering FAIR data in the life sciences. *F1000 Research.* [Version 1; peer review: 1 approved, 1 approved with reservations], *9,* 495. https://doi.org/10.12688/f1000research.23889.1

Shiva, V. (2016). *Seed Sovereignty, Food Security*. North Atlantic Books.

Soto Laveaga, G. (2009). *Jungle Laboratories: Mexican Peasants, National Projects, and The Making of The Pill*. Duke University Press.

Sperschneider, J. (2019). Machine learning in plant-pathogen interactions: Empowering biological predictions from field scale to genome scale. *The New Phytologist, 228*(1), 35–41. https://doi. org/10.1111/nph.15771

Strasser, B. (2019). *Collecting experiments: Making big data biology*. University of Chicago Press.

Tardieu, F., Cabrera-Bosquet, L., Pridmore, T., & Bennett, M. (2017). Plant phenomics, from sensors to knowledge. *Current Biology, 27*, R770–R783. https://doi.org/10.1016/j.cub.2017. 05.055

Wang, H., Cimen, E., Singh, N., & Buckler, E. (2020). Deep learning for plant genomics and crop improvements. *Current Opinion in Plant Biology, 54*, 34–41. https://doi.org/10.1016/j.pbi. 2019.12.010

Welch, E., Louafi, S., Carroll, S.R. et al. (2021). *Post COVID-19 implications on genetic diversity and genomics research & innovation: A call for governance and research capacity*. White Paper (released 6 July 2021). http://www.fao.org/3/cb5573en/cb5573en.pdf

Wilkinson, M. D., Dumontier, M., Aalbersberg, I. J., et al. (2016). The FAIR guiding principles for scientific data management and stewardship. *Nature Scientific Data, 3*, 160018. https://doi.org/ 10.1038/sdata.2016.18

Williamson, H. . F., Brettschneider, J., Caccamo, M., et al. (2021). Data management challenges for artificial intelligence in plant and agricultural research. *F1000*. [Version 1; peer review: 1 approved with reservations]. https://doi.org/10.12688/f1000research.52204.1

Part I
Challenges from/for the Field: Data Linkage Across Crops, Seeds and Field Experiments

Preface

The first part of the book interrogates the ways in which the biological features of plants and their environments inform the data systems devised to capture, communicate and analyze such features. A key concern is how to document the extensive variability that exists not only among plant specimens and varieties, but also among types of soils, climatic conditions, agricultural systems and human uses of plants. Responsible data linkage involves explicitly asking how different kinds of variability feature in data systems, how the success of such systems is assessed, and who is involved in – and affected by – decisions around how data infrastructures document plants and their environments. The authors in this part consider several aspects of data linkage, including: how scientific disagreements on how to set up a data infrastructure are managed, and with which implications; how data coming from a wide variety of breeding initiatives are harmonized, and what happens when harmonization fails; how socio-political conditions affect which seeds come to be defined and valued as objects of analysis, which in turn determines the characteristics of related data collections; how the apparatus devised to extract and manage data intersect with breeding practices on the ground; whether data management methods focused on digital sequence information can fruitfully serve broader phenotype and environmental datasets; and whether and how data systems can and should pay more attention to marginal environments where uniform crop varieties do not perform consistently, rather than prioritizing data collection on selected high-yield varieties.

Theory-Ladenness as a Problem for Plant Data Linkage

Gregory Radick

Abstract This paper draws upon the history of scientific studies of inheritance in Mendel's best-remembered model organism, the garden pea, as a source of two parables – one pessimistic, the other optimistic – on the challenges of data linkage in plants. The moral of the pessimistic parable, from the era of the biometrician-Mendelian controversy, is that the problem of theory-ladenness in data sets can be a major stumbling block to making new uses of old data. The moral of the optimistic parable, from the long-run history of studies at the John Innes Centre of aberrant or "rogue" pea varieties, is that an excellent guarantor of the continued value of old data sets is the availability of the relevant physical materials – in the first instance, the plant seeds.

1 Introduction

Proposals that point the way forward are nowadays routinely called "roadmaps." But on Richard Harrison and Mario Caccamo's showing elsewhere in this volume, the data-world of the future for agricultural plants in Britain is more handily visualized with an image akin to the map of the London underground (Harrison & Caccamo, 2022, Fig. 7). Instead of tube stations we see different kinds of data – genomics data, environmental/simulation data, phenomics data, plant breeding/trial data, Recommended Lists data, Distinctness Uniformity Stability data, Value for Cultivable Use data, Official Seed Testing Station data – plus a range of activities and systems where those data may be integrated and acted upon: the seed certification scheme; the growing and evaluating of certified seeds on farms; the collecting of national statistics bearing on productivity, performance, and environmental impact; and the tracking of seeds, and the profits accruing from innovations in their development, through a distributed ledger system. Looping between these are brightly

G. Radick (✉)
School of Philosophy, Religion and History of Science, University of Leeds, Leeds, UK
e-mail: G.M.Radick@leeds.ac.uk

H. F. Williamson, S. Leonelli (eds.), *Towards Responsible Plant Data Linkage: Data Challenges for Agricultural Research and Development*,
https://doi.org/10.1007/978-3-031-13276-6_2

coloured one-way arrows, mostly solid, occasionally dotted, with the caption below the image spelling out the envisaged benefits. Genomics data, for example, will feed into the determination of how distinct, uniform and stable a variety is (these being the standard criteria for the award of intellectual property rights to the breeder) as well as how valuable it is for cultivatable use, in a way that helps integrate the data generated from these exercises and so increases their value for seed certification.

Here is an ideal of frictionless movement between various kinds of plant data across time – an ideal also encapsulated in phrases such as "historical data mining." Plant data on such a vision is like oil: a valuable resource that only needs tapping to become potentially useful. Between ideal and reality there are, of course, impediments. But nothing here suggests that these are other than infrastructural, as when data are locked away in filing cabinets, or in journals that no one has yet digitized, or on floppy disks written in an outmoded computer language, readable on machines that no one runs anymore or – almost as bad – that are run by firms charging exorbitant fees for the service. These are, in principle, soluble problems, some of them technical, others social, still others as much technical as social. Solve these problems, open up access to the data, and the data will start to flow along the mapped-out channels, to the good of future food security and the knowledge that will underpin it.

I want to suggest in what follows that there may be another class of impediments worth being reflective about: *intellectual* ones. I will dwell in particular on what, for historians and philosophers of science, is an especially conspicuous candidate: the problem that data are, in the canonical phrase, "theory-laden." The basic thought here is that, in important ways, the categories used in classifying observations, and the choices made about which observations to file under which categories, can reflect background theory (see, e.g., Hanson, 1958). By way of making this abstract issue concrete, I'm going to offer two stories, both involving that exemplary Mendelian plant, the garden pea, *Pisum sativum*. Because I intend to draw morals from these stories, I'm calling them "parables." The moral from my first story will be pessimistic: the problem of data theory-ladenness needs to be taken seriously. But the moral from my second story is optimistic: one way to overcome the problem of data theory-ladenness is to retain access to the seeds of the plants featuring in historic data. At this point my chapter will intersect with other chapters in the volume, notably those by Helen Curry, Courtney Fullilove and Richard Ostler on the seed banks that are sometimes labelled – in splendidly theory-laden manner – "germplasm collections." As we shall see, the fact that seeds are more than just containers for genomes can be consequential.

2 The Pessimistic Parable

"The rediscovery of Mendelian genetics ushered in an agricultural revolution. For the first time, varieties that combined performance characteristics were systematically developed, based upon the principles of heredity and the genetic control of characters." So begins the abstract originally circulated with the chapter by Harrison and Caccamo. They are based at the National Institute of Agricultural Botany (NIAB) in Cambridge, and, in the role they assign to revolutionary Mendelism in the making of modern agricultural success, keep faith with NIAB's foundational vision. In the *Memoranda on the Establishment of a National Institute of Agricultural Botany* published in November 1918, A. B. Bruce, superintending inspector for the Board of Agriculture and Fisheries, wrote:

> The undoubted success of plant breeding work at Cambridge is due primarily to the fact that in recent years an entirely new science has been built up as the result of the discoveries made by the monk Mendel in the early sixties. At the time these discoveries were overlooked, and it is only in the last ten years or so that they have received proper recognition. Without going into scientific details, the effect of Mendel's and subsequent work can be summed up by saying that it is now possible to *make* a new plant possessing valuable economic qualities. Just as an architect in building a house has at his disposal different kinds of building materials, so the modern plant breeder can make a new plant out of, as it were, the fragments of another. It will readily be recognised what a powerful weapon this new discovery has placed in the hands of the agricultural botanist. . . . Now we no longer require to wait for nature to act; we can deliberately set about manufacturing what we require.

Bruce proceeded to illustrate with examples from the work of the leading exponent of Mendelian breeding, Rowland Biffen, recently installed as Director of the new Plant Breeding Institute, also in Cambridge. Biffen's first great success was "Little Joss," a high-yield, rust-resistant variety of wheat created by crossing a high-yield but rust-susceptible English variety with a rust-resistant but low-yield Russian variety. Little Joss, Bruce reported, "has now been on the market for nearly ten years, and so far has shown no tendency to revert either to the low yielding character of one of its parents, or to the liability to Rust of the other." More recently, Bruce went on, Biffen had introduced other new varieties of wheat, among them "Yeoman," a Mendelian synthesis of English high-yield with the superior baking quality associated until then with Canadian varieties (Bruce, 1918: 12–13).

When Bruce sang the praises of Mendelian breeding, it had been 18 years since Mendel's "Experiments on Plant Hybrids" (Mendel, 1866) had become an unexpected sensation among botanical hybridzers. By the time of Little Joss's release in 1908, a new science of heredity elaborated around Mendel's paper – the science later known as "genetics," but at this time mainly known as "Mendelism" – had taken off internationally, thanks above all to the efforts of William Bateson and his students at Cambridge, Biffen not least. From then until now, Mendelian principles have been fundamental to the organization of knowledge of heredity. Around the world, at every level of education, the standard point of entry into a scientific understanding of heredity is Mendel and his peas, in a form that Bateson first made teachable (Radick, in press). One key to Mendel's success, students learn, was his focusing in on traits that come in distinct either/or versions: seed colour in the pea as either yellow or green; seed shape as either round or wrinkled; and so on. Another was his assiduity in ensuring that his parental stocks were pure and so "true-breeding," i.e., the yellow-seeded stocks only ever produced yellow seeds, and the green-seeded stocks only every produced green seeds. Yet another was the care he took in ensuring uniformity in the treatment of the large number of plants he dealt with. Thus did Mendel get the data which enabled him to discover what had eluded his predecessors: dominance and recessiveness; the 3:1 ratio of dominant to recessive plants in the second generation of hybrids; and, crucially, the production by hybrid plants of gametes that were not themselves hybrid but were pure for one or the other of the trait-versions (see, e.g., Campbell, 1993: 258–67).

All of that is familiar, indeed foundational. Much less familiar is a scorching critique of all of that from W. F. R. Weldon (1860–1906), who at the time of the Mendelian rediscovery was Linacre Professor of Zoology at Oxford. On a Weldonian perspective, what you gain in control via Mendelian breeding experiments you lose in generality. Yes, if you assiduously expunge from your parental stocks all the variability except for the single either/or difference that interests you, and you then carry out your experimental breeding under uniform environmental conditions, you may well get, at least roughly, the patterns that led Mendel to infer what he did. But take different decisions about what to focus on, what to exclude, and which environment to impose, and you could well find yourself examining different patterns, which could in turn lead you to different, even opposite, conclusions (Weldon, 1904–1905).

In the garden pea, for example, is it really the case that yellowness is dominant to greenness, across the board? And what about the Mendelian corollary that if a seed is green, it cannot harbour any yellow-making factor? Yes, in the particular purified varieties that Mendel worked with, those conclusions seemed to hold. But when Weldon surveyed the world of commercial pea breeding, he found enormous variability – a continuous spectrum of colours stretching between yellow and green, a smooth gradation from extreme roundedness to extreme wrinkledness – as well as longer-range inheritance patterns that, under Mendelian theory, were impossible and so invisible (Weldon, 1902). For Weldon, all of this heterogeneity in traits and their inheritance was not just intelligible but, in a modest way, predictable, given what experimental embryologists had learned in recent years about the role of context in conditioning development. In Weldon's view, the twentieth century deserved a science of heredity that took this heterogeneity, and the multiple, interacting causes that brought it about, as its subject matter – whereas Mendelism was set to treat it as a nuisance, and Mendelians, in line with their training, to categorize actual variability within the capacious categories that elementary Mendelism favoured (Radick, 2016).

For the most part Weldon's perspective on the theory-ladenness of Mendelian observations remains locked away in unpublished letters and manuscripts. The exception is the well-known suspicion that Mendel's pea data are "too good to be true": that is, the numbers he reported are improbably close to the ratios predicted by his theory, given the number of trials he did. Nowadays this suspicion is associated with the mathematical geneticist Ronald Fisher, who published a classic paper about it in the 1930s. But the discovery was Weldon's, made in 1901 and published in the same 1902 paper where he also published photographic plates showing the variability he had found in pea-seed colours and shapes. The suspicion became an object of public hand-wringing and finger-wagging over the possibility that Mendel was guilty of fraud only from the mid-1960s (Radick, 2022). Since then there has emerged a small industry devoted to examining the case (Franklin et al., 2008). Amidst the tremendous ingenuity and technicality, the larger lesson that Weldon drew has mostly been lost: the problem stemmed not from Mendel's character but from his categories – binary categories upon which Mendel erected a theory of heredity which ignored context as a source of variability, and which in turn directed him to classify traits according to his either/or scheme (Radick, 2015). When confronted with a trait not unambiguously belonging to one side or the other of an either/or classification, he probably judged it to belong on whichever side made for tidier ratios. (It's been shown in a classroom experiment with students that if you give them three instead of two categories to work with in classifying pea seeds – say, "yellow," "green," "ambiguous" – they will use all three categories; Root-Bernstein, 1983.)

So: whenever we are dealing with historic plant data from the post-1900 period, we need at least to consider whether what is reported is not just what any competent observer would have reported, but is – sometimes subtly and sometimes unsubtly – inflected with Mendelian expectations, and/or with Mendelism's legacy for intellectual property rights: the insistence on distinctness, uniformity and stability (see Berry, forthcoming; Kochupillai & Köninger, forthcoming).

3   The Optimistic Parable

Given all the natural heterogeneity actively controlled for in a Mendelian experimental garden or laboratory, one might predict a "return of the repressed" once the products of Mendelian experimental breeding enter the wider world. A related prediction is that, when the repressed does *not* return, it's thanks to some combination or other of two sorts of remedy. One is selection. By selecting lineages in which Mendelian traits of interest get expressed most fully and reliably, across the broadest range of environments, the skilled breeder gradually builds up, and builds in, whatever internal context best buffers trait expression in the new breed against the slings and arrows of outrageous fortune. The other remedy is, in some form or other, to extend the controls beyond the limits of the experimental space.

For all Biffen's promotion and, indeed, self-promotion as *the* Mendelian breeder, he relied heavily on selection, as Berris Charnley has noted (Charnley, 2011: 144–5). With Little Joss, it worked a treat. But with Yeoman, selection proved insufficient to ensure the stability of the released variety. Farmers who grew the seed eventually found a noticeable proportion of "rogue" plants – that is, plants departing from the advertised type, in the direction of older, lesser wheat stocks. By the early 1920s, Yeoman's rogue problem had become so bad that Biffen was being quoted in *Nature* as saying "the sooner Yeoman is off the market the better." Biffen placed the blame on an external source: the threshing machines that travelled from farm to farm, contaminating Yeoman-planted fields with seed from older stocks. By the time a successor breed, Yeoman II, was released in 1924, a new, NIAB-run distribution system was in place, with the seed sold in sealed sacks bearing NIAB's emblem. As an anti-contamination effort, it was a modest step. But we do well to see in it the start of the larger-scale control efforts to come, in the form of the fertilizers, pesticides, and herbicides sold along with post-Biffenian seeds and required in order to make them flourish as advertised (Charnley & Radick, 2013).

None of this would have surprised Weldon. He had a lively sense of the commercial value of selection in creating breeds that gave farmers what they'd paid for whatever the vicissitudes of environment (Radick, in press). He also knew how badly breeders often struggled with rogues when attempting to establish varieties sufficiently differentiated from starting stocks to count as new. In the 1902 paper discussed above, he even documented persistent controversies among pea breeders due to rogue troubles (Weldon, 1902: 246–50; Charnley, 2013; Charnley & Radick, 2013: 229).

Problematic for breeders, rogue peas plants are nevertheless the stars of my optimistic parable. Perhaps Weldon's most attentive reader was Bateson, whose *Mendel's Principles of Heredity: A Defence* (1902) is at heart an extended take-down of Weldon's paper (Bateson, 1902). Addressing breeders at a New York hybridization conference a few months after the book had come out, Bateson trumpeted Mendelism's solution of the rogue problem as one of its greatest attractions for his audience. According to Bateson, once it was understood that a plant showing a dominant trait could be either homozygous or heterozygous, and care was taken to ensure breeding from homozygotes only, the tragedy of rogues would disappear. But Bateson well new that the kinds of roguish returns that fascinated the likes of Weldon were not the absent-for-a-generation recessive traits featuring in Mendelian analyses but the absent-for-many-generations atavisms which Mendelian analyses, with their indifference to ancestors beyond the true-breeding parents, did not even register, let alone explain (Bateson, 1904; Radick, 2013).

In the heat of battle with Weldon, Bateson declared Mendelism victorious over rogues. When that victory was secure, however, Bateson allowed that maybe there was indeed more to be learned about rogues. During the 1910s, when he directed the newly founded John Innes Institute, the study of rogue peas became a major research project, conducted in collaboration with Caroline Pellew. Bateson and Pellew became convinced that though some rogues could be explained away as due to contamination or heterozygosity, not all could. As they put it in a 1915 paper:

> The term "rogue" is applied by English seed growers to any plants in a crop which do not come true to the variety sown.... When peas are grown for seed on a commercial scale it will be readily understood that untrue plants are introduced in various ways, mixture, crossing by insects, and the persistent recurrence of a recessive form being the most obvious sources of such plants.... but the facts preclude the supposition that the special rogues with which we are here concerned are introduced either by mixture or crossing, nor can they be regarded as recessives coming from a heterozygote in the ordinary sense.

When Bateson and Pellew crossed these "special rogues" with normal peas, the hybrids all showed the rogue phenotype (indicated that rogueishness was dominant). On Mendelian expectations, the self-fertilizing of these hybrids should have produced offspring showing a 3-to-1, rogue-to-normal ratio. Instead, however, all of the offspring showed the rogue phenotype (Bateson & Pellew, 1915, quotation on 13–4; Charnley, 2011: 114–20).

Bateson and Pellew never got to the bottom of what was behind the rogueish characters and inheritance patterns of the pea plants they collected. But their research was well regarded, so much so that in 1922–3 Bateson served as an expert witness in a court case on whether a pea breeder was liable for the extreme proportion of rogues in some seed bought from him (Radick, 2013).

The rogue pea data Bateson and Pellew reported have remained accessible from their day to ours. What has kept their data tantalizing is not just the gradual emergence of a body of theorizing and technique suitable for investigating such cases (Le Goff et al., 2021: 38), but the prospect of getting beneath the data by working with similar-looking and similar-behaving pea plants as these have come to the attention of breeders. That was true in the 1960s and 1970s, when two John Innes researchers in succession, Kenneth Dodds and Peter Matthews, had a go – but with little to show for it (Matthews, 1973). And it was true in the 2010s, when the agronomist-geneticist José Leitão, based at the Laboratory of Genomics and Genetic Improvement in Faro, Portugal, became intrigued (Anon, 2021). What piqued Leitão's interest was the resemblance he noticed to similar inheritance patterns in other plants known to be due not to genetic differences but to epigenetic ones – that is, to differences in the immediate biochemical environment of the DNA sequence. He honed in on pea seeds held at the GermPlasm Resource Unit of the John Innes Centre (as it is now called) from two lines: a non-rogue variety, called Onward; and a rogue variety derived from Onward and showing the same off-type characters which Bateson and Pellew had studied (known as "rabbit ears," because the narrow, pointed leaflets and stipules give the plants a rabbity aspect). Analysis of DNA in the two lines revealed them to be highly similar genetically. Epigenetically, however, they were different, with Leitão's team identifying a number of methyl groups present in the epigenome of the rogue line but absent from the non-rogue line (Santo et al., 2017).

Are the epigenetic differences *responsible* for the differences in character? The answer remains elusive. Leitão's team managed to carry out expression studies on fourteen pairs of genome segments, methylated (from the rogue line) and unmethylated (from the non-rogue line) – but no significant differences in gene expression were found. In their paper Leitão and his colleagues suggested that perhaps resolution lies with analysis of larger segments of genome/epigenome:

additional studies are needed to unveil the biological consequences of the identified differential methylation. For the moment, we can only speculate that the observed alterations in DNA methylation, and eventual modifications in chromatin conformation, probably spread over larger genomic regions encompassing the identified sequences, and eventually affect the expression of other, surrounding, genes. (Santo et al., 2017: 6)

It is early days for the study of the molecular epigenetics of rogue peas (see too Pereira & Leitão, 2021). But they already look not just "non-Mendelian" but potentially Weldonian, in that the key to understanding them may turn out to lie in differences in internal context of the sort routinely stripped out in the course of

Mendelian standardization. And that key will have been found because investigators had access not merely to historic data but to plausible surrogates for historic plant material.

4   Conclusion

So, seeds matter, not just for all the familiar reasons, but for what access to them can do for anyone wishing to make new uses of old plant data. To say that is not, of course, to say that *only* seeds matter, as though contexts for DNA are interesting up to the seed-coat barrier but not beyond it. Undoubtedly, my second parable would be more fully illustrative of the moral that I wish to draw from it had the John Innes Centre looked after its seeds *in situ* rather than *ex-situ*; had the soil and climatic conditions under which the rogue pea seeds studied by Bateson and Pellew proved somehow indispensable for the expression of the rogue phenotype; and had, over the decades, the seeds and the conditions alike been subjects of rigorous monitoring programs, enabling detection, and remedy, of any deviations. Even so, the rogue pea phenotype's depending not on genes but on extra-genetic context suffices to under-score the point that, when it comes to dealing with the theory-ladenness of old data, the greater our access to the original materials that generated that data, or to plausible surrogates, in all their contextualized complexity, the better, because the less beholden we are to old conceptual choices that we might now want to question.

How much better? On the one hand, as have seen, ours is a scientific agriculture that grants to the systematic study of phenotypic plasticity not just a name ("phenomics") but a place on the data-linkage map of the future. Context looks well catered for already, thank you. On the other hand, that map is one where all the data generated and integrated so frictionlessly support the development of plant varieties which are distinct, uniform and stable. As another contributor to this volume, Mrinalini Kochupillai, has emphasized, the commercial promotion of varieties meeting these criteria has been a disaster for global crop biodiversity, with knock-on effects for human health and for the environment, not least because chemical "inputs" are typically part of the package that farmers buy when they abandon local landraces for commercial varieties (Kochupillai & Köninger, forth-coming). There is room, then, even in the age of phenomics, for taking a much more expansive view of what our duties are when it comes to the contexts in which the genes in our seeds have their effects: duties of conservation, curiosity and care.

Let me end with a story that I learned from Kochupillai's brilliant 2016 book *Promoting Sustainable Innovations in Plant Varieties*. There she wrote about Albert Howard, a Cambridge-trained agricultural botanist from just before the Bateson-Biffen era (Kochupillai, 2016: 84, 90). Howard went on to become a scientific student of traditional agriculture in India. In the counterfactual history that no one has written in which world agriculture in the twentieth century went organic rather than chemical, Howard is the Norman Borlaug figure. According to Kochupillai, Howard reported that Indian farmers in the 1930s were getting sugarcane yields that, she says, have not been surpassed even today. That is an interesting datum. But what would be more interesting still would be an attempt to recreate that feat, using seeds descended from those varieties in use in the 1930s as well as the "green manure" that Howard wrote about, with the seeds planted and the manure applied in the soil types and climatic conditions where the sugarcane that he observed was grown. The fields growing those seeds under those conditions would be true grounds for optimism.

Acknowledgments Many thanks to Sabina Leonelli, Hugh Williamson and Helen Curry for helpful comments on an early draft; to Tina Barsby, then CEO at NIAB, for sending me a copy of the foundational pamphlet containing Bruce's remarks; to Mike Ambrose, then at the John Innes Centre, for alerting me to Santo et al. (2017); to José Leitão for illuminating discussion of his research; and to an anonymous reviewer for helpful comments on a late draft.

References

Anonymous. (2021). *The mystery of the Rabbit-Eared Rogues*. John Innes Centre. https://www.jic. ac.uk. Accessed 10 Nov 2021.

Bateson, W. (1902). *Mendel's principles of heredity: A defence*. Cambridge University Press.

Bateson, W. (1904). Practical aspects of the new discoveries in heredity. *Proceedings: International Conference on Plant Breeding and Hybridization. Memoirs of the Horticultural Society of New York, 1*, 1–9.

Bateson, W., & Pellew, C. (1915). On the genetics of 'rogues' among culinary peas (*Pisum sativum*). *Journal of Genetics, 5*, 13–36. plus plates.

Berry, D. J. (Forthcoming). Scientific activism and IP: How UK-based agroecology and plant synthetic biologists have challenged the status quo. In *Designing nature: Essays on intellectual property law*. Oxford University Press.

Bruce, A. B. (1918). The economic results of plant breeding. In L. Weaver et al. (Eds.), *Memoranda on the establishment of a National Institute of Agricultural Botany* (pp. 12–15).

Campbell, N. A. (1993). *Biology* (3rd ed.). Benjamin/Cummings.

Charnley, B. (2011). *Agricultural science, plant breeding and the emergence of a Mendelian system in Britain, 1880–1930*. PhD dissertation, University of Leeds.

Charnley, B. (2013). Seeds without patents: Science and morality in British plant breeding in the long nineteenth century. *Revue économique, 64*, 69–88.

Charnley, B., & Radick, G. (2013). Intellectual property, plant breeding and the making of Mendelian genetics. In C. MacLeod & G. Radick (Eds.), *Owning and disowning invention: Intellectual property and identity in the technosciences in Britain, 1870–1930* (Special issue of *Studies in History and Philosophy of Science, 44*) (pp. 222–233).

Franklin, A., Edwards, A. W. F., Fairbanks, D. J., Hartl, D. L., & Seidenfeld, T. (2008). *Ending the Mendel-Fisher controversy*. University of Pittsburgh Press.

Hanson, N. R. (1958). *Patterns of discovery: An inquiry into the conceptual foundations of science*. Cambridge University Press.

Harrison, R., & Caccamo, M. (2022). Managing data in breeding, selection and in practice: A hundred year problem that requires a rapid solution. In S. Leonelli & H. F. Williamson (Eds.), *Towards responsible plant data linkage: Data challenges for agricultural research and development* (ch. 3). Springer.

Kochupillai, M. (2016). *Promoting sustainable innovations in plant varieties*. Springer.

Kochupillai, M., & Köninger, J. (Forthcoming). Cast into the stones of international law: A critique of the UPOV standards and the underlying welfare and scientific assumptions they globalize. In

A. Metzger & H. G. Ruse-Khan (Eds.), *Intellectual property ordering beyond borders*. Cambridge University Press.

Le Goff, A., Allard, P., & Landecker, H. (2021). Heritable changeability: Epimutation and the legacy of negative definition in epigenetic concepts. *Studies in History and Philosophy of Science, 86*, 35–46.

Matthews, P. (1973). *Genetic studies on spontaneous and induced rogues in Pisum sativum*. PhD thesis, School of Biological Sciences, University of East Anglia.

Mendel, G. (1866). Versuche über Pflanzen-Hybriden. *Verhandlungen des naturforschenden Vereines in Brünn* 4, second part (*Abhandlungen*): 3–47. An outstanding translation and commentary by Staffan Müller-Wille and Kersten Hall is freely available from the British Society for the History of Science at http://www.bshs.org.uk/bshs-translations/mendel

Pereira, R., & Leitão, J. M. (2021). A non-rogue mutant line induced by ENU mutagenesis in paramutated rogue peas (*Pisum sativum* L.) is still sensitive to the rogue paramutation. *Genes, 12*, no. 1680.

Radick, G. (2013). The professor and the pea: Lives and afterlives of William Bateson's campaign for the utility of Mendelism. In C. MacLeod & G. Radick (Eds.), *Owning and disowning invention: Intellectual property and identity in the technosciences in Britain, 1870–1930* (Special issue of *Studies in History and Philosophy of Science, 44*) (pp. 280–291).

Radick, G. (2015). Beyond the 'Mendel-Fisher controversy': Worries about fraudulent data should give way to broader critiques of Mendel's legacy. *Science, 350*(9 October), 159–160.

Radick, G. (2016). Presidential address: Experimenting with the scientific past. *British Journal for the History of Science, 49*, 153–172.

Radick, G. (2022). Mendel the fraud? A social history of truth in genetics. *Studies in History and Philosophy of Science, 93*, 39–46.

Radick, G. (in press). *Disputed inheritance: The battle over Mendel and the future of biology*. University of Chicago Press.

Root-Bernstein, R. S. (1983). Mendel and methodology. *History of Science, 21*, 275–295.

Santo, T. E., Pereira, R. J., & Leitão, J. M. (2017). The pea (*Pisum sativum* L.) rogue paramutation is accompanied by alterations in the methylation pattern of specific genomic sequences. *Epigenomes, 2017*(1), 6.

Weldon, W. F. R. (1902). Mendel's laws of alternative inheritance in peas. *Biometrika, 1*, 228–254.

Weldon, W. F. R. (1904–1905). *Theory of inheritance*. Unpublished MS. Pearson/5/2/10/4, PP. Papers of Karl Pearson, UCL.

Managing Data in Breeding, Selection and in Practice: A Hundred Year Problem That Requires a Rapid Solution

Richard J. Harrison and Mario Caccamo

Abstract Following the rediscovery of Mendelian genetics, food supply pressures and the rapid expansion of crop varieties with defined performance characteristics, international systems were set up throughout the 20 C to regulate the trade of seed, the protection of intellectual property and the sale of productive varieties of key agricultural crops. These systems are a highly connected but largely linear set of processes. System changes are slow to be adopted due to the cascade of effects that structural alteration would have globally. Multi-omic technologies and the subsequent proliferation of data types used within modern breeding, offer the possibility to gain deeper insights into the performance characteristics of varieties. Current integration of data, standards and ownership structures limit their applications for wider purposes, both private and public. We explore how data within and between breeding programmes and the varietal approval and monitoring processes could be made FAIR. We examine what role expanded or aligned programmes of data collection and expanded trait evaluation at the point of varietal registration and evaluation, as well as on farm could have in ensuring the best linkage of public and private data to address some of the challenges society faces over the next 30 years with the required, rapid transition to sustainable agricultural systems.

1 Our Modern Food System

Key to developing any future system of data linkage in agriculture is the need to understand the structure of the current system. To fully understand how our current system of varietal registration, approval and certification has arisen and therefore some of the pitfalls and opportunities for more efficient data linkage and improved system function, it is also important to understand which drivers led to their establishment and how this has shaped the extant systems today and the flow of

R. J. Harrison (✉) · M. Caccamo
NIAB, Cambridge, UK
e-mail: richard.harrison@niab.com; mario.caccamo@niab.com

© The Author(s) 2023
H. F. Williamson, S. Leonelli (eds.), *Towards Responsible Plant Data Linkage: Data Challenges for Agricultural Research and Development*,
https://doi.org/10.1007/978-3-031-13276-6_3

37

data between them. It is also important to note that our current set of statutory and advisory systems for the registration and marketing of crops of major agricultural importance is intrinsically linked to the development and fate of the organisations that developed and implemented them. While this may seem unimportant, it serves as an indicator of how often function (or dysfunction) follows from and may be of importance when considering future alternations to both national and international organisations.

The modern food system is one that for a large part of the twentieth century prioritised productivity (Benton & Bailey, 2019). The integration of modern plant breeding, agronomy and mechanisation led to rapid and sustained productivity growth and has in-part allowed humanity across the globe to continue its shift from an agrarian society to an industrial society. Modern agriculture has been (along with modern medicine) a key contributor to the enormous reductions in global poverty and hunger (von der Goltz et al., 2020). The shift to industrialisation has also resulted in the primary energy source for human society to shift from photosynthesis-derived phytomass to fossil fuel, a trend that is broadly present in agriculture, as well as most other sectors of society and the economy. For most nations, there remains a linear relationship between fossil fuel energy usage and GDP, though there is emerging evidence that decoupling fossil fuel usage from GDP is possible (Haberl et al., 2020). Industrialisation and the ability to produce cheap food has led to a significant reduction in infant mortality, which in the short term has driven global population growth and overall global prosperity. Balanced against this is the overall lack of effective integration of the myriad externality costs of industrialised societies, including in the area of agriculture. Modern agriculture and the food system that it supports is currently responsible for between 10% and 30% of primary emissions. For agricultural crops, greenhouse gas emissions are largely due to soil-associated, microbially-driven nitrous oxide (N_2O) emission through excess fertiliser use (and the carbon dioxide (CO_2) used in the production of inorganic fertiliser through the Haber-Bosch process- the so-called input foot-print) and soil disturbance and carbon release in the form of CO_2 through tillage. In submerged cropping systems such as rice, the action of anaerobic methanogenic microbes in waterlogged soils leads to methane (CH_4) production (Smartt et al., 2016). There are also then the opportunity costs of agriculture, for example the loss of land for natural carbon sequestration through conversion to agriculture. There are then of course the onward uses of agricultural commodities, for example animal feed, which then leads to animal-associated CH_4 emissions. Beyond the narrow lens of emissions, it is also clear that through both land use change and the use of chemical controls of pests and diseases there is an overall decrease in the carrying capacity of the environment for many species, primarily due to ecosystem fragmentation and destruction (Dudley & Alexander, 2017). Agriculture is the largest contributor to biodiversity loss (Dudley & Alexander, 2017) and as such the need again to either find effective mitigation or simply to reduce the footprint of agriculture is required if we wish to reverse the current and ever declining viability of the ecosystem.

While all of this may seem unrelated to the use and linkage of data in agriculture, it is not. The regulatory and advisory systems that are in place in much of the world can directly shape the traits that are brought to market and at present in many places prioritise yield advantage (which is a key component of agricultural efficiency) under high input farming systems above other traits that deliver public and/or private goods. These include specific efficiency traits (i.e. nutrient use per unit of production), pest and disease resistance traits but also traits that may contribute to improved soil structure, reduction in nutrient loss to the wider environment, performance in mixed cropping systems etc. In this review we will put forward the case for how ensuring more effective data linkage can play a key role in designing, developing and delivering a sustainable farming system through both enabling access to data and through expansion of publicly available data types and enabling trait measurements for key resilience and efficiency traits alongside productivity traits. Of key importance is also the rapidity and urgency that is required to reform our farming system if we are to meet the joint goals of protecting and securing food production, reducing biodiversity loss and reaching net zero emissions.

2 A Short Historic Perspective on the Current Breeding, Protection and Registration Systems in the UK and Their Reliance on Data Linkages

The National Institute of Agricultural Botany (NIAB) was founded in 1919 as response to the food crisis of 1917–1918, when there was a serious shortage of imported food and a lack of seed, fertiliser and equipment to crop large areas of newly ploughed pasture, required to meet quotas for food crops imposed by governmental bodies. From the outset it was a public-private partnership supported through personal donations and governmental support. It has been an Independent Charitable Trust, though for the first 75 years of its life operated effectively as a government-funded institute until its full privatisation in 1996. Its founding Director was Sir Lawrence Weaver, who was at the time Controller of Supplies at the Food Production Department, which had been set up by the Board of Agriculture and Fisheries to deal with the national crisis during the first world war (Wellington & Silvey, 1996). NIAB is a classic example of an organisation whose form followed its function, built to deliver a specific objective.

Parallel developments to NIAB had also led to the implementation of a seed testing system, which primarily had the responsibility of ensuring that seed testing schemes were put in place for domestic and imported seed. Although long called for and already implemented in other European nations, the national food crisis led to Weaver establishing the official seed testing station for England and Wales which began in London but later became a part of NIAB in Cambridge (Wellington & Silvey, 1996). From 1921 the Official Seed Testing Station (OSTS), became a member of the International Seed Testing Association (Wellington & Silvey,

1996), which to this day continues to ensure common standards for data collection are developed across the world for seed testing.

Aside from NIAB's role in seed testing, its objective was to achieve two aims: 'to promote the improvement of existing varieties of seeds plants and crops in the UK' and 'to aid the introduction or distribution of new varieties' but not to breed new varieties itself. This was at a time at which new wheat varieties, such as 'Little Joss' and 'Yeoman', developed in the light of Mendelian principles of inheritance and resistant to yellow rust (a disease caused by the fungus *Puccinia striiformis. f.sp. tritici*) and with higher grain quality had been recently produced by Rowland Biffin the Plant Breeding Institute (PBI), a slightly older Cambridge-based state institute (and part of the School of Agriculture at Cambridge University) that was also under the purview of Lawrence Weaver. Of note is that neither 'Little Joss' nor 'Yeoman' were reported to outperform varieties at the time for yield parameters, but did show improvements for disease resistance and quality in bread making respectively (Charnley, 2011).

In early years NIAB, as well as providing a role in seed multiplication for state-bred PBI varieties, established voluntary schemes for the approval of seed crops and certification of multiplied seed for use on farm, which led to the improvement of the quality of seed for the national harvest. This led to the development of a broader voluntary seed certification scheme, that evaluated both the in-field performance of seed lots at different stages of the multiplication process and the performance and characteristics of the seed as part of the Official Seed Testing function (Wellington & Silvey, 1996).

This undoubtedly contributed to the success of varieties such as Little Joss and Yeoman, as high-quality seed was always available. Of further interest is the fact that 'Little Joss' was reported to be a good low-input variety, that did well in light soil, enabling economic production in the fact of stiff price competition due to cheap imports and 'Yeoman' was suitable for intensive production with heavy fertiliser input enabling higher yields through altered agronomic practice, again allowing profitable production, this time of high-quality bread flour (Charnley, 2011).

Through detailed measurement and increasingly the use of statistical tests and approaches to experimental design, some developed at nearby Rothamsted Research, detailed observations of plant characteristics could be measured and compared for a range of performance characteristics. This led to great success in 'cleaning up' the practice of duplicate naming of varieties, a dubious practice that had occurred since Victorian times. This was especially successful in the area of potato and cereal varieties where synonymous varieties were reported on an annual basis and could be shown to be statistically indistinguishable (Wellington & Silvey, 1996).

In 1923 NIAB established a system of performance trials, in order to compare new varieties to existing varieties. These comprised of multi-site trials that then gave a ranked estimate of overall and regional varietal performance. These trials evolved over time and led to the establishment of 'Descriptive Lists' (DL) and 'Recommended Lists' (RL), designed to allow farmers to select independently-evaluated varieties. Descriptive Lists fulfilled the function of allowing variety attributes to be documented without providing a ranking. The objective of

Recommended Lists was to fulfil the national requirement to maximise productivity and hence production of UK-grown agricultural and horticultural crops.

In post-war years, having seen the benefits of the voluntary seed certification and process in ensuring quality of both domestic supply and imported seed, new legislation began to be developed, to ensure that analytical standards for declarations of purity, germination and weed content were in place as well as freedom from disease. Furthermore, as by now plant breeding had developed many new varieties, the need for seed purchasers to understand the genetic quality of seeds offered for sale was also a key consideration (Wellington & Silvey, 1996). Parallel developments in the area of intellectual property rights for plant breeders ultimately led to the UK legislation, the 1964 Plant Varieties and Seeds Act, which allowed for the establishment of Plant breeders rights, allowing plant breeders to be granted the right to protect their intellectual property in the same manner as other inventors do, therefore affording legal protection and the right to prevent unauthorised multiplication and sale of seed. Although there was national implementation of this scheme, through the 1964 Act, the specified requirements for a common standard of adoption for plant breeders' rights was internationally agreed in 1961, when the International Convention for the Protection of New Varieties of Plants was agreed in Paris for a Union for the Protection Of Varieties (UPOV). This specified international standards for Distinctness (that the variety can be clearly distinguishable from any other variety whose existence is common knowledge at the time of filing of the application), 'sufficient homogeneity' (Uniformity) and Stability (that the relevant characteristics of the variety do not change over generations) across 13 initial crops over a period of 8 years; collectively known as DUS testing. This allowed reciprocity across member states, meaning breeders could have multi-territory protection of their variety, as defined by a common data standard (Wellington & Silvey, 1996). Crucially the botanical varietal descriptors that were (and still are) used for protection bore little or no resemblance to agronomic performance, meaning that other processes were required for evaluating these characteristics. Therefore, the second part of 1964 act established the official index of varieties and the requirement for required statutory performance trials before seed marketing was allowed, for a range of crops deemed important to national food security, effectively the Recommended List system.

Of linked importance to the granting of plant breeders' rights is the convention of the breeders' exemption for the use of genetic material (registered varieties) in further development of plant material (Würtenberger, 2017). The protection and release of intellectual property for societal advancement is in the common interest-for example as stated in the US Constitution "To promote the progress of science and useful arts." (US Constitution, Art 1, s.6). It is on this principle that within the Plant Variety Protection system, breeders are able to utilise other breeders' material in their own crossing and selection process, following protection; this exemption is estimated to have led to tremendous economic returns since its implementation (Lüttringhaus et al., 2020).

Upon entry into the common market in the early 1970's, EEC directives stated that National Lists and Official certification of seed were needed to meet common market standards and for entry into a Common Catalogue enabling marketing throughout the EEC (Wellington & Silvey, 1996). Once again productivity was at the heart of the agricultural policy. Only seed of high quality, with approved performance and distinct identity could be marketed under EEC policy. This meant that seed of important crops could only be marketed after a variety had been accepted for inclusion in a "National List" (NL) and the seed certified by a member state or third country operating to equivalent standards (a full list of species for which National Listing is essential is provided in Supplementary Table 1). Common grades for seeds at different levels of multiplication were also developed and meant that UK voluntary systems were converted into statutory systems that conformed to OECD standards (Fig. 1).

Of note- farm saved seed was (and is) still permitted to be used outside the certification system. Entry into the common market effectively ceded national sovereignty in defining what could be grown and marketed in any single country for market access. Following Brexit the UK no longer participates in the common catalogue and therefore breeders now have to register their varieties in both the UK through APHA (Animal and Plant Health Agency) and in Europe through the Community Plant Variety Office (CPVO).

In practice the established system of IP protection and National Listing means that parallel evaluation of DUS characteristics, required for Plant Varietal Rights (PVR) to be granted and Value for Cultivatable Use (VCU) trials are required to be carried out in parallel for National Listing to occur. At the time of entry into the common market 'Recommended Listing', remained a government-funded activity for many crops, but nowadays, following widespread reform of near market research and development in the mid 1980's, the RL is wholly industry funded through the statutory levy which is administered by the Agricultural and Horticultural Development Board (AHDB). The National List is now delivered by a combination of breeder-funded trials, delivered both by breeders and by NIAB and government-funded disease resistance trials and operated on a cost recovery basis. Often VCU trials and RL trials are intertwined, though this varies on a crop-by-crop basis.

The UK system of a dual NL and RL leads to a two-tier system that means following National Listing a second non-statutory bar is created, meaning that in

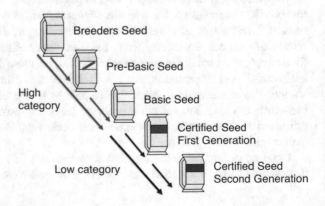

Fig. 1 A simplified overview of a seed certification scheme, ensuring quality standards throughout the propagation chain

Breeders Seed

Pre-Basic Seed

High category

Basic Seed

Certified Seed First Generation

Low category

Certified Seed Second Generation

addition to passing the first statutory approval for marketing and certification, the second advisory tier exists, the Recommended List (see Figs. 2 and 3) to highlight varieties that show a clear improvement in performance to a set of existing varieties. The overall criterion for RL candidate selection- places a strong emphasis of promoting varieties which are 2% or more above a yield target for each market segment of a crop group, though exceptions are possible. Specifically the RL variety selection criteria are stated to be "considered to have the potential to provide a consistent economic benefit to the UK cereals or oilseeds industry"(AHDB, 2020). As can be seen from the evaluated criteria for winter wheat (Table 1) other factors are

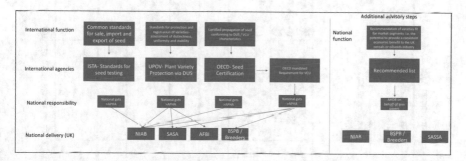

Fig. 2 Overview of the linkage between National and International systems and agencies involved in the registration, regulation and recommendation of crops of agricultural importance

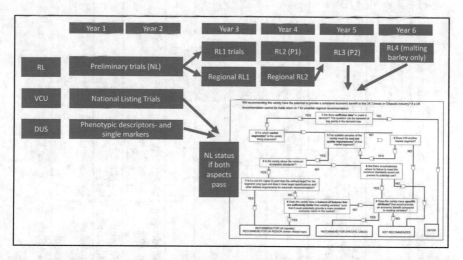

Fig. 3 The National and Recommended List system. DUS and VCU trials (Y1, Y2) define the criteria for entry on to the National list and certification schemes, while subsequent years evaluate the 'best of the best' for inclusion onto the Recommended List. Inset decision tree for the AHDB RL process is reproduced from (AHDB, 2020)

Table 1 The evaluation and criteria for selection of winter wheat lines on the AHDB Recommended List 2020–21

Grain quality	Endosperm texture	Protein content (%)	Protein content (%) – Milling spec	Hagberg Falling Number	Specific weight (kg/hl)	Chopin Alveograph W	Chopin Alveograph P/L
Fungicide-treated grain yield (% treated control)	United Kingdom (10.8 t/ha)	East region (10.7 t/ha)	West region (10.9 t/ha)	North region (11.0 t/ha)			
Untreated grain yield (% treated control)	United Kingdom (10.8 t/ha)						
Agronomic features	Resistance to lodging without PGR (1–9)	Resistance to lodging with PGR (1–9)	Height without PGR (cm)	Ripening (days +/– Skyfall, –ve = earlier)	Resistance to sprouting (1–9)		
Disease resistance	Mildew (1–9)	Yellow rust (1–9)	Brown rust (1–9)	Septoria tritici (1–9)	Eyespot (1–9)	Fusarium ear blight (1–9)	Orange wheat blossom midge
Agronomic features	Lodging % without PGR	Lodging % with PGR	Latest safe-sowing date				
Rotational position	First cereal (11.1 t/ha)	Second and more (9.5 t/ha)					
Sowing date (most trials were sown in October)	Early sown (before 25 Sept) (11.0 t/ha)	Late sown (after 1 Nov) (9.5 t/ha)					
Soil type (about 50% of trials are on medium soils)	Light soils (10.8 t/ha)	Heavy soils (10.9 t/ha)					
Speed of development to growth stage 31 (days +/– average)	Early sown (Sept)	Med sown (Oct)	Late sown (Nov)				

Data obtained from AHDB Recommended Lists for cereals and oilseeds 2021/22 available from www.ahdb.org.uk

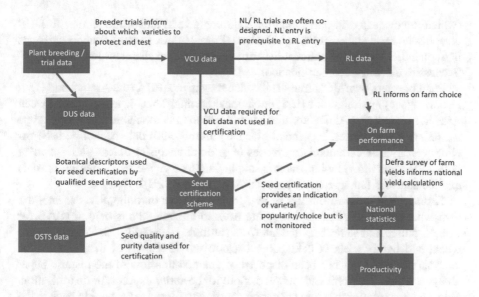

Fig. 4 Current data linkage between nationally applied statutory and advisory varietal registration and trialling processes and their onward linkage to national statistics

considered, but entry onto the list is largely based on the primary results of NL yield trials.

This long and complex history, which is insufficiently summarised here is presented in order to illustrate that the requirement for data standards and data linkage in complex systems is not at all new. The consequence of system of varietal registration, testing and certification for marketing, built largely under policies promoting food production, is that many of the international data standards have been developed by international organisations with functions for seed testing, Plant Variety Rights and Seed Certification schemes (ISTA, UPOV, OECD) an overview of which is presented in Fig. 4. As such the integrity of data stretching back sixty or more years is in-part preserved.

3 Who Owns What Data?

In the current registration and evaluation system, as a result of the many changes in organisational ownership and funding of national listing systems, ownership of the data is disaggregated. For DUS purposes the data is owned by the registrant (i.e. the breeder), and national databases of performance are kept by the bodies that perform the tests (e.g. NIAB) as well as summary data held by the international body UPOV in their "PLUTO" database (https://www.upov.int/pluto/en/). Summary data is made public at the national level to allow seed certification protocols to be administered (which are evaluated in-part based on DUS characters) and while similar to DUS data, has no legal status and is not identical. For example, for certification purposes in Scotland, SASA grow national list varieties to develop character lists for distinctiveness to aid with seed certification.

Plant breeders own VCU trial data, which they give APHA (the Animal and Plant Health Agency) permission to use for National Listing. There is agreement between APHA, breeders and AHDB for the data to be used in combinable and forage crops for RL, or DL purposes. In general practice currently, NL data cannot be used for either research or commercial purposes (e.g. extra analysis) unless permission of both APHA and individual breeders is sought. APHA own the VCU 'matrix' of trials and the analyses (an agreement between APHA and BSPB).

Historic data, prior to transfer of registration systems onto the private sector are somewhat patchy. Electronic data going back to the late 80's is held at NIAB for cereals, pulses and sugar beet and late 70's for herbage and oilseeds. Yield data for wheat and barley trials is held going back to the 1940s. Most of the statistical analysis of these trials has been disposed of apart from a few of the historic paper analyses which were kept and archived, more for posterity than for future utilisation. However, for some crops there are paper records e.g. for sugar beet going back to the 20's. Post-1986 the levy body owns Recommended List data while prior to that, as a government-funded activity delivered through NIAB the data was in public ownership.

Less attention has been given so-far in this overview to the quality of record keeping within breeding programmes, both in the public and private sector. From personal experience, the availability of historic, field and trial data are often patchy and dispersed among paper and digital records of varying quality. It is usual that simple structural problems, such as turnover of staff, the patchiness of digitisation of paper records and the lack of resources to curate and archive data all lead to loss of potentially valuable data. Due to the simplicity of much of the historic data, the issue of data standards is usually not an issue. However, it is often the case that phenotypic descriptors are not necessarily well designed and can be highly subjective. Moreover, the immediate lack of identification of a lack of purpose for some datasets often leads (in the author's own experiences) to short-term decisions being made about the investment in data archiving of the majority of within breeding programme data. This is especially true within breeding programmes when dealing with historic data, as living material may no longer exist and therefore the immediate utility of the data is sometimes not apparent.

4   Data Linkage with Statutory Information: Examples

The revolution in affordable genome sequencing and genotyping technologies has led to the ability to measure genetic variation in plant varieties to a degree and precision that was unthinkable even 10 years ago (Pavan et al., 2020). As such many publicly funded initiatives and collaborative public-private initiatives have led to the widespread availability of DNA sequence data in the public domain. Other multi-omics data is following, but principally it is DNA sequence polymorphism data that is of immediate value. Many studies have shown the value of incorporation of molecular data into both DUS (Cockram et al., 2010, 2012, 2015; Saccomanno et al., 2020) and VCU (Wang et al., 2012) and suggested strategies for deployment (Jamali et al., 2019). For example, in crops where heritability of DUS traits is low, it is of significant benefit to utilise molecular data (Cockram et al., 2010). Within plant variety protection UPOV already have models laid down by the Biochemical and Molecular Techniques (BMT) Working group of UPOV for the use of molecular markers (Jones et al., 2013). However, in the UK use of molecular data or prediction of performance does not yet extend to VCU trial data, nor RL trials, despite some obvious advantages of doing so.

4.1 The Use of Historic Data

Recent work of Fradgley et al. has shown what a valuable resource even relatively simple pedigree information can be in the light of modern genomics (Fig. 5) (Fradgley et al., 2019). Through an analysis of global wheat data, coupled with genetic marker data, they were able to first construct a pedigree of over 2600 wheat varieties and identify signatures of artificial selection across the pedigree and demonstrate that these genomic regions could correspond to genes of known functional importance in key yield components.

Subsequent Genome-Wide Association Studies GWAS using VCU yield data finds around a third of the signatures of selection identified in the pedigree paper to overlap with GWAS hits for yield (White et al., 2021).

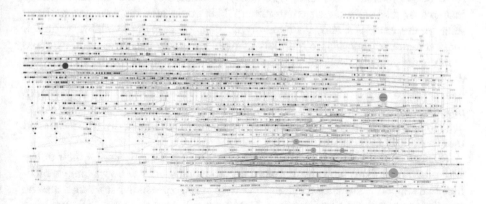

Fig. 5 The pedigree of global wheat, as reconstructed by Fradgley et al. (2019), drawing on Recommended List data, among other sources. (Reproduced without modification under CC-BY 4.0 licence, with permission from the author)

This demonstrates the principle that high-quality data that is publicly available can have scientific value far beyond that originally envisaged through linkage of newer datasets with historic data. Further potential for much more detailed analysis of breeding and selection activity lies within the vast datasets for the National and Recommended lists.

4.2   Data Linkage Between Public and Private Sources and Use Within Breeding Programmes

So far, this review has primarily concentrated on the vast array of data that is generated as part of the varietal listing process, which is largely unknown to the majority of academic researchers working in the area of crop genetics and improvement. However, over the past twenty or so years the generation of breeder-relevant data within the academic sector has grown substantially, especially as the explosion in molecular biology techniques has led to the creation of large datasets. As a general aid, dataset can be separated into two classes. The first is data that informs about the biological function of the crop as a whole- for example a detailed timecourse of gene expression regulation in multiple tissues of a single variety of wheat, grown in a controlled environment. This data and the associated analysis are clearly relevant to breeding a crop improvement- for example in determining specific genes involved in biological process; information which a breeder could use to devise a screen for genetic variation in breeding material. However, it cannot necessarily be integrated directly into a breeding programme or selection scheme. It is likely that raw data and associated metadata is deposited in an archive and that the relevant conclusions will remain available for some time.

The second type of data is that which is breeder-relevant is most likely to take the form of population- level data, potentially associated to some physical genetic resource or relevant growing environment. This could take the form, as we have seen in the examples above, of genotypic data of publicly available varieties, sequencing or resequencing of publicly available diversity panels. A good example of this is the publicly available 'diverse MAGIC' resource which can be used for trait discovery in breeder-relevant material (Fig. 6). The selection of this material was based, again, upon the analysis of important founders of wheat breeding programmes (Scott et al., 2021). As both the dataset and the genetic resource is available, it is possible for the breeder to use this population as a discovery tool and then to directly cross in variation.

It is of course likely that some of these resources are time-limited in their utility and as such not likely to be available as seed beyond a brief window, unless other financial support is provided for their long-term storage. A similar story may also be true for some, but not all of the data. Data from this specific example is available from a variety of sources. Taking this as an example the resources are spread among five different entities: (1) The preprint and ultimately the published paper and its

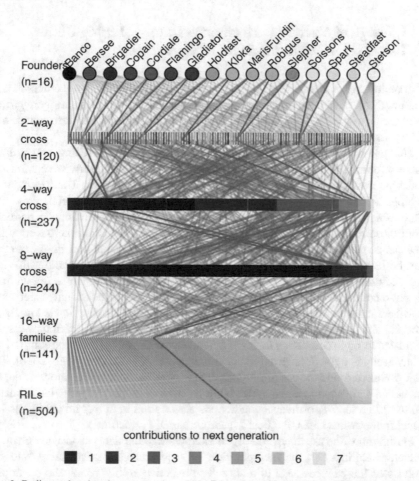

Fig. 6 Pedigree showing the construction of 504 Recombinant Inbred Lines (RILs). One exemplar pedigree is highlighted to show how all 16 founders are intercrossed into each RIL. (Reproduced without modification from Scott et al. (2021) under CC-BY 4.0 licence, with permission from the senior author)

supplementary materials. (2) A laboratory website with files required to associate genetic variation with individual lines (3) A github repository with scripts for analysis (4) The European Nucleotide Archive for sequence data and (5) The location of the physical genetic resource- in this case NIAB in Cambridge. This is typical of modern scientific publication and it is easy to see how breakage or removal of any component of this pipeline and archival system will lead to loss of utility of the resource. Over the medium term (5+) years the risk of this occurring is probably highly likely and is on the whole the current state of affairs across many disciplines, not just crop research.

5   Linkage of Variety Performance Data – Making Better Use of RL Data in the Light of Genomics

With multi-environmental trial (MET) data such as the RL the desired outcome is to measure the performance of varieties on a regional basis as well as simply reporting the outcome of the MET. However, it is also possible to predict the performance of a variety, through the use of mixed models.

This use of use mixed models has gained favour in the breeding community, where information about relatedness of individuals (an all-all pairwise relationship matrix) is combined with performance data gathered across sites. The treatment of incomplete trial block designs, replicates within sites, different trial sites, and relationships between individuals as random effects allows predictions to be made about those random effects. For example, information drawn from across multiple sites can be used to better predict variety performance for yield in a single site that may be lacking in data (Millet et al., 2019). Similarly, treating relationship data (e.g. pedigree) as a random effect allows performance information to be drawn from closely-related individuals to predict the performance of an individual variety in a site where it has not been phenotyped (Millet et al., 2019). Moving beyond pedigree data, the absolute (rather than estimated) relationships between individuals can be calculated through comparing their DNA sequences. The all-all pairwise comparison of any group is termed the genomic relationship matrix (Molenaar et al., 2018). The basic premise around the use of genomic data in combination with phenotypic data is that within a mixed model framework it allows a proxy for phenotype to be estimated and used as a tool. Performance predictions are referred to in this mixed regression model framework as BLUPs (Best Linear Unbiased Predictions).

Furthermore, the use of covariates, for example weather data can lead to enhanced predictive abilities, as the incorporation of other relevant information into the analysis of the trial can lead to a stronger predictive ability which the covariates have a large effect on varietal performance (Gillberg et al., 2019). Similarly incorporating covariates of phenotypic measures into predictions of yield, especially when segregating data into common environmental groups can enable better predictive ability of varietal performance for key yield or resilience components (Ly et al., 2018).

In private breeding programmes there is often similar data to that outlined in public trials programmes. This may take the same form of the data above, but is likely to be held in a local database, or file structure of some kind and never publicly available. This data may have value, in combination with other proprietary data, or through the integration with public data but beyond the provision (as outlined above) of the data for statutory systems the 'internal workings' of breeding programmes are currently rarely revealed. For example, by combining recommended list data with internal genotypic data (present in most modern breeding programmes), better predictions of varietal performance in a given region can be made that are present in the RL itself through the application of mixed-models and BLUP to estimate random effects based on combined RL and private breeder data. This could be improved still further if access to all genotypic data in the trial was possible (Robertsen et al., 2019).

The use of high throughput phenotyping in trials programmes will also add additional insights, especially in combination with environmental covariates and better models of plant development (Zhang et al., 2019). Recent work from Zhou and colleagues illustrated how relatively cheap devices could be deployed for quantification of key developmental (and linked environmental changes) in crops through the use of internet of things devices (Zhou et al., 2017), while the same group also showed how the use of low-altitude, low cost Unmanned Aerial Vehicles (UAV) could be used to determine key yield related traits in wheat (GuoHui et al., 2019). Finally, the use of cameras and machine learning algorithms for seed imaging and analysis could provide key data for the analysis of seed quality (Colmer et al., 2020).

6 Unintended Consequences of the Current System – Do Data Standards Help or Hinder?

It has been known for many years that there is immense value in the historic data captured in NL and RL data, however, lack of adoption of new methods due to lack of national international evolution in standards has hindered progress (Mackay et al., 2011). This is primarily due to the fact that national authorities must now follow international standards for DUS, certification and seed testing, which at their heart are systems based upon botanical characterisation, which although laborious (and sometimes inaccurate) is scalable and relatively low-tech. International bodies must ensure that the broader considerations about equity and implementation of processes around the world are put first. This has the unintended consequence of holding back the application of cutting-edge technology. However, this is not the complete story, as countries that are now signing up to OECD and UPOV schemes are implementing them differently and probably placing more emphasis on data integration at the national level, while countries with established systems find it harder to drive forward change.

As a result of multiple factors, discussed later in this review, developments in scientific research have become more distant from the process, leading to significant divergence in what is technically possible and what is carried out in practice and many are now calling for innovation in VCU and DUS systems (Gilliland et al., 2020; Wang et al., 2016).

A recent study by Yang et al. highlights the weaknesses in the current varietal registration system in the light of new information, specifically in the area of DUS testing (Yang et al., 2020). The study revealed that low heritability traits (i.e. those that are not influenced by segregating variation) are commonly used for DUS and certification purposes, meaning that confidence can be low about assigning varietal identity and in proving distinctness (which requires stable differences to be expressed between varieties). This aspect of the system could be completely avoided through the use of molecular markers. However, if mandated internationally this would increase cost of registration and potentially reduce the accessibility of systems to LMICs, lacking reliable or affordable access to more advanced protocols.

For VCU and recommended list trials, consideration must also be given to the fact that potentially more environmentally sustainable varieties (for example those able to give reliable yields in marginal conditions) are not necessarily able to enter the system due to the fact that many trials occur under near optimum conditions, which are unlikely to be the norm on farm, or may require extremely high levels of inputs and therefore may be less sustainable.

The use of distributed ledger technology (DLT) in certification could lead to less cumbersome processes and could in fact increase adoption of certification systems. The technology does not yet exist to deploy DLT efficiently for genomic data at scale but consideration of this as a useful technology to help maintain some privacy around genomic data, while leveraging benefits may be valuable (Lee et al., 2018; Thiebes et al., 2020).

7   Linking Breeding to Wider Farming Systems and On-Farm Practice

By the last quarter of the twentieth century, the impact of international quality schemes using linked data standards, common markets and highly productive agriculture, due to genetic, agronomic and statutory innovations had led to a fairly centralised, highly regulated, but costly set of agricultural systems. By the beginning of the 1980's this increasing cost and rising waste due to the overproduction caused by the Common Agricultural Policy meant that the last 20 years of the twentieth century were spent attempting to move much of the cost of both the systems (PVR, NL and RL) onto the industry and the cost of the strategic and applied research that NIAB and other institutes did onto industry to reduce government expenditure in what by this time was a system that produced sufficient (even surplus) food (Wellington & Silvey, 1996). In 1986 through the Agriculture Bill arms-length levy bodies were formed following industry consultation. These were tasked with collecting a levy from the industry to commission near market research, development and knowledge exchange. A unified levy body, the AHDB, now commissions RL and DL work and produces the recommendations for industry. More generally over the next 20 years the privatisation or closure of many strategic research organisations led to some fragmentation in the UK between the previously well-connected research institute structure and both the industry it serves and policy makers (Wellington & Silvey, 1996). It can be argued that this step to full cost recovery on near-market research and development has had the unintended consequence of disaggregating knowledge bases, especially at the strategic level, leading to a lack of general oversight of the steps required for system innovation as it effectively separated the science-led decision-making functions from the process-led delivery of the systems.

While past challenges were focussed squarely on productivity, it is now clear that a broader set of considerations are required for our longer-term food and environmental security and that our food system must adapt rapidly to ensure that the joint goals of biodiversity protection, economic production and net zero agriculture are met.

The use of these trials datasets to simulate forward under local and global weather and climate models is crucial for increasing our understanding of how best to adapt crops to changing weather patterns as a result of climate change and contribute both to better recommendations, but better data for breeders to utilise in creating more resilient and lower input varieties.

Moving beyond the trials and listing processes and onto the farm the availability of linked data would allow regional performance estimates to be validated (and improved) through using on-farm data of farm-grown crops and therefore a more dynamic recommendation system could exist. In fact, data from all stages of the registration process should be able to feed back into one another creating more dynamic systems able to update predictions and confidence estimates of predicted performance of varieties all the way back to the breeder.

Linkage of trial data to further agronomic development- where significant differences exist between on farm performance and trial performance, should allow better insights can be made into the causes of these yield gaps. It is highly probable that in many farming systems the performance of varieties in trial does not match the on-farm setting. This could be due to factors such as local pest, disease and weed pressure, specific issues with soil or microenvironment or difference in farming practice. All of these factors could potentially be decomposed at the genetic level (and therefore be subjected to improvement through breeding) through the use of robust and open trial data, if common data standards were used in data capture.

Recalling the original examples of Yeoman and Little Joss, the former suited for high-input, high yield farming, the latter for low-input situations. It is likely the latter would have not success in the current system, despite some potential broader environmental benefits of slightly lower yielding, but much lower input varieties.

Care must be taken to select the appropriate ontologies for trial and registration systems and (just like in the historical examples) be aware of and integrate current international efforts in this area (e.g. https://www.cropontology.org/) (Fig. 7).

8 Potential Systems-Level Solutions That Could Be Achieved Through Improved Data Linkage

In moving from a largely linear set of approval systems, consideration should be given to a more circular or 'systems' approach to improvement ensuring that data at all points in the knowledge chain are utilised and that feedback of data are made possible.

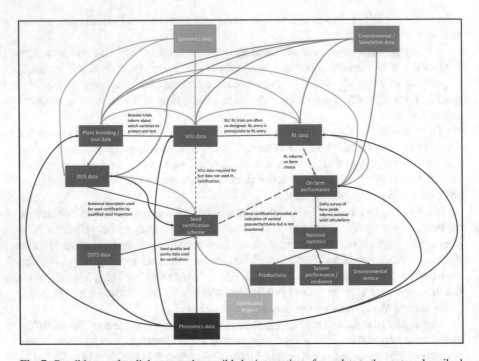

Fig. 7 Possible new data linkages made possible by integration of new data to the system described in Fig. 4. Genomics data (green) would allow better integration of DUS and VCU data, with aligned benefits in seed certification schemes. Furthermore, internal linkages between private breeder data and public data in VCU, RL trials as well as in on farm production could lead to more dynamic recommendations. Integration of environmental and simulation data (orange) would again lead to huge leaps in prediction accuracy as well as paving the way for model-based predictions of crop performance under changing environments as well as greater precision of prediction for farm level growing practice. Distributed ledger systems (yellow) could have an impact on certification systems and potentially offer new methods to track seed thoughout the supply chain in a more accessible way. Phenomics data (red) again impacts all aspects of statutory, advisory and on-farm systems allowing greater linkage between DUS and certification data, as well as providing more phenotypic data to include covariates in yield or other trait predictions. Finally, integration of all methods through the development of common data standards, adopted by statutory and advisory systems would lead to greater power at the national level to assess productivity, environmental service and system-level performance characteristics required to deliver the joint goals of productive, sustainable, net-zero agriculture in tandem with enhanced environmental service

The ability to measure and manage better is made possible through the development of standards and provides a much more coherent dataset upon which policy decisions could be made.

This is crucial to recognise that in enabling the characterisation of and then integration of negative externality costs into the regulation of wider farming system, changes may occur both in the way in which we farm and the performance characteristics of our crop varieties. Careful thought must be given to how the joint considerations of productivity, net zero and biodiversity protection will likely

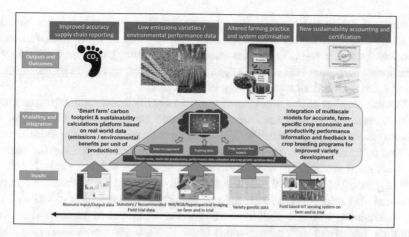

Fig. 8 Examples of possible market opportunities made possible through on-farm data linkages. Underpinned by linked evolution of statutory and advisory systems, these linked datasets could have multiple public good and economic uses, some of which are illustrated in this figure

drive new farming systems which may not solely rely on monoculture. The use of bi-cropping or poly-cropping (where extended phenotypes may emerge-e.g. enhanced biodiversity) are potentially important and trials evaluation systems will likely need to integrate this into the evaluation system. The use of genetics may also change, with variety mixtures (of varying forms), which may be genetically diverse but functionally homogeneous for key traits all being future possibilities that registration systems must deal with.

The ability and desire to derive appropriate metrics at the national level and to have a more holistic view of on-farm performance of varieties and then onward linkage to other data will allow better quantification of life cycle emissions and environmental impacts of primary production. The onward linkage of domain-specific data to metrics about system performance, allowing better estimates of the environmental impact of farming on the environment and both identification and modelling of 'what works' at the policy level, allowing more dynamic implementation of future agricultural policy (Fig. 8).

9 What Structures Are Required for Future System Change and Who Benefits?

Expansion of the national component of statutory systems should be considered to drive forward innovation. This should be considered alongside renewed international engagement with equitable innovation within statutory systems at the global level. In order to maintain market attractiveness for breeders, these enhancements to the national systems should be state funded. This would change data ownership structures to ones where government co-owns data, but provide 'win-win' situations for all points in the supply chain that would likely drive up innovation and overall productivity and sustainability. This may have particular benefit to small breeding companies, that often lack access to genomic resources, but could benefit from the use of them within their own breeding programmes.

The initial pillar of this innovation is principally this need genome sequencing to be a prerequisite for national listing- this should be funded by the government, operate under FAIR principles and be seen as an extension to the current patent system, operating under the principle of publishing and protecting for both the common good and for the benefit of the rights holder (Wilkinson et al., 2016).

Analysis of both varietal data for DUS purposes and DUS data (and onward certification) should evolve in the light of genomics, ensuring that global standards are maintained, but that additional innovation is unlocked through the use of genomes. The use of molecular markers in DUS, VCU and onward RL trials should be made a priority.

However, effort should also be given to drive forward innovation in gathering supporting data, for example phenomic and environmental data of trials that is interoperable used between both public and private functions should be co-funded (much in the same way that pathology data for VCU is funded for the public good). Again, co-benefits could be recognised throughout the food system. An example of success comes from the life science sector where the Pistoia Alliance (www.pistoiaalliance.org) brings together member companies (numbering over 100) from life science companies, technology and service providers, publishers, and academic groups to transform R&D through pre-competitive collaboration. In a fragmented landscape such as agriculture, this may be a new form of collaborative national or international network, that would more rapidly advance linked innovation and statutory innovation than the present separated systems that are currently in operation.

Automated capture and development of data standards for image data and development of standards need to be developed and applied both in registration and advisory systems, but also in on-farm improvement measures; the adoption of common data standards for statutory systems would likely 'cement' a data standard across the industry, even though it could be used on a voluntary basis.

In exchange the use of proprietary data to derive co-benefits that benefit the public good should be requested, enabling DUS and NL data that is currently privately owned to be released into the public domain in some way. Appropriate consideration of the need to retain commercial advantage should be at the forefront of this discussion, but methods are available to ensure that some benefits can be derived without necessarily requiring full data release. The use of trusted intermediaries could be one simple mechanism.

Historically meeting challenges such as these has been done by centralising the challenge and creating form follows function vehicles 'i.e. NIAB' which leverage both public and private investment, but other models could be possible- for an example see the Pistoia alliance in the biomedical sciences, but it is likely that in this case some 're-grouping' of functions are needed to drive change at the required pace. This should include the same three pillars that were there at the inception of our current statutory systems, government and public funders, private enterprise and scientists/academics and be treated as a shared challenge with the power to both design and implement science-based programmes of varietal evaluation and monitoring.

10 Overall Conclusions

There are ample opportunities for improved data linkage to transform our understanding of the changes needed for improved system design. Building depth into reformed long-lived statutory and descriptive systems is likely a good idea as this provides longevity to data and ensures security of data.

Establishment of data standards within statutory and advisory systems often preserves key data linkages but can also ossify and stifle progress and so flexibility is required in any future design and additions and any enhancements or divergence from current international standards by individual nations needs to be supported through public funds to ensure that it is commercially viable to operate and register varieties in the national market. This will lead to a period of duplication, but is required for any broader transformation to occur.

Making any additional data added to statutory and advisory systems open is crucial and should be viewed as a public good. The current system does not allow for FAIR data at this point in time. Making the data standards open and accessible is critical to drive wider adoption of that data. Further standardisation of trials between the public and the private sector is likely a good idea and the use of common ontologies will ensure interoperability.

Extending linkage of data from statutory and advisory systems both into academic research and onto the farm provides great scope for new approaches to measuring and managing system-level properties and assessing the performance and impact of a broader array of crop genetic innovations on farm, with more dynamic feedback into crop breeding programmes.

Function is currently following form, the disaggregation of what was a centralised strategic response has led to 'masters of none' and the inability to drive reform at pace. This can be seen at both the national and international levels. This is not necessarily the fault of any single actor, but a property of the diffuse structure without coordinated oversight.

Current systems emerged out of a need to drive up productivity and were highly centralised efforts, both at the national and then the EU level. The challenge of responding to the joint challenges of economic, resilient and sustainable production, low emissions agriculture and reversing biodiversity decline is too great and swift action is required, likely through a form follows function approach and a re-design of current national systems and organisations but ensuring close cooperation between public and private stakeholders. Plus ça change!

Author Contributions RH devised the overall review with input from MC. RH wrote the first draft of the manuscript. MC provided editorial support and constructive discussion throughout.

Acknowledgements The authors wish to thank Nick Fradgley, Margaret Wallace, Jane Thomas, Stephen Flack, Claire Leaman, Haidee Philpott and Nicola Harrison for information and constructive discussion.

Appendix

Supplementary Table 1 Crop plants requiring UK National Listing

Cereals	
Name	**Common name**
Avena nuda L.	Small naked oat, Hulless oat
Avena sativa L. (includes A. byzantia K. Koch)	Oats and red oat
Hordeum vulgare L.	Barley
Secale cereale L.	Rye
xTriticosecale Wittm. Ex A. Camus Hybrids resulting from the crossing of a species of the genus Triticum and a species of the genus Secale	Triticale
Triticum aestivum L.	Wheat
Triticum durum Desf.	Durum wheat
Triticum spelta L.	Spelt wheat
Zea Mays L. (partim)	Maize (except popcorn and sweetcorn)
Potatoes	
Name	**Common name**
Solanum tuberosum L., including any other tuber-forming species or hybrids of Solanum	Potato
Beet	
Name	**Common name**
Beta vulgaris L.	Sugar beet, fodder beet (including mangel)
Fodder plants: grasses	
Name	**Common name**
Agrostis canina L.	Velvet bent
Agrostis capillaris L.	Brown top
Agrostis gigantea Roth.	Red top
Agrostis stolonifera L.	Creeping bent grass
Arrhenatherm elatius (L.) P. Beauv. ex J. Presl & C. Presl	Tall oatgrass
Bromus catharticus Vahl.	Rescue grass
Bromus sitchensis Trin.	Alaska brome-grass

(continued)

Supplementary Table 1 (continued)

Dactylis glomerata L.	Cocksfoot
Festuca arundinacea Schreber	Tall fescue
Festuca filiformis Pourr.	Fine leaved sheep's fescue
Festuca ovina L.	Sheep's fescue
Festuca pratensis Huds.	Meadow fescue
Festuca rubra L.	Red fescue, Chewings fescue
Festuca trachyphylla (Hack.) Krajina	Hard fescue
Lolium multiflorum Lam.	Italian ryegrass including Westerwold ryegrass
Lolium perenne L.	Perennial ryegrass
Lolium x boucheanum Kunth	Hybrid ryegrass
Phleum nodosum L.	Small timothy
Phleum pratense L.	Timothy
Poa annua L.	Annual meadowgrass
Poa nemoralis L.	Wood meadowgrass
Poa pratensis L.	Smooth-stalked meadowgrass
Poa trivialis L.	Rough-stalked meadowgrass
xFestulolium Asch. & Graebn. Hybrids resulting from the crossing of a species of the genus Festuca with a species of the genus Lolium	Festulolium

Fodder plants: legumes

Name	Common name
Lotus corniculatis L.	Birdsfoot trefoil
Lupinus albus L.	White lupin
Lupinus angustifolius L.	Narrow leaved lupin (previously known as Blue lupin)
Lupinus luteus L.	Yellow lupin
Medicago lupulina L.	Black medick, Trefoil
Medicago sativa L.	Lucerne
Medicago x varia T. Martyn	Sand lucerne
Onobrychis viciifolia Scop.	Sainfoin
Pisum sativum L. (partim)	Field pea
Trifolium hybridum L.	Alsike clover
Trifolium pratense L.	Red clover
Trifolium repens L.	White clover
Vicia faba L. (partim)	Field bean
Vicia pannonica Crantz	Hungarian vetch
Vicia sativa L.	Common vetch
Vicia villosa Roth	Hairy vetch

Other fodder plants

Name	Common name
Brassica napus L. var. napobrassica (L.) Rchb.	Swede
Brassica oleracea L. convar. acephala (DC.) Alef. Var. medullosa Thell. + var. viridis L.	Fodder kale

(continued)

Supplementary Table 1 (continued)

Raphanus sativus L. var. oleiformis Pers.	Fodder radish
Oleaginous and fibrous plants	
Name	**Common name**
Brassica juncea (L.) Czern.	Brown mustard
Brassica napus L. (partim)	Swede rape (including plants commonly known as fodder rape and oilseed rape)
Brassica nigra (L.) W.D.J. Koch	Black mustard
Brassica rapa L. var silvestris (Lam.) Briggs	Turnip rape
Cannabis sativa L.	Hemp
Glycine max (L.) Merr.	Soya bean
Helianthus annuus L.	Sunflower
Linum usitatissimum L.	Flax, Linseed
Sinapis alba L.	White mustard
Vegetable varieties	
Name	**Common name**
Allium cepa L. – Cepa Group.	Onion, Echalion
Allium cepa L. – Aggregatum Group	Shallot
Allium fistulosum L. – all varieties	Japanese bunching onion or Welsh onion
Allium porrum L. – all varieties	Leek
Allium sativum L. – all varieties	Garlic
Allium schoenoprasum L. – all varieties	Chives
Apium graveolens L. – Celery Group and Celeriac Group	No common name
Asparagus officinalis L. – all varieties	Asparagus
Beta vulgaris L. – Garden Beet Group	Beetroot including Cheltenham beet
Beta vulgaris L. – Leaf Beet Group	Spinach beet or Chard
Brassica oleracea L. – Kale Group	No common name
Brassica oleracea L. – Cauliflower Group	No common name
Brassica oleracea L. – Capitata Group	Red cabbage and White cabbage
Brassica oleracea L. – Brussel Sprouts Group	No common name
Brassica oleracea L. – Kohlrabi Group	No common name
Brassica oleracea L. – Savoy Cabbage Group	No common name
Brassica oleracea L. – Broccoli Group	Calabrese type and Sprouting type
Brassica oleracea L. – Palm Kale Group	No common name
Brassica oleracea L. – Tronchuda Group	Portuguese cabbage
Brassica rapa L. – Chinese Cabbage Group	No common name
Brassica rapa L – Vegetable Turnip Group	No common name
Cichorium endivia L.—all varieties	Endive
Cucumis melo L – all varieties	Melon
Cucumis sativus L. – Cucumber Group	No common name
Cucumis sativus L – Gherkin Group	No common name
Cucurbita maxima Duchesne – all varieties	Gourd

(continued)

Supplementary Table 1 (continued)

Cucurbita pepo L. – all varieties	Marrow, including mature pumpkin and scallop squash, or Courgette, including immature scallop squash
Daucus carota L. – all varieties	Carrot and Fodder carrot
Lactuca sativa L. – all varieties	Lettuce
Solanum lycopersicum L – all varieties	Tomato
Petroselinum crispum (Mill.) Nyman ex A. W. Hill – Leaf Parsley Group	No common name
Petroselinum crispum (Mill.) Nyman ex A. W. Hill – Root Parsley Group	No common name
Phaseolus coccineus L. – all varieties	Runner bean
Phaseolus vulgaris L. – Dwarf French Bean Group	No common name
Phaseolus vulgaris L. – Climbing French Bean Group	No common name
Pisum sativum L. – Round Pea Group	No common name
Pisum sativum L. – Wrinkled Pea Group	No common name
Pisum sativum L. – Sugar Pea Group	No common name
Raphanus sativus L. – Radish Group	No common name
Raphanus sativus L. – Black Radish Group	No common name
Rheum rhabarbarum L. – all varieties	Rhubarb
Spinacia oleracea L. – all varieties	Spinach
Vicia faba L. Broad bean – all varieties	Broad bean
Zea mays L. – Sweet Corn Group	No common name
Zea mays L. – Popcorn Group	No common name

References

AHDB. (2020). *Crop Committee Handbook 2020* (Technical report). Agriculture and Horticulture Development Board. https://rl.ahdb.org.uk/media/3911/crop-committee-handbook-2020.pdf

Benton, T. G., & Bailey, R. (2019). The paradox of productivity: Agricultural productivity promotes food system inefficiency. *Glob. Sustain., 2*. https://doi.org/10.1017/sus.2019.3

Cockram, J., White, J., Zuluaga, D. L., Smith, D., Comadran, J., Macaulay, M., Luo, Z., Kearsey, M. J., Werner, P., Harrap, D., Tapsell, C., Liu, H., Hedley, P. E., Stein, N., Schulte, D., Steuernagel, B., Marshall, D. F., Thomas, W. T. B., Ramsay, L., Mackay, I., & O'Sullivan, D. M. (2010). Genome-wide association mapping to candidate polymorphism resolution in the unsequenced barley genome. *Proceedings. National Academy of Sciences. United States of America, 107*, 21611–21616. https://doi.org/10.1073/pnas.1010179107

Cockram, J., Jones, H., Norris, C., & O'Sullivan, D. M. (2012). Evaluation of diagnostic molecular markers for DUS phenotypic assessment in the cereal crop, barley (Hordeum vulgare ssp. vulgare L.). *Theoretical and Applied Genetics, 125*, 1735–1749. https://doi.org/10.1007/s00122-012-1950-3

Cockram, J., Horsnell, R., Soh, E., Norris, C., & O'Sullivan, D. M. (2015). Molecular and phenotypic characterization of the alternative seasonal growth habit and flowering time in barley (Hordeum vulgare ssp. vulgare L.). *Molecular Breeding, 35*, 165. https://doi.org/10.1007/s11032-015-0359-5

Charnley, B. (2011). *Agricultural science, plant breeding and the emergence of a Mendelian system in Britain, 1880–1930* (Doctoral dissertation). University of Leeds.

Colmer, J., O'Neill, C. M., Wells, R., Bostrom, A., Reynolds, D., Websdale, D., Shiralagi, G., Lu, W., Lou, Q., Le Cornu, T., Ball, J., Renema, J., Flores Andaluz, G., Benjamins, R., Penfield, S., & Zhou, J. (2020). SeedGerm: A cost-effective phenotyping platform for automated seed imaging and machine-learning based phenotypic analysis of crop seed germination. *The New Phytologist, 228*, 778–793. https://doi.org/10.1111/nph.16736

Dudley, N., & Alexander, S. (2017). Agriculture and biodiversity: A review. *Biodiversity, 18*, 1–5. https://doi.org/10.1080/14888386.2017.1351892

Fradgley, N., Gardner, K. A., Cockram, J., Elderfield, J., Hickey, J. M., Howell, P., Jackson, R., & Mackay, I. J. (2019). A large-scale pedigree resource of wheat reveals evidence for adaptation and selection by breeders. *PLoS Biology, 17*, e3000071. https://doi.org/10.1371/journal.pbio.3000071

Gillberg, J., Marttinen, P., Mamitsuka, H., & Kaski, S. (2019). Modelling *G* x *E* with historical weather information improves genomic prediction in new environments. *Bioinformatics, 35*(20), 4045–4052. https://doi.org/10.1093/bioinformatics/btz197

Gilliland, T. J., Annicchiarico, P., Julier, B., & Ghesquière, M. (2020). A proposal for enhanced EU herbage VCU and DUS testing procedures. *Grass and Forage Science, 75*, 227–241. https://doi.org/10.1111/gfs.12492

GuoHui, D., Hao, X., MingXing, W., JiaWei, C., Xiue, W., Ji, Z. (2019). Developing cost-effective and low-altitude UAV aerial phenotyping and automated phenotypic analysis to measure key yield-related traits for bread wheat. *Journal of Agricultural Big Data*.

Haberl, H., Widenhofer, D., Virág, D., Kalt, G., Plank, B., Brockway, P., Fishman, T., Hausknost, D., Krausmann, F., & Leon-Gruchalski, B. (2020). A systematic review of the evidence on decoupling of GDP, resource use and GHG emissions, part II: Synthesizing the insights. *Environmental Research Letters., 15*(6), 065003. https://doi.org/10.1088/1748-9326/ab842a

Jamali, S. H., Cockram, J., & Hickey, L. T. (2019). Insights into deployment of DNA markers in plant variety protection and registration. *Theoretical and Applied Genetics, 132*, 1911–1929. https://doi.org/10.1007/s00122-019-03348-7

Lee, S.-J., Cho, G.-Y., Ikeno, F., & Lee, T.-R. (2018). BAQALC: Blockchain applied lossless efficient transmission of DNA sequencing data for next generation medical informatics. *Applied Sciences, 8*, 1471. https://doi.org/10.3390/app8091471

Lüttringhaus, S., Gornott, C., Wittkop, B., Noleppa, S., & Lotze-Campen, H. (2020). The economic impact of exchanging breeding material: Assessing winter wheat production in Germany. *Frontiers in Plant Science, 11*, 601013. https://doi.org/10.3389/fpls.2020.601013

Ly, D., Huet, S., Gauffreteau, A., Rincent, R., Touzy, G., Mini, A., Jannink, J.-L., Cormier, F., Paux, E., Lafarge, S., Le Gouis, J., & Charmet, G. (2018). Whole-genome prediction of reaction norms to environmental stress in bread wheat (Triticum aestivum L.) by genomic random regression. *Field Crops Res., 216*, 32–41. https://doi.org/10.1016/j.fcr.2017.08.020

Mackay, I., Horwell, A., Garner, J., White, J., McKee, J., & Philpott, H. (2011). Reanalyses of the historical series of UK variety trials to quantify the contributions of genetic and environmental factors to trends and variability in yield over time. *Theoretical and Applied Genetics, 122*, 225–238. https://doi.org/10.1007/s00122-010-1438-y

Millet, E. J., Kruijer, W., Coupel-Ledru, A., Alvarez Prado, S., Cabrera-Bosquet, L., Lacube, S., Charcosset, A., Welcker, C., van Eeuwijk, F., & Tardieu, F. (2019). Genomic prediction of maize yield across European environmental conditions. *Nature Genetics, 51*, 952–956. https://doi.org/10.1038/s41588-019-0414-y

Molenaar, H., Boehm, R., & Piepho, H.-P. (2018). Phenotypic selection in ornamental breeding: it's better to have the blups than to have the blues. *Frontiers in Plant Science, 9*, 1511. https://doi.org/10.3389/fpls.2018.01511

Pavan, S., Delvento, C., Ricciardi, L., Lotti, C., Ciani. E., & D'Agostino, N. (2020). *Recommendations for choosing the genotyping method and best practices for quality control in crop genome-wide association studies.*

Robertsen, C., Hjortshøj, R., & Janss, L. (2019). Genomic selection in cereal breeding. *Agronomy, 9*, 95. https://doi.org/10.3390/agronomy9020095

Saccomanno, B., Wallace, M., & O'Sullivan, D. M. (2020). Use of genetic markers for the detection of off-types for DUS phenotypic traits in the inbreeding crop, barley. *Molecular Breeding, 40*, 13. https://doi.org/10.1007/s11032-019-1088-y

Scott, M. F., Fradgley, N., Bentley, A. R., Brabbs, T., Corke, F., Gardner, K. A., Horsnell, R., Howell, P., Ladejobi, O., Mackay, I. J., Mott, R., & Cockram, J. (2021). Limited haplotype diversity underlies polygenic trait architecture across 70 years of wheat breeding. *Genome Biology, 22*, 137. https://doi.org/10.1101/2020.09.15.296533

Smartt, A.D., Brye, K.R., Norman, R.J. (2016). Methane Emissions from Rice production in the United States — A review of controlling factors and summary of research. In: Moya, B. L., Pous, J. (Eds.), Greenhouse gases. INTECH. https://doi.org/10.5772/62025

Thiebes, S., Schlesner, M., Brors, B., & Sunyaev, A. (2020). Distributed ledger technology in genomics: A call for Europe. *European Journal of Human Genetics, 28*, 139–140. https://doi.org/10.1038/s41431-019-0512-4

von der Goltz, J., Dar, A., Fishman, R., Mueller, N. D., Barnwal, P., & McCord, G. C. (2020). Health impacts of the green revolution: Evidence from 600,000 births across the developing world. *Journal of Health Economics, 74*, 102373. https://doi.org/10.1016/j.jhealeco.2020.102373

Wang, J., Cogan, N. O. I., & Forster, J. W. (2016). Prospects for applications of genomic tools in registration testing and seed certification of ryegrass varieties. *Plant Breeding, 135*(4), 405–412. https://doi.org/10.1111/pbr.12388

Wellington, P. S., & Silvey, V. (1996). *Crop and seed improvement.* Henry Ling Ltd.

White, J., Sharma, R., Cockram, J. B., Balding, D., & Mackay, I. (2021). Genome-wide association mapping of Hagberg falling number, protein content, specific weight and grain yield in UK wheat. *Crop Science.*

Wilkinson, M. D., Dumontier, M., Aalbersberg, I. J. J., Appleton, G., Axton, M., Baak, A., Blomberg, N., Boiten, J.-W., da Silva Santos, L. B., Bourne, P. E., Bouwman, J., Brookes, A. J., Clark, T., Crosas, M., Dillo, I., Dumon, O., Edmunds, S., Evelo, C. T., Finkers, R., Gonzalez-Beltran, A., & Mons, B. (2016). The FAIR guiding principles for scientific data management and stewardship. *Sci. Data, 3*, 160018. https://doi.org/10.1038/sdata.2016.18

Würtenberger, G. (2017). Protection of plant innovations. In *Research handbook on intellectual property and the life sciences* (pp. 121–131). Edward Elgar Publishing. https://doi.org/10.4337/9781783479450.00015

Yang, C. J., Russell, J., Ramsay, L., Thomas, W., Powell, W., & Mackay, I. (2020). Overcoming barriers to the registration of new varieties. *BioRxiv.* https://doi.org/10.1101/2020.10.08.331892

Zhang, T., Su, J., Liu, C., & Chen, W.-H. (2019). Bayesian calibration of AquaCrop model for winter wheat by assimilating UAV multi-spectral images. *Computers and Electronics in Agriculture, 167*, 105052. https://doi.org/10.1016/j.compag.2019.105052

Zhou, J., Reynolds, D., Websdale, D., Le Cornu, T., Gonzalez-Navarro, O., Lister, C., Orford, S., Laycock, S., Finlayson, G., Stitt, T., Clark, M. D., Bevan, M. W., & Griffiths, S. (2017). CropQuant: An automated and scalable field phenotyping platform for crop monitoring and trait measurements to facilitate breeding and digital agriculture. *BioRxiv.* https://doi.org/10.1101/161547

Baladi Seeds in the oPt: Populations as Objects of Preservation and Units of Analysis

Courtney Fullilove and Abdallah Alimari

Abstract This essay argues that shortcomings in our approaches to global agriculture and its data infrastructures are attributable in part to a constricted application of population concepts derived from biological sciences in the context of international development. Using Palestine as a case study, this chapter examines the category of *baladi* seeds as a community-generated characterization of population, and one which arguably defies reduction to data. Drawing on quantitative research on farmer participation in informal seed production for wheat in the occupied Palestinian territories (oPt) and oral histories of farmers in the West Bank, this chapter analyzes the relation between participatory plant breeding initiatives, heritage narratives, and international agricultural research in rendering *baladi* seeds legible for archiving. It considers the multiple technological practices through which these institutions characterize and manage access to cultivated seeds, and how they differently approach problems of standardization, scalability, and variability. Through case studies of national and local seed saving initiatives, it asks, in turn, whether *baladi* seeds can be reduced to data, how they might be reduced to data, and whether they should be reduced to data.

1 Introduction

Fundamentally, data constructs a narrative around seeds, characterizing plants according to genetics, morphology, habitat, and a range of other factors. Yet people express numerous ways of living through seeds, in priorities and concepts imperfectly reduced by data schema. This chapter argues that shortcomings in our approaches to global agriculture and its data infrastructures are attributable in part

C. Fullilove (✉)
Wesleyan University, Middletown, CT, USA
e-mail: cfullilove@wesleyan.edu

A. Alimari
National Agricultural Research Centre, Jenin, West Bank, Palestine

© The Author(s) 2023
H. F. Williamson, S. Leonelli (eds.), *Towards Responsible Plant Data Linkage: Data Challenges for Agricultural Research and Development*,
https://doi.org/10.1007/978-3-031-13276-6_4

to a constricted application of population concepts derived from biological sciences in the context of international development. Data infrastructures reflect the priorities of the institutions the produce them, as well as the social and political contexts in which those institutions operate. As a result, data mirror the inequalities and exclusions of the societies in which they are embedded. In historical terms, the imperial/colonial framework of plant science provided the categories from which twenty and twenty-first-century data infrastructures are derived. These infrastructures have simultaneously enabled and restricted our ability to imagine alternative agricultures. Towards exploring these alternatives, this paper explores how multiple institutions and communities of practice define the *population* as an object of conservation, research, and development.

As a fundamental object of data infrastructures, biodiversity has multiple genealogies. As a term, it is commonly used to encompass species, genes, and ecosystems. It was deployed in the 1970s by conservation biologists concerned with species extinction, but also, in agricultural research, by breeders concerned with securing access to global plant genetic material for improved varieties. Beloved by proponents of community sovereignty, the concept of biodiversity is nevertheless trafficked by national governments seeking rhetorical and political tools for control of territory and natural resources, which are documented as biological populations in need of protection. Institutions dedicated to international development inherit this muddle of values and priorities; and so it is little surprise that their databases reflect the complexity and confusion of historical approaches to biodiversity preservation.

The concept of population applies to with cultivated plants quite differently than other flora and fauna, inasmuch as crops are human social and technological productions as well as natural objects. The diverse social and technological styles of agricultural production, and their variable relation to data concepts rooted in population biology, are the subject of this chapter. Within the field of agrobiodiversity preservation, data scientists often classify certain domains of research and production as "informal," where informal is a synonym for community. This identification runs the risk of ignoring diverse local institutional approaches to agricultural practice, which take shape in the absence of, and in opposition to, formal networks of seed production, distribution, and conservation.

Using Palestine as a case study, this chapter examines the category of *baladi* seeds as a community-generated characterization of population, and one which arguably defies reduction to common databases. Through case studies of national and local seed saving initiatives, it asks, in turn, whether *baladi* can be reduced to data, how it might be reduced to data, and whether it should be reduced to data. It considers the multiple technological practices through which institutions characterize and manage access to cultivated seeds, and how these differently approach problems of standardization, scalability, and variability.

Ultimately, this paper identifies a series of social and political considerations that trouble efforts to harmonize data produced in the context of international development. It does not propose universal technical solutions to these problems, because it holds that social and political solutions must precede and direct technological ones. This is a sobering insistence from a place where conflict seems intractable, and where power is alternately sapped by occupation and a bloat of international development agencies complicit in neoliberal development strategies. Agrobiodiversity preservation in the West Bank takes shape against the backdrop of Israeli occupation, which hobbles commercial agricultural development and intensifies dependence on Israeli imports of seeds and finished agricultural products. There is a necessary and relentless focus on access to land and water, amplified by the occupation and the construction of the separation wall snaking the West Bank. Moreover, in a post-Oslo Palestine, local NGOs acquiesce to a multitude of donor priorities and fall in line with their inconsistently expressed requirements. The result is an overlapping array of projects in pursuit of community empowerment, national sovereignty, and neoliberal development. Palestine is an illuminating case study not in spite but because of these tendentious questions of occupation and marginalization, and the ways in which rhetorics of food security and food sovereignty face off or muddle together. These are the world's problems, expressed pointedly in the extremity of the occupied West Bank.

2 "Population" as Unit of Analysis and a Target of Preservation

Ex situ gene banks remain the most prominent conservation strategy for cultivated crops and their wild relatives; but critics have charged that they are insufficient in multiple respects, severing the relation of plant genetic material not simply to its environs, but also to the farmers who have stewarded it. In response, agronomists and breeders have designed in *in situ* conservation strategies aiming to foster on-farm preservation. While both approaches to conservation have created spaces for sustainable agricultural improvement, they have often reified categories of "landrace" and "heirloom" that mark locally adapted seeds as stable artifacts of past agricultural practices, to be collected and preserved in static form. The concept of landrace presupposes a regional ecotype, locally adapted variety, or traditional variety of a domesticated species of plant or animal, generally distinguished by its isolation from other populations of the species. It is typically opposed to a cultivar, produced by selective breeding and maintained by propagation. Practice suggests a more fluid relation between on-farm and ex-situ improved varieties. The landrace concept has been called into question in part because of the hard line it draws between laboratory-based breeding and farmer selection (Berg, 2009). In addition, many "heirloom" seeds are a previous century's commercial varieties, suggesting the ways in which agricultural knowledge is characterized by mobility rather than stasis, and transaction rather than withholding.

This muddle derives from the imperfect application of the population concept to diversity in cultivated plants. In the simplest terms, a population is "all coexisting individuals of the same species living in the same area at the same time." The primacy of the population concept derives from a historical focus on species as the primary unit of analysis, beginning in the natural sciences of the eighteenth century. The species unit has remained fundamental to the twentieth-century disciplines of population genetics and community ecology, as well as their integration in the new population biology of the 1950s. These sciences of the "New Synthesis" were in turn prerequisite to the founding of conservation biology as a "science of crisis" in the 1970s, and subsequent attempts to mark populations for conservation and restoration (Simberloff, 1988; Soulé, 1985). In international agricultural research, population concepts derive from agronomy and conservation biology. Their primary orientation remains toward species protection, codified in the structure of gene banks according to Linnaean binomials. This static taxonomic practice, fundamental to historical plant database design, remains dominant in all formal agricultural research. These continuities obscure the fact that agriculture itself is one of the greatest disruptors of ecosystems. The large-scale farming of the nineteenth and twentieth centuries have intensified this disruption. Even so, agricultural improvement relies on the introduction new genetic material well adapted to local conditions; and thus an imperative to preserve species richness is a focus of modern conservation policies.

Efforts to map species were aspects of a European imperial project to identify nature in an original state, and to justify colonial management of resource stocks (Davis, 2009, 2015; Drayton, 2000; Grove, 1995). The quest for useful plants provided the machinery of imperialism and colonization through European botanic gardens. Forged against fears of colonial degeneracy and the pursuit of valuable natural resources, these scientific projects provided the foundation for nineteenth and twentieth-century models of development rooted in concepts of social evolution and economic growth. In the wake of imperial collecting projects, European and American governments continued to sponsor extensive natural history expeditions (Anker, 2001; Pauly, 2007). The heirs of European botanic gardens oversaw the institutionalization of new sciences of the environment, with institutions such as the New York Botanic Garden incubating the discipline of ecology (Kingsland, 1995; Mitman, 1992). The coalescence of ecology as a discipline in the early twentieth century brought new attention to the study of how organisms live in their environments, and intensified the development of a "baseline concept" in conservation efforts (Alagona et al., 2012). The interwar period, in turn, saw the international development of mathematical models in population growth and dynamics, competition, and predation, inspiring new approaches to the study of biology and population genetics (Huneman, 2019).

By the 1940s, the biologist E.O. Wilson, Ronald Fisher, and others contributed the insights of population genetics to an institutional and intellectual movement ultimately celebrated as a new Darwinian synthesis. Population biologists of the 1950s linked the driving questions of community ecology and population genetics through theorizations of ecological niche and island biogeography, and, crucially, through the application of mathematical modeling to the history of life on earth. As with any synthesis, this one concealed divisions (Huneman, 2019; Kingsland, 1995). Inter and intra-disciplinary debates regarding the relative merits of experimental and laboratory work, theory and practice, and modeling vs. field study are not unique to population ecology; and, indeed, we see them echoed in contemporary discussions of the application of big data to a range of practices, including agriculture and agro-biodiversity. Mathematical models produced striking insights, and yet they seemed to bely the messiness, complexity, and fundamental uncertainty of the life they aimed to characterize. In the field, it is never so simple.

The new synthesis echoed the timelessness of Linnaean natural history rather than the changeability pursued by Charles Darwin and others (Huneman, 2019; Kingsland, 1995). That is: the twentieth century pursued the fixity of the 18th, in denial of the intervening century's messy confrontation with evolution. Crucially, this confrontation was enacted not simply in the theory of natural selection, or in the social Darwinism of Herbert Spencer, but in the agricultural lands of the Maghreb, the Americas (including the American South), and the East Indies. These were the fields of Euro-American colonial expansion, converted for global commodity export. By the mid-twentieth century, they were the sites of agricultural modernization projects. By the 1970s, they were the hosts of a network of Consultative Group on International Agricultural Research (CGIAR) centers for research on food security, rural poverty, and sustainable development.

CGIAR was founded against the backdrop of international agricultural modernization. From the 1950s to the 1970s, the US and Europe competed to establish themselves as dominant exporters of food, then of agricultural inputs, based on a model of input-intensive industrial agriculture. The export of high-yielding seeds and agricultural methods developed by American agronomists aimed to usher in a "Green Revolution," averting the Red alternative of Communism by increasing rural prosperity. Global conservation strategies developed to match these agendas, ori-ented at first toward state control of natural resources, and then toward an international order that recognized the sovereignty of member states over others. Aiming to build on the alleged successes of the Green Revolution, the Food and Agriculture Organization of the United Nations supported programs of agricultural modernization and the free exchange of germplasm between countries for the use of breeders.

CGIAR's mandates for food security and sustainable development included large-scale programs for agro-biodiversity preservation, the most notable of which was a network of international gene banks to amass landraces, and, later, wild relatives of target grains and legumes. When international agricultural research centers turned their attention to biodiversity loss, it was to argue that public and private breeders should have access to global plant genetic resources: moving seed stocks out of the field and into banks from which they could circulate to countries with the capital to pursue research (Curry, 2017; Fenzi & Bonneuil, 2016; Flitner, 2003; Fullilove, 2017; Saraiva, 2013). Today, international research organizations govern the free transfer of global germplasm through standard material transfer agreements (SMTAs) defined by the International Treaty on Plant Genetic Resources (2001). (The Middle East and North Africa is served by the International Center for Agricultural Research in the Dry Areas (ICARDA), headquartered in Syria until 2010, and now in Lebanon.)

Historical arguments for conservation are often nostalgic; and the past provides an imperfect guide to the future at best (Alagona et al., 2012; Cronon, 1992, 1993). In spite of the fashion for heritage seeds and landraces untainted by modern breeding methods, the quest for origins is in many ways misguided, masking the fluidity of agro-biodiversity. These shortcomings suggest the ways in which a focus on species fails to characterize biological diversity, within and beyond the practice of agriculture.

Moreover, arguably human beings have been under-theorized in most studies of populations, defined as ecosystems managers rather than objects of study. The application of entomologist Paul Ehrlich's (1968) population studies to human beings and subsequent discussions of the planet's "carrying capacity" are the exception that prove the rule. These alarmist scenarios of a global overpopulation crisis provided the basis for imperatives of yield that have governed international debates about "food security" from the post-World War II period to the present day.

Alternative approaches to preserving agro-biodiversity have the potential to elevate social and political considerations. Agroecological approaches favor a focus on ecosystem over species, toward polycultural models of production. Agro-ecological approaches have applied practices such as nutrient cycling and intercropping to modern agriculture, drawing on techniques developed over millennia of agricultural practice and applied by farmers across the world (Altieri, 1995). Intellectually, agroecologists are indebted to these millennia of farmers. More narrowly, the discipline draws on concepts of ecological succession and landscape formulated by community and population ecologists such as Frederic Clements and Henry Gleason. Since the 1980s, agroecological approaches have been popularized by agronomists such as Miguel Altieri not simply for their ecological aspects, but also for their social and political implications. These implications are made explicit by the global food sovereignty movement Via Campesina, which promotes agro-ecological methods as an expression of traditional peasant farming.

3   *Baladi* Seeds in Occupied Territory

In recent decades, international agricultural researchers have endeavored to include farmer knowledge in data infrastructures and plant breeding projects. Perhaps ironically, agrarian knowledge provides both the source and the target of their innovations. In Palestine, which provides the case study for this paper, collectors seek local varieties, drawing on the knowledge of local farmers to identify *baladi* seeds (literally "my country," and connoting local and traditional production, native to place) (Nadar, 2018). In common usage, one could regard *"baladi"* as a synonym for local, and it connotes a similar array of associated, yet contested values: community, tradition, ownership, and stewardship, to cite a few examples. In a biological context, *"baladi"* refers generally to a population comprised of numerous heterogeneous lines with their own individual characteristics. In wheat, for example, characteristics might encompass resistance to drought, pests, and rusts, as well as traits related to gluten content and yield (Nadar, 2018). Collectors render *baladi* populations legible for archiving through morphological analysis, physical multiplication, and multiple documentation processes. Thus, even as it shelters and generalizes enormous diversity, the population remains the the object of preservation and the principal unit of analysis.

But *baladi* seeds are differently characterized in projects that seek to express community values of taste, appearance, and texture as primary. That is, *baladi* seeds may stabilize through other means than line characteristics, such as the stories woven around them. Overlapping oral, literary, and documentary practices do not, however, have the same status as data. That is, only certain markings are viable representations of agrarian knowledge in international research and development. The remainder of the essay explores some of these alternative characterizations of population through a survey of four institutions pursuing agro-biodiversity preservation projects in the occupied Palestinian territories (oPt).

In spite of the challenges posed by climate change and occupation, Palestine has one of the highest concentrations of agrobiodiversity in the world, consisting of wild pulses, grains, woody plants, and trees that humans began to modify and domesticate about 12,000 years ago (Tesdell et al., 2020). It is a center of diversity for the crops of the Neolithic (wheat, barley, bitter vetch, chickpea, lentil, flax, and oat) as well as numerous legume species and tree crops (Tesdell et al., 2020; Zohary & Feinbrun-Dothan, 1966). In scientific terms, drylands such as Palestine's are a focus of twenty-first-century breeding research because they host plants and crop varieties adapted to drought, salinity, and high temperatures. These qualities make them objects of interest in the face of global climate change. Seeds form the basis of new research into drought resistant wheat varieties, and through "pre-breeding" can introduce genetic material into parental germplasm used in the production of new seed varieties (e.g. Buerstmayr et al., 2012).

Agricultural science in Palestine took shape against the background of European colonial policy after World War I, Israeli national development after World War II, and Israeli occupation of the West Bank after 1967. Each facilitated governance by circumscribing and cataloguing practices of cultivation in the language of Euro-American agricultural science (Tesdell, 2013). In the early decades of the twentieth century, international wheat breeding initiatives, and the focus on Palestine as a site of domestication, helped remake drylands as targets of colonization (Tesdell, 2017). Although the nascent state of Israel (1946) stood apart from the Cold War on hunger in the third world, it followed a very similar trajectory to other colonized territories in categorizing local agriculture. Policy discourses about local land use mythologized some agricultural practices and degraded others, using historical legal and scientific pretexts to justify intervention (Tesdell, 2013: 79, Salzmann, 2018). A primary theme was that Palestinian agriculture was degraded, backward, primitive, and that the landscape was wasted and barren. The Ottoman-Israeli legal apparatus was used to mark lands as uncultivated, thereby claiming them for the new state of Israel (Tesdell, 2013: 84; Cohen, 1993; Tyler, 2014). These fictions facilitated occupation, governance, and the cultivation of dependency. In June 1967, after brief but decisive conflicts with the surrounding Arab states, Israel occupied the West Bank, along with Gaza, Sinai and the Golan Heights. In the West Bank, Israel supported policies of agricultural modernization intended to bind Palestinian farmers to the Israeli state technical apparatus (Tesdell, 2013: 86). As local production declined and

Palestinians entered the Israeli wage labor market, Palestine effectively became "a captive market for finished Israeli goods" (Abu-Sada, 2009).

While Israeli occupation took on distinctive forms, it shares features with the neoliberal, globalized food system derived from European imperial geopolitics: specifically, as Philipp Salzmann has characterized it, land grabbing, or "accumulation by dispossession. .. within the corporate food regime." Israeli pretexts for land dispossession resembled those used in other settler colonies: displacing current inhabitants, characterizing territory as uncultivated, and casting peasant agricultural practices as primitive and unproductive. While the market replaced the state as the "primary guarantor of food security" after the 1970s, it remained sponsored and enabled by dominant states (Salzmann, 2018). The signing of the Oslo accords in 1993 left Palestine under the twin control of Israel and international finance institutions, marking a moment of neoliberal restructuring and defeat for a nationalist project of liberation (Salzmann, 2018; Samara, 2000). Specifically, the division of the West Bank into Areas A, B, and C, with Area C under full Israeli administrative control, normalized dispossession of Palestinian territory. This reordering paved the way for incursions of Israeli agribusiness and further contributed to the marginalization of rural communities based on peasant agriculture. The World Trade Organization (WTO) and International Monetary Fund (IMF) further institutionalized asymmetries in power between states inherited from their imperial pasts, dictating loan conditions to the governing Palestinian Authority (PA) (Holt-Giménez & Shattuck, 2011; McMichael, 2009; Salzmann, 2018). The PA's rural policy, outlined in the 2008 Palestinian Reform and Development Plan (PRDP), adopted a market vision for agriculture supported by international lenders (Salzmann, 2018). The PA's acquiescence to neoliberal structural adjustment policies also hobbled community development initiatives and economies of resistance that had flourished during the first Intifada (1987–1993) (Kuttab, 2018).

The depoliticized development practice that took shape catered to donors rather than to communities. Palestine has received some twenty-four billion dollars of assistance since 1993 (Kuttab, 2018: 76). In 2008, the Agricultural Project Information System, managed by the Palestinian Ministry of Agriculture with assistance from the FAO, included some 170 international non-governmental, local nongovernmental and community-based organizations, UN agencies, and donors that represent the agricultural sector of West Bank and Gaza Strip (FAO, 2008). The PA remains subservient to the priorities of international actors and donors.

In this climate, international organizations have taken up the mantle of European colonial governments in shaping institutions and regulations to organize natural resources in occupied territory. International development agencies prioritize market potential for an expanded agricultural sector liberated from the impediments of occupation. UNCTAD emphasizes that Palestinian agricultural yields are 43% of Israel's and half of Jordan's. It recommends support required to develop Palestinian agricultural infrastructure, support farmer cooperatives, and stabilize production and transportation costs. The implicit goal is to increase the productivity of the Palestinian agricultural sector for the purposes of trade and development (UNCTAD, 2015).

In practice, agrobiodiversity and rural development projects cross formal and informal domains: CGIAR-funded agricultural research promoted by the Ministry of Agriculture, Palestinian NGO-directed community seed banks supported by international aid, and volunteer-based community organizations oriented toward Palestinian heritage and sovereignty. These various overlapping informal projects skate under the radar, contributing to a patchwork of data infrastructures and undocumented practices. This institutional drift, which is in many respects the product of a post-Oslo development landscape, creates a Swiss cheese of data infrastructures, which in turn masks a Swiss cheese of development and conservation priorities.

4   International Agricultural Research

Institutions dedicated to scientific research interface distinctively with Palestinian agriculture, even as they remain oriented toward market agriculture. In recent decades, biodiversity preservation advocates have emphasized that *ex situ* conservation of seeds in genebanks must be complemented by *in situ* conservation of traditional farming systems, which are often confined to drylands and mountainous areas not extensively cultivated for commercial purposes. While strategies for *in situ* preservation have been drafted by research funded by the Global Environment Facility (GEF)/United Nations Development Program (UNDP), these programs have retained a primarily development-oriented perspective (Freeman et al., 2005). Such research emphasizes the adaptability of landraces to harsh conditions and low input agriculture, and it recommends the pursuit of increased yields through participatory breeding, water harvesting, conservation agriculture, and integrated pest management. These values also make biodiversity loss in arid regions an object of concern for international research. The International Center for Agricultural Research in the Dry Areas (ICARDA), established in 1977 in Tal Hayda, Syria, is the CGIAR center for the Middle East and North Africa, with a broader mandate to promote agriculture in non-tropical dry areas. Its stated mandate is to improve the livelihoods of resource-poor farmers in dry areas through delivery of its research output, working within national agricultural resource systems and directly with farmers. A GEF-funded, ICARDA-coordinated project on "conservation and sustainable use of dryland agrobiodiversity in Jordan, Lebanon, Palestinian Authority and Syria" conducted eco-geographic and botanic surveys in 75 monitoring areas from 2000–2004, identifying threats to wild relatives of field crops, forage legumes and fruit trees. While overgrazing, wood cutting, poverty, and weak environmental protection laws posed threats in the region as a whole, the project identified the "political situation in the West Bank" as a primary threat to biodiversity (Amri et al., 2008).

In Palestine, ICARDA's capacity building projects are channeled through the National Agricultural Research Center (NARC) in Jenin (Northern West Bank). These include its participatory plant breeding programs and the establishment of a gene bank targeting traditional varieties of cereals, legumes, and forage crops.

NARC, which receives variable funding from the UNDP/GEF, the Ministry of Agriculture, the government of the Netherlands, and numerous international NGOs, pursues a range of research in the fields of rainfed agriculture, wastewater conversion, drought-tolerant crops, and informal seed production. ICARDA's participatory plant breeding program, inaugurated in Syria in 1996, has been replicated in 11 countries, including NARC's implementation in the West Bank (Nadar, 2018).

NARC's Genetic Resource Unit (GRU) was the fruit of 2011 funding to support field crop landraces, ultimately resulting in the establishment of the gene bank in 2013. Its objectives are to support agrobiodiversity imperiled by climate change, drought, and disease. Consistent with CGIAR imperatives, the GRU regards crops and their wild relatives as material for breeding improved varieties. Also consistent with the CGIAR centers, its primary clients are researchers and institutions rather than individual farmers or the general public. The GRU provides the infrastructure for a national genebank, with all associated documentation practices. Collection targets ecologically and culturally precise regions: for example, central highland villages outside of Bethlehem and Hebron, known for their drought resistant wheat varieties. Researchers interview farmers about *baladi* varieties, with a focus on elders. Notably the same crop variety may have different names depending on the community in which the crop is grown, creating challenges in documentation. Collected seeds are studied for crop variations characteristic of *baladi* seeds. Accessions are multiplied, dried, catalogued, and split into short, medium, and long-term storage, with label noting GPS coordinates, scientific name, local name, date stored, and viability term (Nadar, 2018). NARC's data standards are consistent with ICARDA's and utilize passport descriptors.

The presence or absence of a national gene bank has major implications for the representation and claims to Palestinian flora. In the absence of an internationally recognized genebank, plants are collected in the Israeli National Genebank, often with different naming conventions. In addition, continued dependence on international donors and the Palestine National Authority's Ministry of Agriculture (MoA) budgets imperils national agricultural research and agrobiodiversity preservation. Without adequate funding for collection and preservation, the gene bank may be underutilized. As an outcome of these conditions, and of continued lack of state recognition, Palestinian flora remain represented by proxy in the state of Israel and its scientific institutions. This condition exacerbates geopolitical inequalities in the production of scientific knowledge.

NARC's goals conform to the MoA's for "increasing agricultural production and productivity and improving livelihoods of the farmers" (MoA, 2013). The MoA strategy prioritizes food security in the West Bank with a fundamental market orientation. The primary targets of its research are grains and legumes, including wheat, chickpeas, and faba beans. NARC shares 70 dunums of land in Beit Qad with a Jenin farmers' association, which benefits from partnership in participatory plant breeding trials and other projects. Farmers in the Jenin region, which is regarded as a breadbasket of Palestine, are comparatively well-served by both NARC and the MoA. This position of privilege may also translate into commitments to small-scale agriculture. Many of these farmers express a commitment to *baladi* seed production, articulated as commitments to Palestinian land inheritance and persistence on the land, as well as preferences for rainfed (*ba'al*) crops requiring less water and fertilizer. They express pride in their relationship with NARC in their production of seeds and knowledge for other farmers, including traditional varieties. They note, for example the superior qualities of smell, color, and flavor of *baladi* wheat varieties for traditional Palestinian cuisine in opposition to hybrid wheat stocks imported from Israel (Nadar, 2018).

Among NARC's projects are surveys regarding participation in community, or informal, seed production, which it identifies as a target for increasing food security by decreasing imports and developing varieties well suited to dry conditions and rainfed agriculture (Istaitih et al., 2020). Community seed production also offers a strategy to reduce dependence on Israeli imports, shore up land claims, and resist the archiving of Palestinian flora as an Israeli national project. To determine farmer participation in informal seed production of wheat in Palestine, NARC conducted surveys of 145 farmers from major seed production sites in Palestine" (Istaitih et al., 2020). The survey aimed to ascertain who participated in these programs and why, finding that "farmers' participation in seed production was significantly influenced by a range of factors, including seed source, planting date, rainfall and productivity, membership in agricultural association, technology adoption, capacity building, frequency of extension contact, and net returns and profit. The most important reason that the participants wanted to participate in the seed production was the access to improved input, the increase in rate of net returns, increase in profit and decreased production costs (Istaitih et al., 2020).

Toward that end, NARC works with agricultural cooperatives to distribute seed and promote best practices. In effect, agricultural cooperatives, liaising with NARC, become middlemen to community seed producers. These practices promote a range of technologies identified as appropriate to the region, including promotion of high yielding and drought tolerant forage crops/species and wild relatives, waste water conversion, scaling up of established water harvesting techniques, and best practices for cultivating medicinal and aromatic plants. The characterization of community production as informal, however valued and however accurate, underscores the market orientation of national agricultural research. Rural development and biodiversity preservation efforts at a national level bridge community empowerment, the PA's market orientation, and movements for international recognition in governance and R&D.

5   Heritage Narratives

In contradiction to a neoliberal development model stand an array of community institutions, universities, and Palestinian NGOs committed to community prosperity through local stewardship of natural resources.

Independent of the Ministry of Agriculture's sponsorship, Palestinian NGOs have led the charge in organizing farmers. This arrangement stems from the founding of multiple agricultural relief committees in the 1980s, primarily in the context of the first intifada (1987–1993). Now nearly forty years on, the politics of Palestinian agricultural NGOs bear faint resemblance to their founding moments, especially as international aid has altered the profile of each organization. Yet some features remain. The agricultural NGOs focus on control of natural resources, including land and water, as a means to achieve both self-sufficiency and territorial sovereignty. They pursue the cultivation of land, primarily through rainfed agriculture, both to diminish dependence on Israeli goods and to forestall Israeli confiscation of uninhabited lands. Like NARC, these NGOs support agricultural modernization to enable competition with Israeli producers; and they organize agricultural cooperatives throughout the West Bank. Cultivating land makes it inaccessible for settlement or protective restrictions applied to nature preserves (Abu-Sada, 2009: 416).

NGOs provide an example of community biodiversity management (CBM) approaches prioritizing community-driven, participatory approaches to biodiversity management and local and subsistence farming (Nadar, 2018; Subedi et al. 2006; Thijssen et al. 2013). Community seed banks have been widely established in Asia, Latin America, and Africa beginning in the 1980s. The Middle East has lagged behind these trends. Advocates hold that CBM empowers farming communities to promote conservation and use of local biodiversity, further aided by partnerships with institutions of research and development (Boef, 2013). Community based seed banks are not necessarily incompatible with national strategies. Several Palestinian NGOs (Union of Agricultural Work Communities [UAWC], Palestinian Agricultural Relief Committees [PARC], Applied Research Institute of Jerusalem [ARIJ]) and universities [Al Najah, Al-Quds, Al-Azhar]) have an MOI with the Ministry of Agriculture and NARC in an attempt to avoid duplication of efforts (Zayed, 2020).

In practice, Palestinian NGOs constantly renegotiate their relationship to international networks of science and capital. NGOs nevertheless prioritize community needs rather than national or international ones. Community seed banks also operate with a shorter time-scale, serving the farmers of the present rather than those of a hypothetical future. To some extent, this frees them from the arts of abstraction and renders immediate their pursuit of drought-resistant crops for which there is local demand: Battir eggplant, Dura serpent squash, and tomato. Moreover, overt orientation toward food sovereignty injects human social life into any ecological equation.

Nor is biodiversity preservation confined to national governmental and nongovernmental organizations. At the university level, nearly every Palestinian University has departments and centers dedicated to water resource management and sustainable land use. At the community level, Um Selim farm, the Bethlehem Farmers' Association, Bustana Community Supported Agriculture (CSA), and the Heirloom Seed Library in Battir, among others, prioritize the preservation of local seed varieties. These are grassroots organizations operating independently of institutionalized research and development.

Perhaps the most well publicized of these is the Palestinian Heirloom Seed Library directed by Vivien Sansour. In 2018, Al Jazeera dubbed Vivien Sansour "the Seed Queen of Palestine" (Anonymous, 2018). Four years earlier, Sansour founded the seed library and the associated organization El Beir Arts and Seeds, crowd-sourcing via Facebook and combing markets to find traditional seeds being cultivated in Palestine. Sansour prioritizes "seed revival" rather than seed saving, enlisting farmers in her project through a mix of enthusiasm and persistence. As a sequel to the Seed Library, she founded the Traveling Kitchen, which hitches itself to a truck and travels from town to town, cooking up the produce of the library. In this practice, she builds on her previous work at Canaan, an olive and almond Fair Trade company selling for export (Nadar, 2018).

The Seed Library is small, enrolling 20 farmers and 4 threatened varieties, including *Jadu'i* watermelon, white (Sahour) cucumber, and *Abu Samra* ("father of the dark one") wheat, also sometimes called *Kahla* ("dark eyes"), or *Haba Soda* ("dark seeds"). (The variety is named for its long, dark awns and dark seeds.) Sansour also raises awareness of wild plants used in Palestinian cuisine. The Israeli government now classifies Akkoub, a thistle commonly used in Palestinian cuisine, as an endangered species. With collection prohibited by law, foraging has become a site of conflict, with Palestinian youth detained by the Israel Defense Forces (IDF). Sansour and others hold that the expansion of Israeli settlements in the West Bank is a primary driver in the plant's endangerment. She adds settlements to Israeli agribusiness and climate change as threats to traditional agriculture.

The Seed Library begins by identifying *baladi* seeds with the aid of local farmers. Seeds are catalogued with crop name and year collected, placed in jars, and shelved in the Seed Library headquarters in Beit Sahour. The library retains seeds for less than a year. The priority is to distribute seeds to farmers, who may "check out" seeds and replenish the stock at the end of the season. As of 2018, the Library housed 40 varieties of *baladi* seeds. Along with the seeds, the Library records stories associated with the crops. It also pursues alternative food networks linking producers and consumers outside of regular market structures (Nadar, 2018).

Sansour is a master storyteller. Putting aside the value of biodiversity and climate change resistance, she links seeds to stories of identity and belonging. Perhaps deliberately choosing a genetic metaphor, she offers that "seeds carry the DNA of who we are, our culture, the work of our ancestors" (Nadar, 2018). She aims to promote agriculture by telling the stories that have made it durable. Hers is a strategy to combat knowledge loss and retain the oral tradition of agriculture, which precedes written language, never mind digital infrastructures.

Sansour has well-trafficked stories about Battir's ancient terracing and aqueduct system (now a UNESCO World Heritage site), and the havoc wreaked by the snaking of the Green Line across its borders. (Only concerted effort by Palestinians and Israelis preserved the village and surrounds from demolition for the construction of the Separation Wall tracking the Green Line.) She celebrates Battir's *baladi* seeds as continuously cultivated since the second c. BCE. El Beir sponsors terrace revival in the zone of the wall, including the planting of *mulukhiya* for stew. Sansour tells stories about how women remember giving birth among the *Jadu'i* watermelons, and about the Quality Street chocolate tins full of seeds kept in every grandmother's drawer. She talks about cooking purple carrot stuffed with pine nuts and rice (Anonymous, 2018). These stories remind us that women are the majority of the world's farmers, and that seeds belong to communities. Fundamentally, her work is about reminding people of their worth and their sense of belonging, in opposition to the imperatives of the market, the constraints of international aid, and the imposition of neoliberal approaches to development in the Middle East and North Africa.

Perhaps it is no surprise that Sansour's interest in agriculture was provoked by her experience in Chiapas, Mexico, where she helped build a cistern for highland coffee growers. During a break, an elder served the team papaya grown from his great grandfather's seeds, inspiring Sansour to consider her own agricultural heritage in Palestine. Chiapas, the site of the anti-NAFTA Zapatista rebellion in 1994, became a center of anti-globalization and indigenous activism in the 1990s. Efforts by the International Cooperative Biodiversity Groups (ICBG) and prominent ethnobotanists to collect Mayan medicinal plants and knowledge collapsed amid charges that ICBG's commercial partners included numerous transnational pharmaceutical and agrochemical companies (RAFI, 1999; Berlin & Berlin, 2003, 2004; Hayden, 2003). The specter of Chiapas haunts twenty-first-century biodiversity collecting enterprises, suggesting that the preservation has always been coupled with exploitation. The proper relation between smallholder agriculture and commercial monoculture remains unresolved, as do broader questions of how local communities should interface with international markets.

Sansour and her collaborators reject the globalizing force of the market. They embrace participatory as opposed to technocratic projects, drawing on the knowledge and engagement of communities. This approach echoes those which elevate traditional environmental knowledge and local knowledge, and promote community involvement in restoration projects. In doing so, they may (or may not) acknowledge the variability of collective memory, preferences, and values in dictating the meanings of the past. (Alagona et al., 2012; Eliott, 2008). These allowances, however consciously made, create a conundrum for the collection and transcription of data, inasmuch as they call into question the very idea of data as "the givens." One farmer who grows hybrid and traditional varieties was asked about the benefits of the latter, to which he replied: "hybrids can give me high yields, but it's not consistent. Hybrids breakdown the soil, eventually stops producing. *Baladi* is consistent, and you get to preserve your country, conserve your tradition. *Baladi* is timeless" (Nadar, 2018). Here, the data is the story, and it is meaningful in the act of retelling.

But there is no such thing as a timeless seed.

6 Agroecosystems

If the vocabulary of taxonomy and genomics universalizes targets of preservation, agroecology prioritizes locality. Through the research group Makaneyyat, which he co-founded in 2015, geographer Omar Tesdell and his collaborators aim to conserve local agrobiodiversity and develop perennial agroecosystems that support polycultures (Tesdell et al., 2020). They see this as a strategy to combat the massive decline of agriculture in Palestine, including wheat, barley, and pulses like lentil and chickpea. They attribute this shift to the transition to wage labor in the Israeli economy, as well as Israeli policies restricting access to land, water, and markets. Olive production provides the single counter-example to a trajectory of agricultural decline. Yet olives themselves were formerly components of polycultures (e.g. olive-grape-wheat), which historically promoted soil health and community subsistence. Tesdell and team aim to design perennial polycultures within existing olive groves with the objectives of improving biodiversity, reducing tillage, rebuilding soils, and providing resilience for climate change.

The founding of Makaneyyat resulted in the compilation of large and diverse data sets consisting of archival sources, aerial images, interviews, and fieldwork. Makaneyyat's methods encompass the frameworks of landscape ecology, geography, and ethnobotany, and history. The research group uses an open-source agroecological research engine, "which allows researchers to manipulate, filter, visualize, and store agroecological data in order to drive their own investigations" (Tesdell et al., 2020). Given the interdisciplinary nature of the work, the archive consists of genetic material, geodata, and ethnobotanical information. In addition to polyculture design, its work encompasses *in-situ* and *ex-situ* conservation of wild food plants and crops, as well as digitization of existing primary source floras of Palestine. A priority of the project is on the "lived experience and knowledge of local farmers and foragers to coproduce knowledge that is relevant to Palestinian communities and agroecosystems." Makaneyyat draws inspiration from Wes Jackson's Land Institute in Salina, Kansas, which aims to develop new perennial grain crops adapted to the prairie's native ecosystem. Makaneyyat collaborates with the Land Institute on research design, candidate selection, and community-based approaches, including "the use of open science models to build climate adaptation into agriculture," developing perennial grain culture oriented toward ecosystem stability (Tesdell et al., 2020).

Acknowledging the historical wealth and complexity of Palestinian agriculture, Makaneyyat and others are less concerned with restoration than attempts to create ecosystems for the future. These projects martial historical evidence and claims about the past to attempt renewed production and partial reconstruction of agrarian landscapes transformed by colonization and occupation. But this makes them ill-fitted to the data regime constructed under the auspices of the International Treaty for Plant Genetic Resources for Food and Agriculture, which aims to protect farmers' rights by insisting that their knowledge and labor belongs to the world (FAO, 2009: v).

Drawing on social ecological and agroecological approaches, Makaneyyat attends to the social as well as biophysical aspects of agriculture in its approach to community and socioeconomic relations. By extension, its approach acknowledges the value of historical agricultural inputs (here, seeds and crop wild relatives); but it regards them as components of a social system rather than isolated variables, and it considers the social system rather than the inputs to be its primary focus. It explicitly rejects the "agrilogistical" approach of annual grain culture, which holds people apart from nature and demands intensive disturbance of local ecosystems. It finds patriarchy and settler colonialism implicated in the broader shift to input-intensive and carbon exploitative agriculture of scale (Streit Krug & Tesdell, 2020).

This overtly utopian outlook on perenniality and diversity, oriented clearly toward social relations rather than technological solutions, sits awkwardly in an international system targeting seeds as determinative inputs of successful agricultural systems. It also scales poorly, focused as it is on the precise needs and configurations of local communities. That is, its "landscape-scale agricultural intervention" makes plant knowledge a social process rather than an object to be universalized as data.

Makaneyyat at odds with related projects constructed on values and infrastructures of improvement, development, capitalism, and settler colonialism; but it does not exist in isolation from them, nor in simple opposition. Rather, Makaneyyat draws widely on available technologies, leveraging global datasets as well as local observations to build knowledge about the Palestinian landscape. Its architects regard gene editing not as a bogeyman but as a path to domestication of new crops for degraded and stressed environments (Van Tassel et al., 2020). It seeks to ally itself with open-source software and seed movements through the adoption of open-source data management and storage tools. Arguably these are ideological commitments as well as practical ones; but they nevertheless set the stage for future projects that take collaborative labor as the basis of agricultural knowledge rather than an object to be secured or commodified. In policy terms, openness and transparency provide the technical basis for just social relations to devise novel agroecosystems.

7 Conclusions

As Alfred Lotka and his heirs brought mathematics to ecology, so has big data come to biodiversity. But it is not clear where to go from here. Since their 2016 publication, the FAIR data principles of findability, accessibility, interoperability, and reusability have been endorsed by a broad range of stakeholders in digital data (Jacobsen et al., 2020). This wide acceptance signals the ascendance of Open Science as an international movement, and of FAIR in the age of Big Data (Mons et al., 2017).

Good data stewardship, however, is incomplete without a more broadly based ethics and practices of stewardship to undergird it. That is, these are not merely questions of implementation, but rather questions about the quality and scope of the data itself. In summary, "the givens" itself never was.

FAIR provides regulations for access to data, but, ironically, these may further limit the scope of the data itself, disqualifying legacy collections or sources badly suited to reduction. Too often, open source data projects manage to excel at the stated principles of access and usability while still failing to make room for other forms of qualitative data, narrative sources, and archival material. Scholars oriented towards "communities of practice" have pursued alternative formulations for data collection, but they remain hamstrung by inability to scale up to the level of big data (Louafi et al., this volume). In response to the limits of FAIR, the International

Indigenous Data Sovereignty Interest Group of the Research Data Alliance has developed 'CARE Principles for Indigenous Data Governance' (Collective Benefit, Authority to Control, Responsibility, and Ethics) (Carroll et al., 2020). These guidelines were the products of consultations with Indigenous Peoples, scholars, non-profit organizations, and governments, and authored by a network of Indigenous data sovereignty networks. Like FAIR, CARE specifies principles with little guide to application. But their publication is a clear demand to orient conversations about international data sharing away from the funding bodies who endorse FAIR, and toward the communities they have too often failed to engage.

NARC, Makaneyyat, and El Beir provide different approaches to interfacing with international agricultural research data infrastructures and the Open Source data platforms that have gained traction in the past decade. Their varied paths suggest that there is no technological end run around the social and political problems posed by international agricultural development and biodiversity preservation initiatives, but rather an approach to technology as a set of diverse material practices associated with particular communities. By representing the values of communities in their data collection and preservation, these institutions aim to counter exclusions reproduced in the legacies of imperial science and governance.

Between science, policy, and community are the spaces to be cultivated. In Palestine, it's often difficult to get farmers on board with agricultural development programs for many reasons: because these programs should pay, and they don't. Because they should make things easier, and they can't. Because although it may be an ecologically appropriate technology, no one wants to manually dig hundreds of curved swales in the ground to channel rainwater: because, whether or not it was traditional agricultural practice in the region, it is back-breaking labor, and the returns don't justify it. Because no one wants to grow vegetables they can't sell amid a glut of lower priced Israeli products. In place of market incentives, there are fervent political commitments to Palestinian sovereignty and heritage, with farming deployed to prevent land seizures and to promote solidarity and community. These very commitments mitigate participation in neo-colonial and neoliberal international development projects. Data won't be FAIR until it speaks to the needs and desires of these communities so pervasively excluded from the benefits of international development.

References

Abu-Sada, C. (2009). Cultivating dependence: Palestinian agriculture under the Israeli occupation. In A. Ophier, M. Givoni, & S. Hanafi (Eds.), *The power of inclusive exclusion: Anatomy of Israeli rule in the occupied Palestinian territories*. Zone Books.

Alagona, P. S., Sandlos, J., & Wiersma, Y. F. (2012). Past imperfect: Using historical ecology and baseline data for conservation and restoration projects in North America. *Environmental Philosophy, 9*(1), 49–70.

Altieri, M. (1995). *Agroecology: The science of sustainable agriculture*. Westview Press.

Amri, A., Monzer, M., Al-Oqla, A., Atawneh, N., Shehadeh, A., & Konopka, J. (2008). Status and threats to natural habitats and crop wild relatives in selected areas in West Asia region. *Proceedings of the International Conference on Promoting Community-Driven In Situ Conservation of Dryland Biodiversity.* 18-21 April 2005, ICARDA, Aleppo, Syria.

Anker, P. (2001). *Imperial ecology: Environmental order in the British empire, 1895–1945.* Harvard University Press.

Anonymous. (2018) *The Seed Queen of Palestine.* Al Jazeera News, December 10. https://www.aljazeera.com/program/witness/2018/12/10/the-seed-queen-of-palestine

Berg, T. (2009). Landraces and folk varieties: A conceptual reappraisal of terminology. *Euphytica, 166*(3), 423–430.

Berlin, B., & Berlin, E. A. (2003). NGOs and the process research: The Maya ICBG project in Chiapas, Mexico. *International Social Science Journal, 55*(178), 629–638.

Berlin, B., & Berlin, E. A. (2004). Community autonomy and the Maya ICBG project in Chiapas, Mexico: How a bioprospecting project that should have succeeded failed. *Human Organization, 63*(4), 472–486.

Buerstmayr, M., Huber, K., Heckmann, J., Steiner, B., Nelson, J. C., & Buerstmayr, H. (2012). Mapping of QTL for Fusarium head blight resistance and morphological and developmental traits in three backcross populations derived from Triticum dicoccum × Triticum durum. *Theoretical and Applied Genetics, 125*(8), 1751–1765.

Carroll, S. R., Garba, I., Figueora-Rodriguez, O. L., Holbrook, J., Lovett, R., Materechera, S., Parsons, M., Raseroka, K., Rodriguez-Lonebear, D., Rowe, R., Sara, R., Walker, J. D., Anderson, J., & Hudson, M. (2020). The CARE principles for indigenous data governance. *Data Science Journal, 19*, 43. https://doi.org/10.5334/dsj-2020-043

Cronon, W. (1992). A place for stories: Nature, history, and narrative. *The Journal of American History, 78*(4), 1347–1376.

Cronon, W. (1993). The uses of environmental history. *Environmental History Review, 17*(3), 1–22.

Curry, H. (2017). From working collections to the World Germplasm Project: Agricultural modernization and genetic conservation at the Rockefeller Foundation. *History and Philosophy of the Life Sciences, 39*(5).

Davis, D. K. (2009). Historical political ecology: On the importance of looking back to move forward. *Geoforum, 40*(3), 285–286. https://doi.org/10.1016/j.geoforum.2009.01.001

Davis, D. K. (2015). Historical approaches to political ecology. In T. Perreault, G. Bridge, & J. McCarthy (Eds.), *The Routledge handbook of political ecology.* Routledge.

Drayton, R. H. (2000). *Nature's government: Science, imperial Britain, and the "improvement" of the world.* Yale University Press.

Ehrlich, P. (1968). *The population bomb.* Ballantine Books.

Eliott, R. (2008). *Faking nature: The ethics of environmental restoration.* Routledge.

FAO. (2009). *International treaty on plant genetic resources.* FAO. https://www.fao.org/3/i0510e/i0510e.pdf

FAO. (2008). *Agricultural projects in the West Bank and Gaza Strip 2008. Agricultural Projects Information System (APIS) report.* FAO. https://www.un.org/unispal/document/auto-insert-203615/

Fenzi, M., & Bonneuil, C. (2016). From "genetic resources" to "ecosystems services": A century of science and global policies for crop diversity conservation. *Culture, Agriculture, Food and Environment, 38*(2), 72–83. https://doi.org/10.1111/cuag.12072

Flitner, M. (2003). Genetic geographies: A historical comparison of agrarian modernization and eugenic thought in Germany, the Soviet Union, and the United States. *Geoforum, 34*(2), 175–185.

Freeman, J., Mehdi, S., & Duwayri, M. (2005). *Conservation and sustainable use of dryland agrobiodiversity, Jordan/Lebanon/Syria/Palestinian authority, terminal evaluation final report.* December 2005, United Nations Development Program Evaluation Resource Center. https://erc.undp.org/evaluation/documents/download/849.

Fullilove, C. (2017). *The profit of the earth: The global seeds of American agriculture*. University of Chicago Press.

Grove, R. (1995). *Green imperialism: Colonial expansion, tropical Island Edens, and the origins of environmentalism, 1600–1860*. Cambridge University Press.

Hayden, C. (2003). *When nature goes public: The making and unmaking of bioprospecting in Mexico*. Princeton University Press.

Holt-Giménez, E., & Shattuck, A. (2011). Food crises, food regimes and food movements: Rumblings or feform or tides of transformation? *Journal of Peasant Studies, 38*, 109–144. https://doi.org/10.1080/03066150.2010.538578

Huneman, P. (2019). How the modern synthesis came to ecology. *Journal of the History of Biology., 52*, 635–686.

Istaitih, Y., Alimari, A., & Jarrar, S. (2020). Determinants of farmers' participation in informal seed production for wheat in Palestine. *Research on Crops, 21*(1), 186–176. https://doi.org/10.31830/2348-7542.2020.028

Jacobsen, A., de Miranda Azevedo, R., Juty, N., et al. (2020). FAIR principles: Interpretations and implementation considerations. *Data Intelligence, 2*(1–2), 10–29. https://doi.org/10.1162/dint_r_00024

Kingsland, S. (1995). *Modeling nature: Episodes in the history of population ecology*. University of Chicago Press.

Kuttab, E. (2018). Alternative development: A response to neo-liberal de-development from a gender perspective. *Journal für Entwicklunspolitik, 34*(1), 62–90. https://doi.org/10.20446/JEP-2414-3197-34-1-62

McMichael, P. (2009). A food regime analysis of the 'world food crisis'. *Agriculture and Human Values, 26*, 281–295. https://doi.org/10.1007/s10460-009 9218-5

Mitman, G. (1992). *The state of nature: Ecology, community, and American social thought, 1900–1950*. University of Chicago Press.

Mons, B., Neylon, C., Velterop, J., Dumontier, M., da Silva Santos, L. O. B., & Wilkinson, M. D. (2017). Cloudy, increasingly FAIR; revisiting the FAIR Data guiding principles for the European Open Science Cloud. *Information Services and Use, 37*, 49–56.

Nadar, D. (2018) *How community saving of traditional seeds promotes agrobiodiversity and farmer autonomy in the West Bank, Palestine*. M.A. thesis, American University of Rome.

Pauly, P. J. (2007). *Fruits and plains: The horticultural transformation of America*. Harvard University Press.

Rural Advancement Foundation International (RAFI). (1999). *Biopiracy Project in Chiapas, Mexico Denounced by Mayan Indigenous Groups*. News Release, December 1. www.rafi.org

Salzmann, P. (2018). A food Regime's perspective on Palestine: Neoliberalism and the question of land and food sovereignty within the context of occupation. *Journal für Entwicklunspolitik, 34*(1), 14–34. https://doi.org/10.20446/JEP-2414-3197-34-1-14

Samara, A. (2000). Globalization, the Palestinian economy, and the "peace process". *Journal of Palestine Studies, 29*(2), 20–34. https://doi.org/10.2307/2676534

Saraiva, T. (2013). Breeding Europe: Crop diversity, gene banks, and commoners. In N. Disco & E. Kranakis (Eds.), *Cosmopolitan commons: Sharing resources and risks across Borders*. MIT Press.

Simberloff, D. (1988). The contribution of population and community biology to conservation science. *Annual Review of Ecology and Systematics, 19*, 473–511.

Soulé, M. E. (1985). What is conservation biology? *Bioscience, 35*(11), 727–734.

Streit Krug, A., & Tesdell, O. I. (2020). A social perennial vision: Transdisciplinary inquiry for the future of diverse, perennial grain agriculture. *Plants, People, Planet, 3*(4), 355–362. https://doi.org/10.1002/ppp3.10175

Tesdell, O. (2013). *Shadow spaces: Territory, sovereignty, and the question of Palestinian cultivation*. PhD dissertation, University of Minnesota. https://hdl.handle.net/11299/174898

Tesdell, O. (2017). Wild wheat to productive drylands: Global scientific practice and the agroecological remaking of Palestine. *Geoforum, 78*, 43–51.

Tesdell, O., Othman, Y., Dowani, Y., Khraishi, S., Deeik, M., Muaddi, F., Schlautman, B., Streit Krug, A., & Van Tassel, D. (2020). Envisioning perennial agroecosystems in Palestine. *Journal of Arid Environments, 175*, 104086. https://doi.org/10.1016/j.jaridenv.2019.104085

United Nations Conference on Trade and Development (UNCTAD). (2015). *The besieged Palestinian agricultural sector*. United Nations. https://unctad.org/system/files/official-document/gdsapp2015d1_en.pdf

Van Tassel, D. L., Tesdell, O., Schlautman, B., Rubin, M. J., DeHaan, L. R., Crews, T. E., & A Streit Krug. (2020). New food crop domestication in the age of gene editing: Genetic, agronomic and cultural change remain co-evolutionarily entangled. *Frontiers in Plant Science, 11*, 789. https://doi.org/10.3389/fpls.2020.00789

Zayed, D. (2020). *Interview by Courtney Fullilove at UAWC Local Seed Bank*. Ramallah.

Zohary, M., & Feinbrun-Dothan, N. (1966). *Flora Palaestina*. Israel Academy of Sciences and Humanities.

Data Management in Multi-disciplinary African RTB Crop Breeding Programs

Afolabi Agbona, Prasad Peteti, Béla Teeken, Olamide Olaosebikan,
Abolore Bello, Elizabeth Parkes, Ismail Rabbi, Lukas Mueller,
Chiedozie Egesi, and Peter Kulakow

Abstract Quality phenotype and genotype data are important for the success of a breeding program. Like most programs, African breeding programs generate large multi-disciplinary phenotypic and genotypic datasets from several locations, that must be carefully managed through the use of an appropriate database management system (DBMS) in order to generate reliable and accurate information for breeding-decisions. A DBMS is essential in data collection, storage, retrieval, validation, curation and analysis in plant breeding programs to enhance the ultimate goal of increasing genetic gain. The International Institute of Tropical Agriculture (IITA), working on the roots, tubers and banana (RTB) crops like cassava, yam, banana and plantain has deployed a FAIR-compliant (Findable, Accessible, Interoperable, Reusable) database; BREEDBASE. The functionalities of this database in data management and analysis have been instrumental in achieving breeding goals. Standard Operating Procedures (SOP) for each breeding process have been developed to allow a cognitive walkthrough for users. This has further helped to increase the usage and enhance the acceptability of the system. The wide acceptability gained among breeders in global cassava research programs has resulted in improvements in the precision and quality of genotype and phenotype data, and subsequent improvement in achievement of breeding program goals. Several innovative gender responsive approaches and initiatives have identified users and their preferences which have informed improved customer and product profiles. A remaining bottleneck is the

A. Agbona (✉)
International Institute of Tropical Agriculture (IITA), Ibadan, Oyo State, Nigeria

Molecular & Environmental Plant Sciences, Texas A & M University, College Station, TX, USA
e-mail: a.afolabi@cgiar.org

P. Peteti · B. Teeken · O. Olaosebikan · A. Bello · E. Parkes · I. Rabbi · C. Egesi · P. Kulakow
International Institute of Tropical Agriculture (IITA), Ibadan, Oyo State, Nigeria

L. Mueller
Boyce Thompson Institute for Plant Research, Ithaca, NY, USA

H. F. Williamson, S. Leonelli (eds.), *Towards Responsible Plant Data Linkage: Data Challenges for Agricultural Research and Development*,
https://doi.org/10.1007/978-3-031-13276-6_5

effective linking of data on preferences and social information of crop users with technical breeding data to make this process more effective.

Keywords BREEDBASE · Genotyping · Phenotyping · Database · Ontology · Customer and product profiling

1 Introduction

An appropriate database management system (DBMS) is essential for any organization to generate reliable and accurate information that will guide decision-making (Meiryani, 2019). Similarly, African breeding programs need a well-structured database management system, to carefully manage the large multi-disciplinary phenotypic, genotypic, and social datasets generated from multiple breeding programs across diverse regions and agroecosystems comprising genetic resource collections, plant breeding trials, on-farm trials, processor evaluations, and consumer testing. A DBMS is a software designed to define, manipulate, retrieve and manage data in a database. It generally manipulates the data itself, the data format, field names, record structure and file structure (Jaekel, 2013). A DBMS is essential in data collection, storage, retrieval, validation, curation and analysis in plant breeding programs to enhance the ultimate goal of increasing genetic gain, and informing breeding efforts at each advancement stage.

The International Institute of Tropical Agriculture (IITA), working on roots, tubers and banana (RTB) crops like cassava, yam, banana and plantain have deployed the use of a Findable, Accessible, Interoperable, Reusable web-based database system; BREEDBASE (https://www.BREEDBASE.org) which was borne out of the NextGen Cassava Project (https://www.nextgencassava.org). The NextGen project seeks to modernize cassava breeding using cutting-edge tools for efficient delivery of improved cassava that satisfies end user needs to farmers in sub-Saharan Africa. The ultimate vision is to improve genetic gain and to deliver cassava varieties with increased yield and disease resistance and other highly preferred traits into the hands of these farmers.

BREEDBASE is an open source, open access, web-based, breeding software available to the scientific community. It can accommodate phenotypic, genotypic, and environmental data collection, storage and analysis tools. It also includes support of the PhenoApps and the breeding API (BrAPI; Selby et al., 2019)) allowing tool integration among the breeding community (Simoes et al., 2019). The functionalities of these databases in data management and analyses have been instrumental in achieving key breeding goals such as monitoring and improving genetic gain, as well as increasing adoption of varieties with end-user preferred traits based on improving data quality and data analysis.

BREEDBASE capabilities include employment of:

- a user-friendly ontology (https://www.cropontology.org): The ontology is the controlled vocabulary that describes the trait phenotyped in a crop. Traits are grouped into classes (Morphological, Physiological, Quality, etc.) and each has its associated method and unit of measurement defined. In simple terms, it is a dictionary of traits.

- statistical analyses: Data analysis tools that can help breeders make inferences from their dataset. Some of the tools include mixed models for single trial analysis, Genome Wide Association Study, stability analysis, and simple descriptive statistics.
- interfaces with BrAPI (Breeding API, https://brapi.org/): BrAPI is a RESTful (Representational State Tranfer) web service Application Programming Interface (API) that helps to simplify integration and exchange of data across system and databases. Through its interface, exchange of phenotypic and genotypic data is possible.
- barcode-based data collection using the PhenoApps (http://phenoapps.org/): PhenoApps are a suite of barcode-enabled phenotyping tools. Tools include *Fieldbook* for phenotype data collection, *Coordinate* for genotype tissue sample collection and tracking, and *Inventory* for weighing samples without the need for data transcription.

BREEDBASE instances such as cassavabase.org can be launched for any crops using the Docker solution for ease of deployment.

2 Structure of BREEDBASE

BREEDBASE is developed using an open-source data schema (CHADO) and other software (PERL, JavaScript frameworks like JQuery, D3, Bootstrap). The CHADO database schema is a widely used database schema for model organism databases (Jung et al., 2011). It is a modular, ontology based, flexible design that can easily be implemented (Figs. 1 and 2).

All the source codes and database schemas used in the development of BREEDBASE are open source and available for download at https://github.com/solgenomics (Tecle et al., 2014) (Fig. 3).

Cassavabase was the first and is the most widely used instance of BREEDBASE and currently accommodates 1164 users from 22 breeding programs. It holds 459,000 accessions from 4070 phenotyping trials (of which 2642 are from the IITA cassava breeding program), 365 phenotyping traits collected for 34,000 genotypes, and over 15.3 million phenotypes representing 962,000 plots from 436 locations with approximately 19,000 images linked to plot information (Fig. 4).

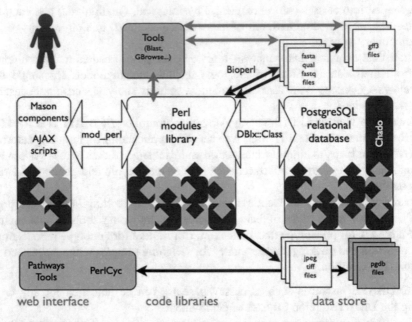

Fig. 1 BREEDBASE database schema. (Photo credit: Mueller's Lab in BTI, Ithaca)

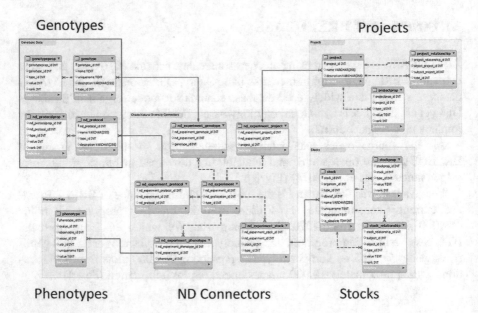

Fig. 2 BREEDBASE entity relationship schema. (Photo credit: Mueller's Lab in BTI, Ithaca)

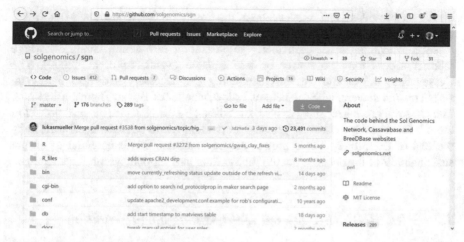

Fig. 3 BREEDBASE code available on solgenomics.net

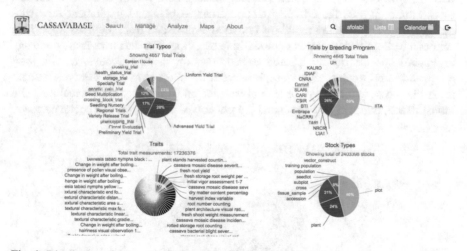

Fig. 4 Distribution of data currently hosted in Cassavabase

2.1 Implementation of the Cassavabase Mirror Site for Data Sustainability

The Cassavabase mirror site https://iita-mirror.cassavabase.org has been hosted at IITA in Ibadan, Nigeria since 2016. A mirror site is a replica of a website or network node. The concept applies to network services, and have different URLs than the original site, but a mirror site has identical or near-identical content (Glushko, 2014) as the primary database. Similarly, the Cassavabase mirror site is currently used as the replica of the main site hosted at the Boyce Thomson Institute Ithaca, NY USA (https://cassavabase.org/). The software and databases are updated every week. The mirror site provides a real-time backup of the original site, it reduces network traffic, improves access speed, and ensures availability of the original site for technical reasons. Mirror sites are particularly important in developing countries, where internet access may be slower or less reliable (Sekikawa et al., 2000). The Cassavabase mirror site was installed using the necessary dedicated hardware, software, and servers, with the help of Lukas Mueller of Boyce Thomson Institute. The Cassavabase mirror site was also established by the NextGen cassava project to build local capacity to host the primary production site to be maintained at IITA by the end of the project. For a project driven database, this is essential to ensure sustainability of the database should the funding and support system for it change.

2.2   BREEDBASE-Centered Data Management Workflow

IITA cassava breeding implements a data management workflow centered around Cassavabase. It uses the DBMS functionalities to plan and implement new trial designs, drawing from the wealth of knowledge provided from previous data. This creates a well-guided approach to modernize breeding. The database easily connects to PhenoApps for quality assured data collection using barcodes, and is then integrated into Dropbox (http://dropbox.com) for short-term data storage and also to enable access to the data prior to final curation and uploading to Cassavabase. The three different types of data generated are phenotypic, genotypic and social datasets.

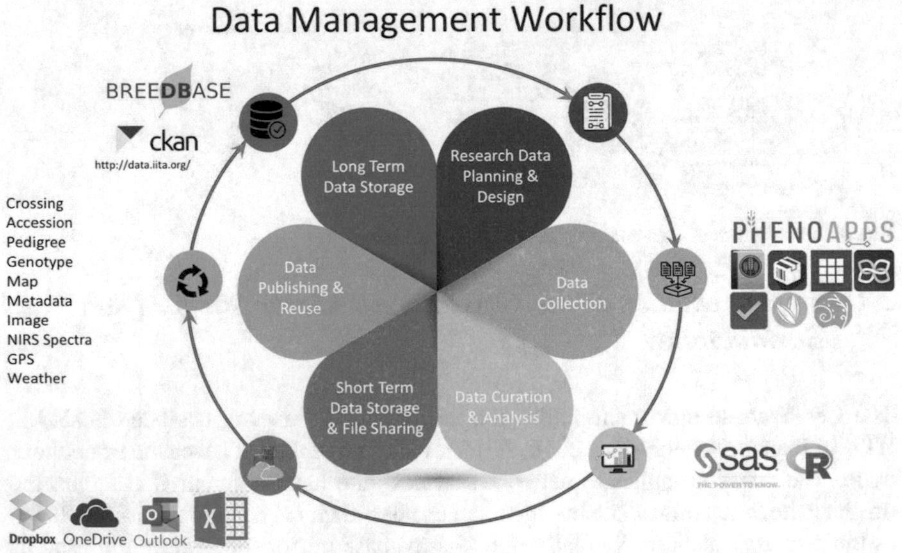

Fig. 5 Cassava breeding data management workflow

Data analysis is carried out using in-house scripts developed in R (R Core Team, 2018) purposefully for plant breeding trial data processing. Efforts are in place to enhance the phenotypic analytical capacity of the database to ensure maximal usage. However, there are often challenges uploading social data from surveys and farmers' trials into Cassavabase due to routine variations in variable terms used from one study to another oweing to the descriptions provided by the respondents who are mostly processors, marketers and farmers. Complimentary Knowledge Archive Network (CKAN), an open-source data portal (http://data.iita.org/) is adopted for storing these datasets. BrAPI calls can then be used to source data from Cassavabase into CKAN (Fig. 5). Data linkage with other crop breeding applications can also achieved using BrAPI. This reduces unnecessary duplication of tasks and more efficient use of resources.

3 Application of the Cassava Trait Ontology

The cassava trait ontology enhances the interoperability and effectiveness of data exchange between databases by providing standard concepts (including breeder, farmer and end-user terms) to describe the phenotypic information stored in those databases. The cassava ontology workspace within the database (https://cassavabase.org/search/traits) currently describes 365 variables terms and 206 post-composed terms representing important trait groups for several characteristics captured including traits from recent surveys with end-users for crop improvement (agronomic, biotic and abiotic stress, morphological, physiological and quality traits). The cassava ontology has been migrated from solgenomics: (https://github.com/solgenomics/cassava/tree/master/ontology) to the Planteome repository (https://github.com/Planteome/ibp-cassava-traits) and issues are being tracked via https://github.com/Planteome/variable-issue-submission/issues. Making changes to the cassava trait ontology file involves a sequence of validation process. When a new trait is to be added, a request can be made using a submission form available for the entire RTB community on https://submit.rtbbase.org/. The request is then posted as an issue on the Planteome repository (https://github.com/Planteome).

4 Product Profiles and Customer Profiles

Product profiles and customer profiles deal with an important question in breeding: for whom are we breeding (which users) and what are the breeding products needed to achieve the targeted outcomes of increasing genetic gain in farmers' fields and improving livelihoods and stimulating gender equity of cassava and empowerment of users along the value chain? To know for whom breeding is conducted it is necessary to prioritize the segment of people targeted. Donors often stress the need to improve livelihoods of small-scale farmers and other value chain actors, so the first step is to clearly define the target groups to be able to include as many cassava users as possible in a socially inclusively manner. Tools to improve social inclusion in product profile development were developed under the gender and breeding initiative led by The CGIAR (Consultative Group on International Agricultural Research) Research Project on Roots Tubers and Bananas program in cooperation with IITA, the Alliance of Bioversity and CIAT (International Center for Tropical Agriculture) and CIP (International Potato Center) with support from the CGIAR Excellence in Breeding Platform. The result of this was the creation of the Gender Plus tools (https://www.cgiar.org/innovations/g-tools-for-gender-responsive-breeding/) (Ashby & Polar, 2021a, b, c; Orr et al., 2021a, b; Polar et al., 2021). Information to inform customer and product profile needs sustained cooperation between socio-economists, anthropologists, gender specialists, and food scientists. Information on customer and product profiles is not static but subject to continual change as a result of socio-economic and variety preference dynamics. The customer and product profile information informs breeding on the number of pipelines needed, the number of preferred traits to monitor and the specific traits to prioritize and use throughout the selection stages. This demands a proper documentation and prioritization of traits of which the results and sources should be systematically documented and integrated within the DMBS system, independently from specific project funding to assure continuity and advancement. The Gender Plus (G+) tools partly addressed the structural integration of value chain actor information segregated by gender and other social factors, as well as information on the relative importance of the different food products made from cassava. Redesigning of the CGIAR customer and product profile tools is ongoing and cassava breeding has played an important exemplary role in this demand-led stage gate (Cooper, 1990; Kotch, 2018; Ragot et al., 2018) breeding effort, as cassava programs held much of the required information on user preferences accounting for the intersections of value chain actors, social segments such as gender groups, poverty, food security status and socio- cultural regions (Polar et al., 2021; Polar & Ashby, 2021; Teeken et al., 2021). Such information needs proper investigation and systematic alignment with the whole breeding process.

Currently cassava breeding programs in West Africa and East Africa have identified 4 provisional product profiles that are heavily Nigeria and Uganda focused because of the 10-year efforts funded through Nextgen cassava and complemented by RTBFoods project funding. The provisional product profiles are:

- *Cassava for food security*: Focus on cassava to be used for fermented food products gari-eba and fufu (major cassava food products in Nigeria) produced and processed by smallholder farmers and processors.
- *Cassava for the fresh market*: Cassava that can be boiled and eaten and/or pounded, important for food security among cassava farming households and specifically for the northern half of Nigeria where cassava is a secondary crop.

This cassava can also be dried into cassava chips that can be milled to create cassava flour which is most common in Northern Nigeria and Uganda.

- *Biofortied cassava for improved nutrition*: Cassava that is biofortified and aimed at increasing nutritional security especially for nutritionally insecure social segments that do not have access to sufficient other sources of vitamin A such as vegetables.
- *Cassava for industry*: Cassava that can be used as starch and ethanol sources for processed food products (such as composite foods) and non-food products.

These profiles are currently end product focused. In the course of the new initiatives on customer and product profiling these classifications can change through including profiles specifically focused on certain user segments that need extra attention from a social inclusion perspective.

The matrix of issues to consider includes prioritizing locations and customer (user) segments, inputs from value chain actors, ecological and cultural regionality, as well as the demand led stage gate breeding focus on variety replacement and commercialization of seed delivery, and finally the intersection of all these domains with gender. This complex matrix needs to be filled as completely as possible in order to informed the minimum number of product profiles needed and the composite of traits needed to be prioritized within each of them. This will assure that maximal impact can be cost effectively achieved with regards to the different development goals' impact areas.

Fig. 6 Marie Angelique Laporte, Elizabeth Arnaud from Crop Ontology and Afolabi Agbona visited BTI to develop strategies for storing farmer/processor related traits in BREEDBASE in November 2019

4.1 The Need to Harmonize Datasets Generated Within the Breeding Program to Fully Optimize Adoption of Product Design and Development Strategy

Continuous interaction between the crop ontology, database management, and breeders will be an important step going forward. This used to happen through the crop ontology workshop but this may not have all stakeholders present. In 2019, an RTB (Roots, Tubers and Bananas) workshop (Fig. 6) was held at the Boyce Thompson Institute where the crop ontology group including the curators met with the database group to discuss a common approach for RTB ontology quality content for agronomic, quality, gender-sensitive and Participatory Varietal Selection (PVS) traits and variables.

5 Application of BREEDBASE for Quality Control

Plant phenotypic data comprises information that can be analyzed as datasets individually or combined with existing datasets and reanalyzed. The correct interpretation, comparability, replicability and interoperability of these data is only possible provided the collected data are equipped with an adequate set of useful metadata. The metadata contains information needed to understand and effectively use the data. Thus, metadata is receiving increasing attention across a broad spectrum, to help interpret phenotypic data to achieve the goals of the scientific community. The rows and columns of numeric and textual observations contained within a data set are frequently referred to as raw data. Raw data are usually considered valuable if they can be used within the scientific framework of the study that generated the data. Interpreting and using raw data to investigate a study's underlying theoretical or conceptual model(s) requires an understanding of the types of variables measured. The measurement units, the data quality, the conditions under which the variables were measured, and other relevant facts are all needed and are provided in the metadata. Information is then generated from the combination of raw data and metadata. BREEDBASE collects metadata such as study name, study description, study year, location, date of planting, date of harvesting, plot length, plot width, number of replications, number of blocks, plant stands per plot, field size, unit spacing etc. Metadata like plot length, plot width and field size are very important when estimating yield.

Other useful data collected by BREEDBASE and linked to the metadata include weather data, GPS, plot images, and crossing information. The collected weather data are available on https://weather.rtbbase.org/. Weather data collected include temperature, rainfall, light intensity and day length. Plot level GPS data has also been collected using the handheld Garmin 20x device, for recent trials from Ibadan and Ubiaja in Nigeria. This information is also available on Cassavabase. Plans are under consideration to install a Real Time Kinematic (RTK) positioning system, which would help to collect more accurate GPS data. More than 19,000 plot images, linked to traits described in the ontology have been uploaded, and image analysis like root necrosis, whitefly counts, etc. can be performed on these images. The crossing information, which includes cross name, female parent, male parent, cross type, number of flowers, number of fruits, and number of seeds are properly managed in Cassavabase. Crossing block activity can include more than 10,000 crosses, carried out over a 3-month time period, by a large team of specialized field technicians. We have implemented a new PhenoApps tool; Intercross, which makes it easier to track and collect the crossing information using barcodes that increases the quality of the crossing data. This information can be linked to the seedling nursery trial to provide proper pedigree linkages. Information on task or gender specific benefits of some traits also informed trait selection in breeding complementing phenotype and genotype information. Breeders are informed of the social or gender implications of some traits when selected or prioritized.

6 Integrating Feedback from Social Data for Enhancing Decision-Making in the Breeding Pipeline

The choice of parents fed back into the breeding pipeline is guided by selections from the advanced and late testing stages as well as the feedback from demand creation trials (DCTs), farmers' trial evaluations, surveys, participatory processing, and other methods that provide direct feedback from end users. Demand creation trials are mostly used for variety promotion. It enables processors to choose the most suitable variety for their needs. Variety demand by diverse end users in differentiated cassava markets is determined by communication with the end-users, demand creation trials and inferences from cassava variety adoption studies. Cassavabase contains most of the DCT data from recent years across different locations such as Ago-Owu, Ikenne, Ilorin, Abuja, etc. This provides a decision-making tool for processors to access production and processing value of varieties to generate demand.

Fig. 7 Social engagement with farmers and processors. (Photo courtesy of Béla Teeken)

In order to complement quality data collection efforts to inform breeding or crop improvement initiatives, the survey team constantly or periodically engages farmers, processors and other food chain actors/end-users in the evaluation of new and existing (crop-cassava) varieties alongside local and commonly grown popular varieties, to determine their trait preferences at the production, processing and utilization stages, market valuation of traits and benefits or gains accrued utilizing these improved crop varieties (Fig. 7). Using approaches such as mixed methods to collect quantitative and qualitative data on gender and social aspects (Teeken et al., 2018), informative datasets and inferences have been generated over the years using the Tricot triadic comparison citizen science technology in large scale participatory variety selection and consumer testing (van Etten et al., 2019, 2020; Moyo et al., 2021). The online tablet-based 1000Minds (www.1000minds.com) survey, that is a pairwise comparison between options of equal monetary value using the PAPRIKA method (Potentially All Pairwise Rankings of all possible Alternatives), Mother-Baby trials (Teeken et al., 2021), RTBfoods project (https://rtbfoods.cirad.fr/) methodologies, participatory processing and consumer testing activities (Ndjouenkeu et al., 2021; Forsythe et al., 2021; Teeken et al., 2021; Amah et al., 2021), have all been managed in the DBMS. Although scalable protocols and ways to systematically and more effectively connect these data to to breeding and food science data to inform breeding need further development which is fortunately a mandate of the new Market Intelligence and Product Profiling initiative of the One CGIAR (CGIAR, 2021). Survey results have already significantly informed further improvement to the cassava ontology in order to integrate farmer's and other users' descriptions as traits. The results have also informed gender responsive product profiles for food (especially related to processibllity and food product quality), industry and biofortification. Solutions that have been adopted by some farmers through training platforms like Tricot include standardized cassava spacing and the slant planting which farmers have observed to give more and better or increased yield per plot.

It is imperative to understand the way farmers and processors describe a trait as well as the value placed on such a trait. This will inform the ontology of such trait. Scalable approaches such 1000minds and Tricot are promising in providing such user information because they focus on centralized data management (e.g. www.ClimMob.net for the Tricot approach) and scaling and aim to standardize procedures to generate user feedback in such a way that it becomes an integrated part of the breeding data allowing optimal data integrationLinking of ClimMob and Breedbase is one of the main objectives of the developers of Tricot and ClimMob (van Etten et al., 2020). There is a need for controlled and unified trait description before the social data collected during field surveys or participatory varietal selection can be useful to breeders. The crop ontology will provide a controlled vocabulary set for economically important traits (Shrestha et al., 2010) described by farmers, processors and other value chain stakeholders. Such traits will not be useful or meaningful to breeders until an interdisciplinary study is conducted by breeders, social scientist and food scientists, in which food scientists translate verbal trait descriptions collected by social scientists into measurable traits for breeders' use.

Language barriers, the translation/interpretation of verbal/raw field data collected by social scientists, the adoption of a unified concept for inquiry during surveys as well as the design of an appropriate template that can accommodate social science and food science data are part of the bottle-neck limiting integration of feedback of social data for decision making in breeding programs.

This calls for adoption or incorporation of gender responsive studies to identify regional and cultural differences/similarities in description of traits preferred by men and women (Olaosebikan et al., 2018, 2019; Teeken et al., 2018). This was one of the objectives of the GREAT (Gender Responsive Researcher Equipped with Agricultural Transformation) program organized in sub-Saharan Africa to train researchers that can transform African Agriculture through conducting innovative socially inclusive and end-user-oriented studies. The program trained multidisciplinary teams to structurally integrate gender into the technical and bio-physical sciences.

Trained experts (data curators, application developers, data analysts) or scientists (in breeding, social science, food science) will work together to validate trait descriptions and process traits to a form that can easily be comprehended by breeders before such traits can be used in breeding programs. With regards to food product quality traits this is currently happening within the RTB foods project (https://rtbfoods.cirad.fr/) where the social scientist presents the crop characteristics preferred by users, ranked in order of importance based on survey, participatory processing and consumer testing, to food scientists and breeders for a further translation into operational traits and to determine the first two additional traits to focus on. E.g. the product profile for the gari and eba cassava food products currently addresses colour and food product texture, with Standard Operation Procedures (SOP) being developed to measure these informed by evaluating good and less good varieties as processed and evaluated by users. This multidisciplinary process has happened but will be formalized into a RTB foods product profile document. The focus crops here are Cassava, Cooking banana, Sweet potato, Potato and Yam. Product profiles are formulated per country and for the different major food products made from the crops. With regards to cassava a current MSc research, using ground penetrating radar technology, is looking at how the current Nextgen varieties of cassava perform with regards to early maturity, which will reveal if special selection for early maturity will be necessary.

Decision making in breeding cannot be reached without synthesis of social and food scientists' data collected at different points of contact with farmers and consumers. Social data collected on the emotional, hedonic and organoleptic descriptions of trait preferences of raw food crops as well as culinary food product characteristics will be tested in the laboratory by food scientists to confirm social data. Food science transforms social data into measurable data that can be quantified and presented to the breeder for incorporation into breeding objectives and development of market driven or demand-led breeding to meet the need of stakeholders in crop value chains, as championed by the Excellence in Breeding Platform.

More so, it is necessary to develop Product Profiles for crops which will assist breeders in shaping breeding objectives that will be beneficial to end-users. This is achievable through concerted efforts of social, physical and bio-physical scientists by engaging in an interdisciplinary project that will lead to development of SOPs which can be adopted across the CGIAR's international research centers as well as national partners in sub-Saharan Africa. This will enhance the development of agriculture as well as the reliability of data collected since the centers and partner institutes will be using the same approaches and methods of operation from the field through the laboratory to data storing methods. This will also foster relationships within and between the institutes.

Near Infrared Spectroscopy (NIRS) offers possibilities to link root and food characteristics (Alamu et al., 2021). This requires following rigorous SOPs for NIRS scans on fresh roots from breeding trials, on the intermediate food products (if applicable) as well as on the final food products as prepared in food science labs following processing and preparation steps that have been externally validated by participatory processing of contrasting clones (new clones and processor preferred clones) in the working environment of the users (village cottage processing, or larger scale using mechanized processing depending on the product profile). The clones to be evaluated with the users in their own environment are preferably grown in the same trail and close to the production source of the users, which often implies an on-farm trial or breeding location close to the communities of the processors. This concert of activities can result in finding relationships between food product quality traits and physiochemical characteristics of fresh roots which would allow for earlier selection for food product quality and processability (the amount of drudgery involved in processing) of different varieties. All these activities are currently put in place through the cooperation between the Nextgen cassava and RTB foods projects. Within cassava breeding two proof of concepts are currently identified: the investigation of the discoloration during processing of the root resulting in pale or brownish food products, which is hypothesized to be related to dextrinization, the browning related to caramelization of sugars (non-enzymatic browning) in relation with the presence of polyphenols, (enzymatic browning) in the fresh roots. Another proof of concept is related to the final textural properties, hardness, smoothness, mouldability, adhesiveness and strechability of the dough like food products which are hypothesized to be related to amylase and pectin contents.

7 Promoting BREEDBASE Functionality for Increased Usage

To ensure the adoption of technical solutions in BREEDBASE and the proper integration of social solutions to inform decision-making, Quality Champions were appointed from different breeding programs. They are the "go to" people for quality control (QC) and BREEDBASE-related topics. Their roles include:

- Creating awareness and access to best practices/state-of-the-art techniques in quality management.

Fig. 8 Capacity development across partner stations

- Developing and implementing SOPs, making use of key performance indicators (KPIs) and quality metrics to ensure the correctness of data and other practices involved in breeding.
- Improving efficiency of breeding data collection, curation and storage
- Effecting an increase in the usage of Cassavabase in daily breeding activities
- Training users on QC and data management.

The Quality Champions are also actively involved in promoting digitization of practices; promoting the use of electronic data capture, procuring digital inputs like barcode labels for both phenotype and genotype stocks to improve data quality, among others.

Knowledge sharing is key for the successful implementation of our technical and social solutions. To this effect, regular training of users on the usage of BREEDBASE and PhenoApps tools is essential (Fig. 8).

Every year at IITA Headquarter, we organize biweekly trainings for technicians, supervisors and students for a 2 to 3 months period, usually in the first quarter of each year. During these trainings, we focus on functionalities of Cassavabase, which are mainly useful for the technicians to do their day-to-day breeding activities, such as creating lists, searching for and downloading phenotypic data and layouts, designing barcodes, etc. We also focus on PhenoApps tools like Fieldbook, Coordinate, Intercross, etc. We have also extended this training beyond Nigeria, Uganda and Tanzania to include all IITA regional locations in Southern Africa, East Africa and Central Africa along with national agricultural research systems (NARS) partners. A dedicated community of practice partnership (COPP) has been set up to engage additional NARS breeding staff from 9 countries including Rwanda, DR Congo, Kenya, Cote d'Ivoire, Sierra Leone, Ghana, Zambia, Malawi, and Mozambique. The COPP mainly focuses on expanded use of digital tools, germplasm exchange, support in field management, trial design, data management, use of genotyping for markers and variety identifications and peer visits for knowledge exchange among breeding programs.

8 Fostering Continuous Improvement

To maintain continuity of practice and to further improve the system, we are exploring collaborations with different cassava initiatives ranging from the fundamentals of breeding program optimization, development of improved product profiles and efficiently delivering new products to farmers; to addition of new features, increased training of technical staff and linking traits across the value chain using the cassava trait ontology.

In conclusion, the wide applicability of BREEDBASE has encouraged wide acceptability among breeders in the global cassava research programs. The user-friendliness of the DBMS and availability of SOPs for most breeding processes allows a smooth cognitive walk through for the users. All these solutions have resulted in improvements in precision and quality of phenotypic and genotypic data, thus resulting in the overall improvement of breeding program goals.

A bottleneck identified is the full integration of social and anthropological data related to gendered, regional and socially inclusive trait preferences and other relevant social information that informs the customers and product profiles to focus on a more demand led breeding approach, but also the way breeding is organized as it also should determine stakeholders that are to be represented in product advancement and variety release procedures. The mentioned One Cgiar Market Intelligence and Product Profiling innitiatve (CGIAR, 2021) will be important to allow to tackle this bottleneck. The cassava breeding unit in Ibadan has played a role in setting an example for acquiring such social information and is continuing to contribute to redesigning customer and product profiling procedures that will also be informed by market intelligence tools. This is especially important because public breeding is specifically tasked with creating social impact in the form of poverty alleviation among smallholder value chain actors. The Tricot citizen science scaled participatory variety selection approach offers a platform for learning and variety dissemination and building partnerships with users but is also important to test the external validity of the multilocation breeding trials as it systematically evaluates variety performance under farmer conditions. This is important as the main objective of breeding is to increase genetic gain in farmers' fields. Current initiatives are ongoing to connect Tricot data efficiently to BrAPI and Breedbase through the ClimMob online platform (www.climmob.net) (van Etten et al., 2020; Manners et al., 2022, forthcoming) also including almost real time climate data through 'climatrends' and 'chirps'(https://CRAN.Rproject.org/package=climatrends, and https://CRAN.Rproject.org/package=chirps) (de Sousa & van Etten, 2021; de Sousa et al., 2020). Although public breeding shares demand driven approaches with the private sector, it is the explicit social inclusiveness and focus on sustainable development goals (https://www.un.org/sustainabledevelopment/sustainable-development-goals/) that sets public breeding apart. We are confident that during the years to come this part will be well integrated with the modernized DBMS systems.

References

Alamu, E. O., Nuwamanya, E., Cornet, D., Meghar, K., Adesokan, M., Tran, T., Belalcazar, J., Desfontaines, L., & Davrieux, F. (2021). Near-infrared spectroscopy applications for high-throughput phenotyping for cassava and yam: A review. *International Journal of Food Science and Technology, 56,* 1491–1501. https://doi.org/10.1111/ijfs.14773

Amah, D., Stuart, E., Mignouna, D., Swennen, R., & Teeken, B. (2021). End-user preferences for plantain food products in Nigeria and implications for genetic improvement. *International Journal of Food Science and Technology, 56,* 1148–1159. https://doi.org/10.1111/ijfs.14780

Ashby, J. A., & Polar, V. (2021a). User guide to the G+ product profile query tool (G+PP). In *CGIAR Research Program on Roots, Tubers and Bananas, user guide 2021–2.* Lima, Peru. www.rtb.cgiar.org/gbi. https://hdl.handle.net/10568/113167

Ashby, J. A., & Polar, V. (2021b). User guide to the standard operating procedure for G+ tools (G+SoP). In *CGIAR Research Program on Roots, Tubers and Bananas, User Guide. 2021–3.* International Potato Center. www.rtb.cgiar.org/gbi. https://hdl.handle.net/10568/113166

Ashby, J. A., & Polar, V. (2021c). Description sheet to the gender plus product profile query tool (G+PP). In *CGIAR Research Program on Roots, Tubers and Bananas.* International Potato Center. www.rtb.cgiar.org/gbi. https://hdl.handle.net/10568/113191

CGIAR. (2021). *Market intelligence and product profiling.* Retrieved from: https://storage.googleapis.com/cgiarorg/2021/10/INIT05-Market-Intelligence-and-Product-Profiling.pdf

Cooper, R. G. (1990). Stage-gate systems – A new tool for managing new products. *Business Horizons, 33,* 3.

de Sousa, K., & van Etten, J. (2021). *ClimMobTools: API client for the 'ClimMob' platform* (R package version 0.3.9). https://CRAN.R-project.org/package=ClimMobTools

de Sousa, K., van Etten, J., & Solberg, S. Ø. (2020). *Climatrends: Climate variability indices for ecological modelling* (R package version 0.1.6). https://CRAN.R-project.org/package=climatrends

Forsythe, L., Tufan, H., Bouniol, A., Kleih, U., & Fliedel, G. (2021). An interdisciplinary and participatory methodology to improve user acceptability of root, tuber and banana varieties. *International Journal of Food Science and Technology; Special Issue: Consumers Have Their Say: Assessing Preferred Quality Traits of Roots, Tubers and Cooking Bananas, and Implications for Breeding, 56,* 1115–1123.

Glushko, R. J. (2014). *The discipline of organizing: Core concepts edition.* O'Reilly Media.

Jaekel, T. (2013). The role concept for relational database management systems.. Hong Kong

Jung, S., Menda, N., Redmond, S., Buels, R. M., Friesen, M., Bendana, Y., Sanderson, L. A., Lapp, H., Lee, T., MacCallum, B., Bett, K. E., Cain, S., Clements, D., Mueller, L. A., & Main, D. (2011). The chado natural diversity module: A new generic database schema for large-scale phenotyping and genotyping data. *Database (Oxford), 2011,* bar051.

Kotch, G. P. (2018). *Applying stage-gates to better manage public breeding programs* | Excellenceinbreeding [www document]. https://excellenceinbreeding.org/blog/applying-stage-gates-better-manage-public-breeding-programs. Accessed 12.8.19.

Manners, R., de Sousa, K., Teeken, B., et al. (2022). An agile framework for decentralized on-farm testing supported by citizen science. A review (under review with *Agronomy for Sustainable Development*) (forthcoming).

Meiryani, A. S. (2019). Database management system. *International Journal of Scientific & Technology Research, 8*(06), 309–312.

Moyo, M., Ssali, R., Namanda, S., Nakitto, M., Dery, E. K., Akansake, D., Adjebeng-Danquah, J., van Etten, J., de Sousa, K., Lindqvist-Kreuze, H., Carey, E., & Muzhingi, T. (2021). Consumer preference testing of boiled sweetpotato using crowdsourced citizen science in Ghana and Uganda. *Frontiers in Sustainable Food Systems, 5,* 620363. https://doi.org/10.3389/fsufs.2021.620363

Ndjouenkeu, R., Ngoualem Kegah, F., Teeken, B., Okoye, B., Madu, T., Olaosebikan, O. D., Chijioke, U., Bello, A., Oluwaseun Osunbade, A., Owoade, D., Takam-Tchuente, N. H., Biaton Njeufa, E., Nguiadem Chomdom, I. L., Forsythe, L., Maziya-Dixon, B., & Fliedel, G. (2021).

From cassava to gari: Mapping of quality characteristics and end-user preferences in Cameroon and Nigeria. *International Journal of Food Science and Technology, 56*, 1223–1238. https://doi.org/10.1111/ijfs.14790

Nextgen Cassava. (n.d.). https://www.nextgencassava.org accessed Friday 16 July 2021 & Grant proposal narrarive Phase II: https://www.nextgencassava.org/wp-content/uploads/documents/NextGenRenewal_Proposal%20Narrative_2017%20copy.pdf

Olaosebikan, O., Kulakow, P., Tufan, H., Madu, T., Egesi, C., & Teeken, B. (2018). A case study of cassava trait preferences of men and women farmers in Nigeria: Implications for gender-responsive cassava variety development. In H. A. Tufan, S. Grando, & C. Meola (Eds.), *CGIAR gender and breeding initiative* (pp. 35–43). CIP.

Olaosebikan, O., Abdulrazaq, B., Owoade, D., Ogunade, A., Aina, O., Ilona, P., Muheebwa, A., Teeken, B., Iluebbey, P., Kulakow, P., Bakare, M., & Parkes, E. (2019). Gender-based constraints affecting biofortified cassava production, processing and marketing among men and women adopters in Oyo and Benue States, Nigeria. *Physiological and Molecular Plant Pathology, 105*, 17–27. https://doi.org/10.1016/j.pmpp.2018.11.007

Orr, A., Polar, V., & Ashby, J. A. (2021a). User guide to the G+ customer profile tool (G+ CP). In *CGIAR Research Program on Roots, Tubers and Bananas, user guide 2021–1*. Lima, Peru. www.rtb.cgiar.org/gbi. https://hdl.handle.net/10568/113168

Orr, A., Polar, V., & Ashby, J. A. (2021b). Description sheet to the gender plus customer profile tool (G+CP). In *CGIAR Research Program on Roots, Tubers and Bananas*. International Potato Center. www.rtb.cgiar.org/gbi. https://hdl.handle.net/10568/113190

Polar, V., & Ashby, J. A. (2021). Gender report and template forms for G+ tools. In *CGIAR Research Program on Roots, Tubers and Bananas, report template 2021–1*. Lima, Peru. www.rtb.cgiar.org/gbi. https://hdl.handle.net/10568/113189

Polar, V., Teeken, B., Mwende, J., Marimo, P., Tufan, H. A., Ashby, J. A., Cole, S., Mayanja, S., Okello, J. J., Kulakow, P., & Thiele, G. (2021). Chapter 16: Building demand-led and gender-responsive breeding programs. In G. Thiele (Ed.), *Root, tuber and banana food system innovations*. Springer. https://link.springer.com/content/pdf/10.1007/978-3-030-92022-7.pdf

R Core Team. (2018). *R: A language and environment for statistical computing*. R Foundation for Statistical Computing. https://www.R-project.org/

Ragot, M., Bonierbale, M., & Weltzien, E. (2018). From market demand to breeding decisions: A framework. In *CGIAR gender and breeding initiative working paper 2*. CGIAR Gender and Breeding Initiative.

RTB Foods. (2018). *Project description*. https://rtbfoods.cirad.fr/project/rtbfoods-description. Accessed 1 Apr 2022.

Sekikawa, A., et al. (2000). Internet mirror sites. *The Lancet, 355*(9219), 2000.

Selby, P., et al. (2019). BrAPI – An application programming interface for plant breeding applications. *Bioinformatics, 35*(20), 4147–4155.

Shrestha, R., Arnaud, E., Mauleon, R., Senger, M., Davenport, G. F., Hancock, D., Morrison, N., Bruskiewich, R., & McLaren, G. (2010). Multifunctional crop trait ontology for breeders' data: Field book, annotation, data discovery and semantic enrichment of the literature. *AoB Plants, 2010*, 1–11. https://doi.org/10.1093/aobpla/plq008

Simoes, C., et al. (2019). *Breedbase: A digital ecosystem for plant breeders*. African Plant Breeders Association (APBA) Conference.

Tecle, I. Y., Edwards, J. D., Menda, N., Egesi, C., Rabbi, I. Y., Kulakow, P., Kawuki, R., Jannink, J. L., & Mueller, L. A. (2014). solGS: A web-based tool for genomic selection. *BMC Bioinformatics, 15*, 1–9. https://doi.org/10.1186/s12859-014-0398-7

Teeken, B., Olaosebikan, O., Haleegoah, J., Oladejo, E., Madu, T., Bello, A., Parkes, E., Egesi, C., Kulakow, P., Kirscht, H., & Tufan, H. (2018). Cassava trait preferences of men and women farmers in Nigeria: Implications for breeding. *Economic Botany, 72*, 263–277. https://doi.org/10.1007/s12231-018-9421-7

Teeken, B., Agbona, A., Bello, A., Olaosebikan, O., Alamu, E., Adesokan, M., Awoyale, W., Madu, T., Okoye, B., Chijioke, U., Owoade, D., Okoro, M., Bouniol, A., Dufour, D., Hershey, C., Rabbi, I., Maziya-Dixon, B., Egesi, C., Tufan, H., & Kulakow, P. (2021). Understanding cassava varietal preferences through pairwise ranking of gari-eba and fufu prepared by local farmer–processors. *International Journal of Food Science and Technology, 56*, 1258–1277. https://doi.org/10.1111/ijfs.14862

United Nations. (2021). *The sustainable development goals.* https://www.un.org/sustainabledevelopment/sustainable-development-goals/. Accessed 16 July 2021.

Van Etten, J., Beza, E., Calderer, L., Van Duijvendijk, K., Fadda, C., Fantahun, B., et al. (2019). First experiences with a novel farmer citizen science approach: Crowdsourcing participatory variety selection through on-farm triadic comparisons of technologies (tricot). *Experimental Agriculture, 55*(S1), 275–296. https://doi.org/10.1017/S0014479716000739

Van Etten, J., Abidin, E., Arnaud, D., Brown, E., Carey, E., Laporte, M.-L., López-Noriega, I., Madriz, B., Manners, R., Ortiz-Crespo, B., Quirós, C., de Sousa, K., Teeken, B., Tufan, H. A., Ulzen, J., & Valle-Soto, J. (2020). In www.rtb.cgiar.org (Ed.), *The tricot citizen science approach applied to on-farm variety evaluation: Methodological progress and perspectives* (RTB Working Paper. No. 2021-2). CGIAR Research Program on Roots, Tubers and Bananas (RTB).

Part II
Challenges from/for the Data: Data Linkage Across Standards, Infrastructures and Scales

Preface

The second part of the book shifts the focus from the field and circumstances of data collection to the nature of the data themselves, and how the characteristics of data as technical artefacts affect efforts to develop and link data infrastructures. On the one hand, data infrastructures are supposed to be both interoperable and scalable – in other words, they need to support the ability to connect, compare and cross-analyze large volumes of data generated under a variety of different circumstances. On the other hand, there are legitimate worries that the standardization processes required to facilitate interoperability and scalability may result in the loss of precious system-specific information captured by situated efforts of data collection. This tension is accentuated by the different characteristics of data generated under laboratory conditions, in trials and experiments in the field, and through agronomic governance, trade and regulation. Among the issues explored by contributors to this part are the efforts and skills required to maintain data collections over time, including both rescuing legacy data and ensuring that data produced today are sustainably preserved in the long run; the extent to which data linkage requires taking account of the whole data lifecycle, which in turn means assessing how data may be used in the future; and how a critical perspective on data re-use may inform the development and implementation of data standards.

From Farm to FAIR: The Trials of Linking and Sharing Wheat Research Data

Christopher John Rawlings and Robert P. Davey

Abstract This paper describes progress towards an integrated data framework that supports the sharing of data from the Designing Future Wheat (DFW) strategic research programme funded by the UK BBSRC. DFW is a 5 year project (https://designingfuturewheat.org.uk/) that spans eight research institutes and universities, and aims to deliver pre-breeding germplasm to breeders to improve and increase the genetic diversity of their breeding programs. DFW is committed to making its data open to the wider research community by adopting FAIR data sharing approaches. It is also a good example of a data-intensive strategic research programme which follows a cyclical Field-to-Lab-to-Field approach that is representative of much contemporary and multidisciplinary crop science research. However, even with dedicated funding to develop crop data research infrastructures within DFW, we found that there are many challenges that require pragmatic and flexible ways to enable them to interoperate. We present key DFW data resources as a case study to assess progress and discuss these challenges with a view to developing infrastructure that exposes metadata-rich datasets and that meets FAIR principles.

1 Background to Designing Future Wheat

The Designing Future Wheat (DFW) project is a strategic research programme funded by the UK BBSRC that spans eight research institutes and universities with aims to deliver pre-breeding germplasm to breeders to improve and increase the genetic diversity of their breeding programs. The DFW partners are John Innes Centre, Rothamsted Research, Earlham Institute, the Quadram Institute, the

C. J. Rawlings (✉)
Rothamsted Research, Harpenden, UK
e-mail: chris.rawlings@rothamsted.ac.uk

R. P. Davey
Earlham Institute, Norwich, UK
e-mail: robert.davey@earlham.ac.uk

H. F. Williamson, S. Leonelli (eds.), *Towards Responsible Plant Data Linkage: Data Challenges for Agricultural Research and Development*,
https://doi.org/10.1007/978-3-031-13276-6_6

European Bioinformatics Institute, the National Institute for Agricultural Botany and the Universities of Nottingham and Bristol. DFW was originally funded for 5 years (2017–2021) but has been extended a further year due to the impact of the COVID pandemic (https://designingfuturewheat.org.uk/). DFW builds on the success of an earlier BBSRC-funded cross-institute strategic collaboration – the Wheat Improvement Strategic Programme (WISP) which ran from 2011 to 2017 (http://www.cerealsdb.uk.net/cerealgenomics/WISP/). The genesis of both WISP and DFW were responses by BBSRC to an independent review of its funding of Crop Science by Prof. Chris Gilligan in 2005 which recommended, among other things, that BBSRC should increase and focus its funding on crops rather than solely model species. This was aimed at encouraging better coordination of research in the grasses and small grain cereals research community. A key outcome of the WISP program was to develop high throughput phenotyping techniques for use across large field experiments with thousands of plots; the success of this has resulted in the generation of large datasets, emphasizing the need for efficient data collation, storage and sharing platforms.

In addition to developing pre-breeding germplasm, an important aspect of DFW is an explicit declaration that results are made available free of intellectual property restrictions. This continues the principle adopted in previous BBSRC-funded wheat research initiatives such as WISP. A key output from DFW are panels of novel germplasm (seeds) that can be freely incorporated into other academic research or commercial breeding programmes by crop breeding companies that participate as collaborators, i.e. "pre-competitive breeding", or pre-breeding. Complementing this pre-breeding research in DFW is a wide-ranging and underpinning programme of genetics, genomics and trait biology research, including the generation of new genomics resources. The same principles of openness in the pre-breeding research applies to data from the wider project and DFW is committed to making data open to the whole research community by adopting FAIR data sharing approaches. DFW is therefore a good example of a data-intensive strategic research programme which follows a Field-to-Lab-to-Field approach that is representative of much contemporary and multidisciplinary crop science research.

2 Challenges and Approaches for Data Management

We will highlight a range of challenges and approaches to creating a consistent and reusable integrated strategy for common heterogeneous agricultural datasets. These challenges include the variety of trial designs, difficulties harmonising environmental data across remote sites, keeping up to date with data generation tools, technologies and formats (e.g. sensors, drones), and monitoring research outputs.

2.1 Characterising the Origins of Genetic Material

All crop improvement programmes focus on using diversity inherent in germplasm (seed) collections to access and exploit potentially beneficial traits. Measures of that diversity across clades, species or lines are used to describe and construct core collections. The ancestry and provenance of a seed provides a biological context crucial to integration. A well-managed genebank with readily available and good quality passport information (Food and Agriculture Organization of the United Nations, 2018) and minimal marker sets for establishing variation (e.g. SNPs) that identify germplasm are intrinsic to integration of data from derived samples. Active quality control becomes part of the story to ensure study design and experimentation have a firm basis to power biological interpretation.

In a healthcare research setting, metadata integration is becoming essential to bring together a patient's record with information about a cell line, a disease state, a set of phenotypes, etc., to understand patient characteristics and clinical outcomes. The main driver for standardisation was the development of nomenclature (e.g. Read Codes (digital.nhs.uk/services/terminology-and-classifications/read-codes) and SNOMED (Spicrs et al., 2017)) to harmonise clinical information in electronic medical records. Biomedical ontologies captured terms that may have been specific to a disease or body part, which subsequently necessitated the development of increasingly overlapping domain ontologies, e.g. UBERON (Mungall et al., 2012), to facilitate integration of data which increasingly needed to be at the systems level.

Enabling biological integration through access to consistently described and integratable data sources is not a new concept (Berti-Equille, 2001), but has remained essential due to the "data deluge" and increasingly multi-disciplinary science. Care has to be taken that technical integration is also paired with biological integration, i.e. bringing datasets together can be technically feasible, but whether that is appropriate to a biological question is often down to availability, quality, and richness of metadata of individual datasets (Börnigen et al., 2015).

Genebanks can be integrated into the downstream data management process, leading to FAIRer data infrastructure solutions for crops (Lapatas et al., 2015). Therefore, genebanks and the researchers that use them have a key role to play in making sure germplasm is fit for purpose with a view to designing field trials based on representative genetic diversity, analogous to designing a clinical trial based on representative demographics.

2.2 Describing the Variety of Field Trial Experimental Data

Designing Future Wheat is a crop research programme with a major pre-breeding component conducted in collaboration between UK research institutes and with a consortium of commercial wheat breeders and seed companies. While the design of the pre-breeding trials follow industry standard multi-site germplasm evaluation, the trials at each partner research site are designed to answer more specific questions and are hence more varied. This might comprise singular experimental outcomes, such as assessments of a particular treatment, but each of these experiments is carried out by different research scientists. This data is recorded and held in spreadsheets and/or bespoke databases, each with their differing formats between research organisations. As such, the data produced is not standardised between trials and is therefore not easily linked. To do so requires coordinated effort across the whole programme in order to understand what data is being produced, how it can be put into context using common metadata, and how it might be relevant to other trials and experiments (see Challenges for Data Linkage).

Therefore, an important objective of our project data management strategy has been to bring all the DFW trial datasets into one data repository annotated with the necessary metadata to support both findability and interoperability. We were not able to identify an established metadata standard which adequately describes field experiment design, plot layouts and related spatial information and so we have established our own and incorporated it into our central data management tools. Our data ingestion framework allows field trial data managers to upload their designs and link to trait measurements. This has also resulted in identification of new metadata terms that are required for describing elements of these trials but are not part of the commonly used ontologies (see "Ontology richness and standardisation of trait names").

2.3   Harmonising Environmental and Management Data

Agricultural research is often conducted to evaluate how the interactions between the crop genotype (G), environmental (E) and management (M) factors influence the behaviour of a particular trait (e.g. crop yield or flowering time), often called GxExM studies. Alternatively, a trial may evaluate the response of a particular trait to a treatment (e.g. amount of fertiliser applied) or change in management (e.g. use growth suppressors). The difference between a management or treatment effect depends on the nature of the research question. Management and treatments are aspects of a trial that can be controlled. Environmental factors, on the other hand, are those aspects of a trial that can't be controlled, such as levels of pest infestation, rainfall, temperature or level of sunlight.

When planning field trials and associated measurement protocols, the key decisions will be the selection of the genotypes (varieties), standardisation of trait measurements and any treatments, but also what environmental monitoring and management measures will be used in the trial. In the DFW project, we are working on a single crop (wheat) and our consortium has collaborated for many years. The wheat community has also been developing shared standards for trait measurements for some years (e.g. Dzale Yeumo et al., 2017). Therefore, in many ways we have less of a challenge in this area than if were we building a data sharing infrastructure for multiple crops.

In the case of our project trials, most of the harmonisation of data and methods have been agreed in the trial design stage. Where work is still ongoing is the collection of the associated environment data and management data alongside the core trait measurements and the management information in common forms so that the datasets we hold are complete and satisfy the requirements of re-usability.

2.4   Ontology Richness and Standardisation of Trait Names

Consistent descriptions of experimental processes and measurements of observed entities is vital to ensure cross-compatibility of datasets. The generation of agricultural data comes in many forms and from a wide range of instruments and methods (see Table 1 "*Complexities of data types based on their collection and analysis profiles*" for example). Without some form of internal consensus on standards for data description and sharing, attempting to compare results across high variable spaces such as these is at worst impossible and at best will require a great deal of manual curation. The description of these datasets is dependent on the availability of ontologies that comprehensively cover the domain, based on the richness of the terms within them.

An example of this is how trait measurements are recorded across sites (Pérez-Harguindeguy et al., 2013). Manual scoring of traits within field or greenhouse settings is commonplace and a route to understanding growth, development, and heredity often based on external factors. Traits tend to be semi-descriptive, where a measurement will be inferred but specifics of how that measurement was made are not, e.g. "plant height" is a common trait to measure, but says nothing of the process, constraints, or units when measuring. Therefore, we have produced a trait measurement catalogue which brings together the project-level consensus on terms used when measuring traits, based on the Crop Ontology (Shrestha et al., 2012) http://www.cropontology.org/ontology/CO_321/Wheat. Where traits and measurements are not harmonised with user experience, we feed back these potential additions and improvements to the ontology as appropriate.

2.5   Data Curation Tools and Techniques

Producing FAIR data requires active management through curation. Tools and services that help users meet FAIR data requirements without needing extensive manual intervention are often lacking. When they are provided, they often are geared towards data managers and curators themselves rather than tackling the issue of solving standardisation of datasets at the point where they are generated by a researcher.

Within DFW, using our trait measurement catalogue, users who need to record and submit trait measurements into our centralised data services can use a dedicated

Table 1 Complexities of data types based on their collection and analysis profiles

Data class/origin	Characterisation	Challenge/progress
HTP and UAV imaging	High volume, relatively low complexity	Wide range of different sensors/ instruments. Useful data comes only after bespoke data analysis pipelines – often research projects. Provenance tracking in analysis pipelines unsolved problem
Pangenomic datasets Wheat 10+ www.10 wheatgenomes.com	Medium to high volume, high complexity. Problems with *de novo* vs "lift over" annotations etc	Large polyploid genome sizes. Data processing pipelines compute intensive and in research projects. Visualisation is challenging, but tools are beginning to emerge.
Single cell genomics	High volume, high complexity	Metadata capture and processing across 1000s–100000s cells is complex. Data processing pipelines can be compute intensive, and software development is still required within research projects
Field trial datasets	Low volume, medium complexity "Integrative" already – images, traits, geolocation, unstructured metadata, low standardisation across sites	Solutions vary across sites and with different crops. Community awareness of the importance of standardisation is patchy. Trial / experimental design metadata has no accepted standard.
Diversity set genotyping	Medium volume – and complexity is a factor of the SNP density	Generally well developed standards and data pipelines exist in crop genetics labs.
Epigenetic datasets	Medium to high, high complexity (interpretation). Often need other datasets to contextualise	Similar to crop pan-genome datasets. Data pipelines are research projects in progress.
LTP and "physical" phenotyping -architectural traits	Low throughput, low complexity. Manual technologies lead to human-centric problems	These are the traditional datasets used in crop genetics and plant breeding. Range of standards in place (e.g. for breeders) and some good community agreements in place. In general, problems relate to formalisation of metadata and data quality issues at point of recording. Portable devices offer big improvements in field recording.
Chemical phenotyping	Low/medium volume, low complexity. Quantitative bulked datasets (typically plot based)	Complexity will depend on the compositional analysis required. Issues will be about tying together analyses from multiple instrumentation used

web service to search for consistent terminology *and* measurement properties to use in their collection strategy. For example, users can select a number of traits they wish to measure, and our system will produce an organised spreadsheet that they can fill out and then directly submit into our data repository. This reduces: the need for user support in terms of ontology use; incompatibility of generated datasets; friction experienced when trying to reformat datasets for deposition in repositories.

3   Challenges for Data Linkage

A central challenge of large projects such as DFW is the feasibility of real world management and coordination across large, varied data generation technologies. Researchers are increasingly reliant on a greater number of more varied datasets housed in multiple locations, and the integration of field data with large genomic and phenotypic datasets is needed to push forward the understanding of the relationship between genotype, phenotype and the environment. Approaches need to involve understanding the characterisation of these datasets into estimates of their volume and complexity which, when coupled to the availability of standards and curation tools, often forms the main barriers to data sharing and integration.

There are experimental technologies, e.g. high content phenotype data (phenomics) using state of the art imaging technologies and UAVs, that are being applied to the same germplasm used in the rest of the project for genomics studies. Integration of metadata-rich field trial, trait measurement, imaging and genotyping will offer new ways of predicting GxExM relationships using automated methods such as statistical inference, machine vision and AI.

Additionally, different strategies are needed for integration of each class of data which adds to complications and a large time cost when attempting to standardise – sometimes the plot is the reference unit, sometimes gene name, sometimes genotype, sometimes sample from plot. There are a wide range of different "primary" keys needed and the challenge is getting standardisations and adherence to naming and identifying schemes. This is especially important when retro-fitting technology to an ongoing project that is being run by multiple organisations and may have large legacy datasets.

There are other challenges when developing data management infrastructure for large research consortia, such as DFW. These emerge from the independence of research leaders within the consortium to set their research agenda while remaining within an agreed programmatic or strategic framework. Understandably, these leaders also focus on their science specialisms among the participating research groups and within the programme. Such large consortia therefore create a landscape of research, with research groups functioning as islands of outputs that contribute to the bigger picture. Providing the support to capture all outputs and making necessary connections (usually retrospectively) between them, for example a detailed finding about crop genetics and a trait brought into pre-breeding material, is not immediately possible without manual processes involving curation.

The process of collecting common data about field trials in DFW has highlighted the gap in tooling for data management of phenotyping from field experiments and the need to coordinate and share these data across the consortium to facilitate collaboration. Other data domains have recognised repositories, e.g. for wheat genome and transcriptome sequences, genetic variation (SNP) data, etc. There is also active development of plant phenotype metadata standards (Papoutsoglou et al., 2020) but as yet no dedicated publicly accessible repository. In future there will be a need to integrate other types of data that provide *data waypoints* in the landscape of crop science.

The following three research papers from DFW illustrate the variety of data and where there has been partial success (within the project) to link these data together in the DFW repositories.

In a time course study a high throughput plant phenotyping platform was used to extract plant height information (an important agronomic trait in wheat) using computer vision methods (Lyra et al., 2020). This was an investigation of new statistical genetic methods to address some of the challenges of high throughput datasets from time course studies. The data came from 197 wheat lines grown over two seasons, where 22–26 time points were measured by laser scanning, and plant height was extracted from the subsequent point-cloud data. Statistical genetics analysis permitted identification of persistent and transient QTLs. Genotype data came from a SNP array.

In Shorinola et al. (2019) mutant lines from a wheat tilling population were sequenced. Grain size and root phenotyping methods were used to explore the genetic links between root and grain development. The data processing steps involved extraction of quantitative measures of root growth based on imaging techniques.

In a time-course study of senescence in wheat (Borrill et al., 2019) the experimental focus is on a single wheat variety (bobwhite). The plants were phenotyped for chlorophyll content and grain moisture content. RNA expression data was generated from different tissues and this was used to infer gene regulatory networks. Through integration with public datasets from *Arabidopsis* and the public wheat genome sequence, gene function annotations were transferred to propose the genetic control mechanisms that underlie senescence.

All these papers are based on material that has been grown in a field or greenhouse and demonstrate a large scale generation and reuse of data across the variety of data types and resources:

- Phenotyping (single trait)
- Omics (multiple traits/genes/transcripts)
- Genetics (multiple varieties/QTL)

Our approach in DFW to providing a higher-level view of the data landscape has been to further develop the KnetMiner system. KnetMiner integrates information extracted from public databases and literature resources as well as data services from other wheat information resources to create a comprehensive knowledge graph of wheat information (Hassani-Pak et al., 2021). It has also been possible to interoperate with some of the large-scale DFW data services into this resource (e.g. gene co-expression networks, (Ramírez-González et al., 2018). Furthermore, the KnetMiner team is experimenting with knowledge-level integration by means of "lightweight ontologies", such as Bioschemas (Gray et al., 2017). Knetminer for wheat (knetminer.com/Triticum_aestivum/) therefore provides an open-access wheat resource to DFW participants and the wider community as presented in published papers and datasets in public repositories. However, KnetMiner provides linkage at the knowledge level, and it does not support integration and linkage down to the individual datasets and measurements from those studies.

It is increasingly clear that researchers are producing data with future reusability in mind. However, this does not implicitly make the task of data linkage easier in terms of the complexities of data types and their contexts. Within one experiment, the challenge of linkage is somewhat manageable. To interlink trials and experiments across multiple organisations is very difficult to accomplish in retrospect, even within a single coordinated programme.

4   How to Do It – Data Stewardship Strategy and Infrastructure

As we have described above, in DFW we aim to support integration and sharing of data for our multi-faceted, multi-year, cross-institute programme comprising a portfolio of different experiments, and make research outputs visible both internally and to a range of external stakeholders. To do this, we needed to implement multiple strategic layers of physical, virtual, and coordination infrastructure.

4.1   FAIR Data Sharing within DFW

When the DFW project was being developed, it was agreed that sharing data **within** the project would be essential and that the different work streams would need to coordinate efforts and share best practice. It was also important that data management and sharing was not seen as the sole responsibility of the informatics teams, but one that was shared with the plant scientists too. Coordination was needed to inform the development of the data sharing tools and ensure interoperability between the specialist bioinformatic and genomic data resources that individual partners were developing during the project.

So, the DFW Data Coordination Task Force (DCTF) was established to bring together a multidisciplinary group with representation from all parts of the project including crop breeders, UAV drone phenotyping experts, and software developers. Members were selected from across the programme's plant science teams to act as local experts on the data sharing activities. As such, DCTF members help and encourage individual scientists to come forward with data and get the support they need to annotate it with the necessary metadata and submit to data repositories. Coordination among DCTF members has been achieved through video calls held every 2 months that focus on prioritisation of collaborative elements of the programme, including data resource development that requires input from biologists and statisticians, data generators, data analysis and visualisation experts, and information system software engineers. Regular hackathons have brought developers together with plant scientists to refine data submission and data sharing technologies and to make rapid progress on software and web site interoperability. Hackathons were initially face to face working meetings which led to continued joined-up activities for periods of time after the event. During the COVID-19 pandemic we have used collaboration tools (Microsoft Teams) to run Hackathons with no obvious reduction in their effectiveness in terms of engagement during the meeting or in followup discussions. Hackathon topics have been nominated either by DCTF members or through wider calls for topics from project work-package leaders. More recently, a hackathon to discuss the potential implication of new wheat pangenome data sets (e.g. Walkowiak et al., 2020) for the project was organised with input from the broader wheat and pangenome research community.

Individually, the DCTF members and many of the wider DFW team are active participants in larger national and international communities with shared interests in wheat and wider grass and cereals research and the translation to crop improvement, e.g. the Monogram network (www.monogram.ac.uk), The International Wheat Initiative (www.wheatinitiative.org) and The Wheat Genetic Improvement Network (www.wgin.org.uk). Many of the DCTF are also members of the ELIXIR plant sciences bioinformatics and data infrastructure community (http://elixir-europe.org/communities/plant-sciences). This allows skills, expertise and knowledge to be shared more effectively through a formalised network of peers with a clear remit.

4.2 Compute, Storage, People, Skills

Research Data Management (RDM) lifecycles (Higgins & Others, 2012) are not only concerned with the human-level aspects of collecting, managing, analysing and sharing results, but also the technical aspects where Research Infrastructures (RIs) now play a huge part in the modern high-throughput crop research arena. Traditional research outputs of publications, software and data are increasingly underpinned by a fabric of digital infrastructure, intrinsically woven into how RDM is carried out. UKRI recently produced a landmark report on the status and future recommendations for UK digital infrastructures across the UKRI family of Research Councils and HEIs, and part of this report was concerned with this "ferrying" of data in and out of life science RIs – termed "data stewardship".

As part of DFW, our data stewardship strategy is a formal part of our programme, where we use investments made into digital infrastructure at each of the partner sites in order to facilitate the wider integration of research data through effective coordination. For example, the DFW Data Portal houses a large amount of pre-publication data and acts as a long term storage area for datasets typically not suited to public repositories. This repository comprises an infrastructure of virtual servers and data storage architectures that is provided by CyVerse UK cloud, running within the EI National Capability for e-Infrastructure (Earlham Institute, 2018).

The underpinning infrastructure is invisible to end users via a typical "as-a-Service" architecture (www.intel.co.uk/content/www/uk/en/cloud-computing/as-a-service.html), providing data access APIs, hosted websites, and analytical platforms. This enables us to rapidly develop new tools, share new data, and explore technical solutions collaboratively across the project and beyond.

4.3   Benefiting from Open Source Tools

Our approach relies on open source platforms, both in terms of our own developments but also when using off-the-shelf solutions, e.g. CKAN (ckan.org) for our published outputs, including papers and supplementary datasets. The software and data resources developed within the project (see Table 2) are heavily reliant on open source software (e.g. Neo4J, REACT, PostGreSQL), typically with free academic license agreements. This allows us to maximise the cost-effectiveness of our investments in research, and retain the ability to adapt and modify tools and methods as appropriate. A key challenge is keeping up with new technologies whilst also being able to interoperate in a backwards-compatible manner with previous resources or those developed by other groups. Open source tools give flexibility to try and adapt new ways of working with data and metadata without needing paid software, proprietary codebases or formal licensing agreements.

We did consider open source solutions that exist for technical integration of crop improvement data, e.g. BreeDBase (http://breedbase.org), Germinate (https://germinateplatform.github.io/get-germinate/; Lee et al., 2005), etc. but currently they do not implicitly provide the curation, QC, filtering and importing steps needed to help reach a suitable quality level for biological integration across a wide range of data types. They also assume a mature data and metadata specification. In a project

Table 2 List of main DFW funded data resources to date

CerealsDB	www.cerealsdb.uk.net
KnetMiner	https://knetminer.com/Triticum_aestivum
Wheat Expression Browser	www.wheat-expression.com
Wheat Germplasm Resource	www.seedstor.ac.uk
Ensembl Plants – Wheat	plants.ensembl.org/Triticum_aestivum
DFW Field Trials	grassroots.tools/dfw
DFW Data Portal	opendata.earlham.ac.uk/wheat
DFW Digital Repository	ckan.grassroots.tools

such as DFW which represents a large legacy and heterogeneous data landscape, we needed to build resources that provided a route of least resistance to harmonisation but also included APIs to remain interoperable.

4.4   FAIR Publication Strategies

An important aspect of our commitment to FAIR principles is to ensure that research publications and supplementary data sets are linked and that the data are in usable formats (e.g. not simply referenced in a published PDF). To achieve this we make use of the CKAN digital repository framework and capture all DFW publications and associated data as explicitly listed resources alongside a publication, allowing users to search for and access supplementary data, data files, code on GitHub and other outputs in one place. CKAN is in use by many research institutions, and as such has an active development and support community, and has a fully featured API for integration and programmatic access.

All the data resources and other information are linked from the main project web site (designingfuturewheat.org.uk).

4.5   Meeting Community Obligations

Large research projects have to balance project level obligations and those from stakeholders. For example, DFW reports regularly to funders, researchers, breeders, farmers, and policymakers in food security, nutrition, national farming, etc. These reports contain evidence of our data resources and their use in the community. Openness and transparency is essential to maintain effective communication in a complex landscape of commercial and academic interests. In DFW, FAIRness of our data is a keystone to contribute effectively to the pre-competitive aspects of wheat research in the UK and beyond. We aim to deliver the benefits of publicly funded wheat research with the least barriers to access as possible, from data to seeds to research outputs:

Data Availability is a project-level commitment, which has to be agreed from the beginning and necessary behaviours of all participants reinforced by the project management team. From the outset, the funders of DFW (BBSRC-UKRI) mandated that the project should release data to maximise public benefit. All other data generated in the project is also expected to make it as quickly as possible into the public domain, conforming to FAIR data sharing principles. The stakeholders in the best position to exploit data from the project are other researchers and crop breeders.

Germplasm Availability A key resource being developed by DFW are the Breeders Toolkits – pre-breeding germplasm that are evaluated by a pre-competitive community of wheat breeders associated with the project for potential use in their own (competitive) breeding programmes. After a 2 year embargo, the breeders are also obliged to return their assessment data to the project in order to share with the wider community. The germplasm is publicly available to any user outside the consortium as well (via https://www.seedstor.ac.uk/).

Research output Availability The BBSRC-UKRI have paid close attention throughout the project to the research outputs being generated and the biannual reporting to them includes information on publications, datasets and software outputs. All members of the project are expected to report into a central (Google spreadsheet) on a range of research outputs and this sheet is also made available to BBSRC. Collecting this project-wide information in a simple and transparent way has proved to be extremely useful to show both project progress and commitment to openness. In particular it is the reference source for all research publications for the project for all partners which has provided the basis for a publications portal using CKAN. It has also had the beneficial side-effect of creating confidence in the project with the funder who can easily follow the result of their investment and use this evidence internally and with the government to demonstrate the value of a major investment in UK crop research.

The delivery of these resources does not implicitly improve data linkage in and of itself. Community obligations in this sense are related to the deployment of services and infrastructures that adhere to community standards, such as implementing the Breeding API (BrAPI) (Selby et al., 2019) on top of a data resource or the use of agreed controlled vocabularies, data formats, etc. To ensure these APIs work suitably, the data itself needs to be prepared and described adequately which is not simply a technical task, but a sociological one. This requires community acceptance for the need for standardisation and having resources to comply with agreed standards and protocols, but these efforts are typically not explicitly funded through research grants or programmes. To avoid siloed information and abandoned data warehouses, data linkage requires dedicated funding and resources to bring together both the people and the technology to deliver fit-for-purpose tools and services that demonstrate strong and useful interoperability. Future sustainability then demands openness across these decentralised but interconnected data resources.

5 Conclusions

There are other crop programmes internationally, so DFW is not unique in terms of coordinated efforts to bring about improved access to crop data, e.g. Sol Genomics, MaizeDB, Brassica Information Portal. Other national and international efforts have also been focused on wheat, e.g. Wheat Initiative WheatIS, Triticeae Toolbox (T3), CIMMYT. Indeed, DFW efforts interlink some of these existing resources, e.g. KnetMiner and T3, and uses API standards in some of its outputs to harmonise with the community such as BrAPI (Selby et al., 2019). The geographic and data-centric heterogeneity and diversity of all these resources has influenced and focused

DFW's strategy for data linkage in that we can learn from prior work and also drive the adoption of best practice and standards for FAIR data within the UK's national wheat programme to comply with the consensus of the broader community. The DFW strategy explicitly promotes both openness and FAIRness in an effort to support future wheat data access with the fewest barriers possible.

A strategy for data linkage should be set out at project start and all research outputs would be aligned to that strategy, supported by fully-featured production-level tools to manage datasets FAIRly. However, this is often infeasible due to the challenge of producing a cohesive data strategy without *a priori* oversight of all data types within a programme, and then ensuring compliance with that strategy from day one.

A "catch-22" situation arises where data managers cannot know all data outputs from the start of a project, and will not have all the required data management tools at their disposal, so cannot accurately model data linkage. Furthermore, even if these elements are prepared in advance, there is still the issue of the required manual curation of datasets to ensure that the data is modelled in such a way that biological interpretation is correctly maintained. This curation is often under-resourced compared to the generation of hypotheses, data, and publications. This situation is commonplace – we interact with many stakeholders within research grants, advisory and policy work, and strategic scientific programmes, and throughout all these scenarios we see the same points raised.

A solution would be to strictly control *all* data collection and sharing activities across all researchers, mapped to complete standards that can comprehensively represent all data types. This simply is not feasible for the majority of the data generated from the fast-moving world of omics-intensive agriculture. This also comes at a very large sociological cost, and would likely be rejected as overbearing by the community.

So, a solution of three parts remains:

– Constant ongoing supportive coordination to ensure a data management and linkage strategy is sufficient

   • We suggest the use of coordination committees to ensure inclusion of researchers across a scientific project in order to encourage data standardisation and sharing, and to address societal and technical changes in data management methods

– Proactive, potentially automated, management of well-known "standard" data types and studies

   • Whilst not at the forefront of scientific projects, formalised data management is absolutely required to ensure the automation of routine tasks, in turn promoting effective data reuse by the community at large

– Retrospective application of integrative methods to cater for the new, "known unknown", or less-well-standardised data types

- This is an exciting area for opportunities to develop inventive techniques to add structure to unstructured data, extracting information within scientific publications and datasets, thus future-proofing data for advances in linkage and analysis

Funders are increasingly motivated to look to facilitate these solutions, but key elements that we have highlighted are still lacking in the research landscape, including the adequate resourcing of sociological and technical research that would underpin their data management policies.

To summarise:

- Implementing FAIR principles within a project of the size of DFW is a significant undertaking
- There is a hidden cost of FAIR that is not often taken into consideration by funders and other stakeholders, leaving major FAIR data management tasks to *ad hoc* efforts by research staff rather than dedicated data stewards
- Openness leads to collaborations – in our DFW experience, communities are more willing to engage and share when our strategy for FAIR data is evidenced by open tools.
- Openness ensures that the project and its participants are visible as good collaborators and technical ecosystems benefit from open communication about issues, benefits, and functionality leading to better interoperability and future-proofing.
- Involvement is required across all stakeholders: field experiment managers, farm staff, greenhouse technicians, genebank managers, statisticians, experimentalists, molecular biologists, bioinformaticians, software developers, data managers, project managers and reporting coordinators, PIs, and funders. This is a *"Matrix of Responsibilities"* which takes time and effort to establish in a complex project.
- Increasing multidisciplinarity within a common strategic backdrop for standardisation leads to multiple teams working together to produce multiple datasets more effectively, and with better FAIRness.
- Planning for FAIR in a live research project has to adapt as experiments and methods evolve over time. Initiation often looks very different to milestone delivery so it is still too commonplace that FAIR considerations are only thought about at the time a project has finished and its outputs are being released.
- Transparency of project outputs has benefits across obligations to project-level and a wider stakeholder community.

Acknowledgements The authors gratefully acknowledge support from the UKRI-BBSRC Designing Future Wheat project (Rothamsted BB/P016855/1; Earlham Institute: BBS/E/T/ 000PR9783) as well as the strategic support funding to Rothamsted Research and the Earlham Institute.

The authors wish to thank the members of the Designing Future Wheat Data Coordination Task Force (designingfuturewheat.org.uk/dfw-data-coordination-taskforce/) for their support in the writing of this paper and for their work to make much of the progress described in this paper possible.

References

Berti-Equille, L. (2001). Integration of biological data and quality-driven source negotiation. In *Conceptual modeling—ER 2001* (pp. 256–269). Springer.

Börnigen, D., Moon, Y. S., Rahnavard, G., et al. (2015). A reproducible approach to high-throughput biological data acquisition and integration. *PeerJ, 3*, e791.

Borrill, P., Harrington, S. A., Simmonds, J., & Uauy, C. (2019). Identification of transcription factors regulating senescence in wheat through gene regulatory network modelling. *Plant Physiology, 180*, 1740–1755.

Earlham Institute (2018) *National Capability in e-Infrastructure*. https://www.earlham.ac.uk/national-capability-e-infrastructure. Accessed 23 Feb 2021

Food and Agriculture Organization of the United Nations. (2018). *Genebank standards for plant genetic resources for food and agriculture*. Food & Agriculture Org.

Gray, A. J., Goble, C., & Jimenez, R. C. (2017). *The bioschemas community (2017) bioschemas: From potato salad to protein annotation*. ISWC 2017 Poster Proceedings.

Hassani-Pak, K., Singh, A., Brandizi, M., et al. (2021). KnetMiner: A comprehensive approach for supporting evidence-based gene discovery and complex trait analysis across species. *Plant Biotechnology Journal*. https://doi.org/10.1111/pbi.13583

Higgins, S., & Others. (2012). The lifecycle of data management. *Managing research data*, 17–45.

Lapatas, V., Stefanidakis, M., Jimenez, R. C., et al. (2015). Data integration in biological research: An overview. *Journal of Biological Research, 22*, 9.

Lee, J. M., Davenport, G. F., Marshall, D., Ellis, T. H. E., Ambrose, M. J., Dicks, J., van Hintum, T. J. L., & Flavell, A. J. (2005). GERMINATE. A generic database for integrating genotypic and phenotypic information for plant genetic resource collections. *Plant Physiology, 139*(2), 619631. https://doi.org/10.1104/pp.105.065201

Lyra, D. H., Virlet, N., Sadeghi-Tehran, P., et al. (2020). Functional QTL mapping and genomic prediction of canopy height in wheat measured using a robotic field phenotyping platform. *Journal of Experimental Botany, 71*, 1885–1898.

Mungall, C. J., Torniai, C., Gkoutos, G. V., et al. (2012). Uberon, an integrative multi-species anatomy ontology. *Genome Biology, 13*, R5.

Papoutsoglou, E. A., Faria, D., Arend, D., et al. (2020). Enabling reusability of plant phenomic datasets with MIAPPE 1.1. *The New Phytologist, 227*, 260–273.

Pérez-Harguindeguy, N., Díaz, S., Garnier, E., et al. (2013). New handbook for standardised measurement of plant functional traits worldwide. *Australian Journal of Botany, 61*, 167.

Ramírez-González, R. H., Borrill, P., Lang, D., et al. (2018). The transcriptional landscape of polyploid wheat. *Science, 361*. https://doi.org/10.1126/science.aar6089

Selby, P., Abbeloos, R., Backlund, J. E., et al. (2019). BrAPI—An application programming interface for plant breeding applications. *Bioinformatics, 35*, 4147–4155.

Shrestha, R., Matteis, L., Skofic, M., Portugal, A., McLaren, G., Hyman, G., & Arnaud, E. (2012). Bridging the phenotypic and genetic data useful for integrated breeding through a data annotation using the crop ontology developed by the crop communities of practice. *Frontiers in Physiology, 3*, 326. https://doi.org/10.3389/fphys.2012.00326

Shorinola, O., Kaye, R., Golan, G., et al. (2019). Genetic screening for mutants with altered seminal root numbers in Hexaploid wheat using a high-throughput root phenotyping platform. *G3, 9*, 2799–2809.

Spiers, I., Goulding, J., & Arrowsmith, I. (2017). Clinical terminologies in the NHS: SNOMED CT and dm+d. *British Journal of Pharmacy, 2*. https://doi.org/10.5920/bjpharm.2017.02

Walkowiak, S., Gao, L., Monat, C., et al. (2020). Multiple wheat genomes reveal global variation in modern breeding. *Nature, 588*, 277–283.

Yeumo, E. D., Alaux, M., Arnaud, E., Aubin, S., Baumann, U., Buche, P., Cooper, L., Ćwiek-Kupczyńska, H., Davey, R. P., Fulss, R. A., Jonquet, C., Laporte, M.-A., Larmande, P., Pommier, C., Protonotarios, V., Reverte, C., Shrestha, R., Subirats, I., Venkatesan, A., Whan, A., & Quesneville, H. (2017). Developing data interoperability using standards: A wheat community use case. *F1000Research, 6*, 184314407. https://doi.org/10.12688/f1000research.12234.2

Linking Legacies: Realising the Potential of the Rothamsted Long-Term Agricultural Experiments

Richard Ostler, Nathalie Castells, Margaret Glendining, and Sarah Perryman

Abstract Long-term agricultural experiments are used to test the effects of different farm management practices on agricultural systems over time. The time-series data from these experiments is well suited to understanding factors affecting soil health and sustainable crop production and can play an important role for addressing the food security and environmental challenges facing society from climate change. The data from these experiments is unique and irreplaceable. We know from the Rothamsted experience that the datasets available are valued assets that can be used to address multiple scientific questions, and the reuse and impact of the data can be increased by making the data accessible to the wider community. However, to do this requires active data stewardship. Long-term experiments are also available as research infrastructures, meaning external researchers can generate new datasets, additional to the routine data collected for an experiment. The publication of the FAIR data principles has provided an opportunity for us to re-evaluate what active data stewardship means for realising the potential of the data from our long-term experiments. In this paper we discuss our approach to FAIR data adoption, and the challenges for refactoring and describing existing legacy data and defining meaningful linkages between datasets.

1 Introduction

Long-term agricultural experiments (LTEs) are used to test the effects of different treatments on the sustainability of crop production and resilience of soil health (Dyke, 1974; Johnston & Poulton, 2018). The data collected from LTEs provide irreplaceable time-series while physical samples provide unique snap shots in time for future research. Since the experiments operate over extended periods of time,

R. Ostler (✉) · N. Castells · M. Glendining · S. Perryman
Computational and Analytical Sciences, Rothamsted Research, Hertfordshire, UK
e-mail: richard.ostler@rothamsted.ac.uk; nathalie.castells@rothamsted.ac.uk;
margaret.glendining@rothamsted.ac.uk; sarah.perryman@rothamsted.ac.uk

© The Author(s) 2023
H. F. Williamson, S. Leonelli (eds.), *Towards Responsible Plant Data Linkage: Data Challenges for Agricultural Research and Development*,
https://doi.org/10.1007/978-3-031-13276-6_7

they are well placed to monitor soil processes and responses to climate change. Samples and data can also be subjected to new and improved analytical, statistical and data science techniques, not imaginable or available when collected, such as gene sequencing and machine learning.

Overtime, cumulative additions or omissions of treatment factors can give rise to radically different soil environments between experiment plots resulting in contrasting levels of soil pH, nutrient availability, organic matter or biological activity. Researchers can take advantage of these conditions to use a long-term experiment as a living laboratory. In-field observations and measurements can be made, and, provided they do not interfere with the continuity of the experiment design and data collection, short-term interventions, which take advantage of quantitative differences between plots, may be introduced. Alternatively soils from these plots can be sampled for use as growing media for laboratory and glasshouse experiments.

However, LTEs are costly to run, and experiment managers must demonstrate their cost effectiveness to funders as measured by their continuing impact on science and agriculture. A good return on funder investment can be achieved if an experiment can serve more than one research objective, beyond its original purpose (Johnston & Poulton, 2018), and this can be realised by re-using LTE data and integrating it with other data. For the experiment manager this means a plurality of routine and relevant crop, agronomic and environmental observations are made, detailed experiment management records kept (Dyke, 1974) and appropriate procedures for experiment and sample access are in place. For the data curator the responsibility is to ensure an experiment and its data are sufficiently well stewarded to remain findable, accessible, interoperable and re-usable over time (Wilkinson et al., 2016). However, a major challenge for LTE data curators is a lack of widely adopted standards for managing data across the global LTE community.

The data from long-term experiments should therefore have an important role in understanding agricultural systems and the challenges they face. However, realising this potential means making the data from long-term experiments findable, accessible, and re-usable to the wider scientific community. This potential can be magnified by making the data both machine actionable and interoperable with data from other experiments, both short and long-term, and other datasets. Only then can the data from these unique experiments be used to help address challenges of food security, soil health and climate change adaptation.

This paper uses the Rothamsted Long-term Experiments and Electronic Rothamsted Archive (e-RA) to examine some of the challenges facing experiment data curators who manage LTE data and how publication of the FAIR Data Principles (Wilkinson et al., 2016) has stimulated a reappraisal of LTE data provision using e-RA.

2 Long-Term Experiments at Rothamsted

Between 1843 and 1856 Sir John Bennet Lawes and his scientific collaborator Joseph Henry Gilbert established a series of experiments, including Broadbalk and Park Grass, at the Rothamsted Estate, Hertfordshire, UK (2019). The experiments aimed to test the effects of different organic and inorganic fertilisers on yields for a range of cereal and root crops and hay. Of the nine experiments established between 1843 and 1856, known as the "Classicals", five are still running; these are Broadbalk Wheat, Park Grass Hay, Hoosfield Barley, Exhaustion Land and Garden Clover. Since then at least 40 other long-term experiments, running for at least 10 years, have tested a diverse range of treatment factors including crop rotations, cultivation, manuring, pest and disease control and liming.

2.1   Data and Samples: Lawes and Gilbert's Enduring Legacy

Early on, Lawes and Gilbert recognised the potential value of the data archives and physical sample collections for future scientists. In 1864 Lawes noted the rapid progress in soil science during his own time and speculated what further knowledge future progress could yield (Lawes & Gilbert, 1864). Their foresight in establishing a tradition of data collection and sample archiving has continued to benefit generations of scientists to the present day.

Lawes and Gilbert didn't just keep records and file them away. From as early as 1862 data and results were published, initially as 'Memoranda' (1862) and later as 'Supplements' then 'Yield books'. By 1927 (1928) the Supplements were publishing data and results alongside experiment documentation, recognisable today as structured metadata (Fig. 1), including experiment design; treatment factors and factor levels; plot plans; plot areas; crop varieties; and agronomic events. Later Yield Books added objectives; notes; investigator names; previous cropping; and plot dimensions.

The knowledge about an LTE also extends into an experiment narrative which describes its history in terms of the events and decisions that have shaped an experiment from inception through to either its termination or the present. This narrative provides crucial information for researchers which is often critical to appropriately interpret and re-use experiment data over time.

2.2   The Long-Term Experiments National Capability

Today the ongoing importance of the Rothamsted Long-term Experiments is recognised and funded by a Biotechnology and Biological Sciences Research Council National Capability Grant. The grant objective is to provide long-term

SUGAR BEET.
MANURING.
Nitrochalk as top dressing, applied :—(*a*) early ; (*b*) early and late.
Superphosphate.
Muriate of potash and potash manure salts.

CULTIVATION.
Subsoiling.   Ridging.

Great Harpenden, 1928.
N.W.

	R	F	F	R	R	F	R	F	R	F	F	R	
I	2	6	1	5	10	9	12	11	4	3	7	8	O
II	1	8	5	9	3	7	11	10	6	4	12	2	S
III	6	3	2	11	5	10	4	7	12	8	1	9	O
IV	7	5	9	12	4	8	6	3	2	1	11	10	S
V	12	4	8	3	11	6	5	1	9	10	2	7	S
VI	8	10	11	7	1	12	2	4	3	5	9	6	O
VII	4	2	3	1	9	5	10	6	7	12	8	11	O
VIII	10	7	12	8	2	11	1	9	5	6	3	4	S
IX	3	12	7	4	8	1	9	2	10	11	6	5	S
X	9	1	10	2	6	4	8	12	11	7	5	3	O
XI	5	11	6	10	7	2	3	8	1	9	4	12	O
XII	11	9	4	6	12	3	7	5	8	2	10	1	S

VARIETY : Dippe.
SYSTEM OF REPLICATION : 12×12 Latin Square.
AREA OF PLOT : .014 acre.
TREATMENTS : Muriate of Potash at the rate of 2 cwt. per acre or equivalent Potash Manure Salts (30%). Superphosphate at the rate of 2 cwt. per acre. Top dressing of Nitrochalk at the rate of 2 cwt. per acre, applied early (June 23), and both early and late (July 21). All plots had basal dressing of 10 tons compost in winter, and 2 cwt. per acre Sulphate of Ammonia with other artificials on May 4.
R, F = Pairs of strips one way allotted at random to ridged and flat seed bed.
S, O = Pairs of strips the other way allotted at random to sub-soiling and " not " sub-soiling. The 12 plots of each treatment had 3 allotted to each of the 4 cultivation treatments.
Seed sown May 5 ; roots lifted October 26–November 3.

Fig. 1 An early example of structured field experiment metadata published in the Rothamsted Reports, 1927–1928

experiments and their data for the benefit of the UK bioscience community and to facilitate internationally excellent research in the field of food security and agroecological research. The experiment plots, data and samples are available for scientific research and the grant supports this by funding ongoing maintenance of the experiments, routine data collection and data stewardship, the Rothamsted Electronic Archive (e-RA) and Rothamsted Sample Archive (2019).

2.2.1   e-RA Data Curators

The primary roles of the e-RA Data Curators are data stewardship, data provision and servicing data requests. Data provision includes both appending new experiment data to existing datasets, preserving the backlog of legacy LTE data into accessible research data and maintaining metadata and supplementary documents.

Servicing data requests typically involves advising researchers on the appropriate plots and treatments to use for their research question. It may also include working up new or bespoke datasets for a request. The curators can also help foster new collaborations by match-making researchers with Rothamsted scientists having similar interests. For requests where the curators provide significant support they would expect to be included as co-authors in any publications, otherwise standard acknowledgement text, rather than a citation would be provided plus notification for any publications using the data requested.

A further role of the Curators is to collate impact metrics for reporting to BBSRC. This includes information about data requests, website traffic and data downloads. The information collected for a data access request is detailed, but necessary to show which sectors are using the data, and how and where the LTE National Capability is supporting other BBSRC research.

2.2.2 The Electronic Rothamsted Archive and Data Provision 2013–2020

Development of the electronic Rothamsted Archive, commonly known as e-RA, started in 1991 and its evolution to the launch of a public website in 2013 (referred to as e-RA 2013) is documented by Perryman et al. (2018). In 2021 a new version of e-RA (e-RA 2021) was released, and the changes made are discussed in the following sections.

e-RA provides detailed information about the long-term experiments and meteorological stations, and either direct or request access to LTE and weather data. The site also hosts the Rothamsted Document Archive (e-RADoc) http://www.era.rothamsted.ac.uk/eradoc/, which contains scanned copies of historical documents including the Memoranda, Annual Reports, Yield Books, Guides, Farm Maps and Experiment Plans, making these printed resources available online.

The LTE Data are stored securely in the e-RA database, implemented in Microsoft SQL Server. Prior to e-RA 2021, researchers accessed datasets held in the e-RA database by submitting a data access request agreement, stating the scientific basis for the request and datasets required, to the e-RA Data Curators. The researcher would either be given password access to requested datasets via the online Data Extraction Tool, renamed e-RA Data for e-RA 2021, or the e-RA curators would compile a bespoke dataset. While useful for experienced users, e-RA Data has limited functionality, allowing users to filter, sort, and download subsets of data from defined tabular datasets. It does not allow dataset (table) joins and downloaded data are provided without accompanying metadata such as an identifier, query parameters, experiment name, plot treatment details or column definitions; bespoke datasets extracted by the e-RA curators would be provided with supplementary documentation, but not published with a DOI.

This data access process allowed collection of usage data for impact reporting to funders and to control the release of data as a safeguard against misinterpretation or misrepresentation of the experiments.

Since 2016 aggregated 'Open Access' datasets have been freely available for download. These datasets are published with a DOI and accompanying metadata following the DataCite Schema (Group, 2019) recommendations. Unlike e-RA Data which provides access to annual plot data, the Open Access datasets provide an overview of key findings and trends or changes in the data and are typically averaged over several years or plots. For example the Broadbalk Mean long-term winter grain yields dataset (2017) uses 10 year means for selected plots to illustrate differences between fertilizer treatments and cropping system alongside the introductions of new agricultural technologies.

2.2.3 Data Reuse and Impact

The Long-term experiments are a well-used resource. In the first 5 years of e-RA 2013's public launch, there were approximately 400 requests for long-term experiment data and between 2011 and 2020 an average of 24 publications per year (updated from (Perryman et al., 2018)).

2.2.4 The Rothamsted Sample Archive

The Rothamsted Sample Archive holds over 300,000 soil, grain, herbage, fertiliser and organic manure samples from the long-term experiments, dating back to 1843. The samples are a unique resource freely available to scientists across the globe, and around 15–20 requests are received annually. The Sample Archive has been used to investigate diverse subjects ranging from the effects atmospheric pollution on agriculture (Fan et al., 2008) to evolutionary trends in pesticide resistance (Hawkins et al., 2014) and wheat grain quality traits (Mariem et al., 2020). The Sample Archive is not currently searchable online.

2.3 Sources of Long-Term Experiment Data

Long-term experiment datasets are created and/or added to in one of three ways:

1. Routine data creation by the LTE National Capability
2. Non-routine data creation by researchers external to the LTE National Capability
3. Digital preservation of legacy data by the e-RA Data Curators

2.3.1 Routine Data

The long-term experiments collect routine data for yields and yield traits, management data, soil chemistry and botanical (weed) diversity. Data management is a mature process with data collection and ingestion workflows, analytical methods, and quality assurance documented by internal standard operating procedures (Perryman et al., 2018). However, there is scope for modernisation to better reflect new practices, for example, creating data that is 'born FAIR' rather than making FAIR at a later stage.

2.3.2 Non-routine Data

The long-term experiments and sample archive can be used as a living laboratory resource by researchers external to the National Capability. This provides opportunities for new data creation and together these externally created datasets represent a highly heterogeneous collection, ranging from tabular observations to imagery and sequence data. Non-routine data are generated via three routes including:

1. In field observations and surveys using either manual assessments or sensor technologies (Edwards & Lofty, 1982; Morris, 1992).
2. Soil and vegetation laboratory analyses, using either archived samples or fresh samples collected from experiment plots (Hawkins et al., 2014).
3. Using soil collected from experiment plots as a growing medium for pot and laboratory experiments which are analysed to generate additional datasets (Neal et al., 2020).

Unlike routine data collection, which adds to the LTE time series, non-routine data collection events normally cover a subset of plots and treatment factors for a limited time and are not required to follow prescribed data collection methods.

Before accessing either an experiment or sample archive, researchers are required to submit a scientific justification, however, there is no requirement to provide a data management plan to demonstrate how the data will be collected, managed, and published.

2.3.3 Legacy Data

Rothamsted has conducted many Long-term Experiments over the decades, but much of the data collected is inaccessible or in need of preservation. An ongoing task for the Curators is to mobilise these data, however, this can be a slow process requiring data transcription, and, finding and checking source documents such as experiment plans. If legacy data are not being requested, the potential value of the data may be unclear, and the effort required to recover it difficult to justify.

2.4   Complementary Data: Environmental Monitoring Activities

Environmental monitoring at Rothamsted began in 1853 when Lawes and Gilbert started recording meteorological observations to better understand variations in yield due to weather. Since then the variety, velocity, and volume of additional environmental variables available has been extended through technological innovations and participation in long-term environmental monitoring networks. Together these provide important complementary datasets.

Rothamsted is part of the UK Environmental Change Network (http://www.ecn. ac.uk/) which records biodiversity data and atmospheric, water and soil chemistry data, and the UK Cosmic-ray Soil Moisture Monitoring Network (UK-COSMOS) (https://cosmos.ceh.ac.uk/).

In 1964 Rothamsted established the Rothamsted Insect Survey (https:// insectsurvey.com/), a national network of light traps and later suction traps, for recording moth and aphid distributions. Light traps operate at all four sites and suction traps at Rothamsted and Brooms Barn.

In 2019 permanent soil moisture sensors were added to selected Park Grass plots.

3   Challenges for Long-Term Experiment Data Stewardship

Lawes and Gilbert left a remarkable data legacy but providing continuing access to reusable data remains a challenge. The previous section provided the context for LTEs and in this section we elaborate on the data challenges facing them.

When the Elliot 401 computer was introduced to Rothamsted in 1954 data management entered a new digital age. Since then the technologies and practices for managing and accessing data have evolved rapidly. Just as archive samples can be reanalysed in ways previously unimagined, today data can be published, linked, chunked, and reused, all as a machine actionable resource. But getting data to this state requires specialist data science skills and effort and, just because a computer can link data, it doesn't mean it always should. Understanding how to provide, interpret and integrate LTE data with confidence remains imperative if it is to be used to generate meaningful knowledge continuing impact through re-use.

The FAIR data principles are being widely adopted across research institutions in the agricultural sector and promoted by communities such as Elixir (https://elixir-europe.org/system/files/elixir_statement_on_fair_data_management.pdf) and outputs of the Research Data Alliance Agricultural Data Interest Group such as the Wheat Data Interoperability Group Guidelines (https://ist.blogs.inrae.fr/wdi/) and Agrisemantics Working Group 39 hints guide (Brandon Whitehead and Aubin, 2019). Devare et al. in their chapter highlight the new responsibility of data curators now extends to wider data governance, including active data stewardship to adopt these new standards and guidelines to support wide access and re-use.

Adopting the FAIR principles is a challenge which cannot be ignored, and, in the case of LTEs, raises multiple issues. There are issues of choice and agency ranging from technical decisions around standards adoption to determining how far the responsibility to steward LTE data runs. LTEs can have complex histories and understanding this narrative alongside various sources of variability that affect the interpretation of data is essential. Reducing barriers to data access, interoperability and, ultimately seamless data linkage, while retaining oversight of how data is used for funder reporting is a significant challenge. Further challenges exist for how to understand the potential value of currently inaccessible legacy data then mobilise them and how to ensure externally generated data are retained as part of the experiment narrative alongside routine LTE data.

3.1 Navigating Experiment Narratives

There is a long-held fear, founded in experience (Stroud, 2018), of data misinterpretation and misrepresentation which stems from a view that LTEs are inherently complex and therefore require expert interpretation.

However, while it is true the experiments are complex, this is something scientists can deal with, but only if they have the necessary information to support interpretation, so rather than using this as a reason to raise barriers, the data should be sufficiently well described to withstand and challenge deliberate cherry picking of data to present false narratives. Maintaining generational records for an LTE in terms of the experimental and agronomic decision making, methodological changes and external events that impact interpretation and uses of LTE data forms the LTE narrative. This narrative is not only crucial for using the data, but also provides consistency by giving Curators and researchers a reference point, and ensures knowledge is not lost as LTE staff move on.

For example, Macholdt et al. (2020) in an analysis of yield stability on Broadbalk, used this experiment narrative to explain why certain plot years data are excluded from the analysis. A further example is changes in phosphorous applications on the Broadbalk Wheat and Hoosfield Barley Experiments where phosphorous has been withheld on selected plots since 2001 and 2003 respectively, but for different reasons. On Broadbalk, phosphorous is non-limiting and being withheld as a management decision to allow plots to reduce to more agronomically realistic levels when phosphorous applications will resume. By contrast on Hoosfield Barley, phosphorous is being withheld as a treatment decision to study residual effects on yield. In the Broadbalk case withholding phosphorous should not impact the continuity of data over time, but in the Hoosfield LTE, a new boundary condition is being introduced that does affect this continuity and how data can be analysed.

Since 1906 experiment narratives have been published in a series of 12 Guides, updated on an irregular basis, the most recently published update in 2019 (2019). e-RA 2013 consisted of publicly accessible HTML files with free text descriptions and supplementary files provided for each LTE.

e-RA 2021 has improved the earlier version by adopting the Global Long-term Agricultural Experiments Network (GLTEN) schema to replace free text with structured and consistent experiment descriptions. Launched in 2018, the GLTEN is a community of LTE researchers which aims to improve the visibility and use of these experiments. An early output was the GLTEN schema (https://github.com/GLTEN) which describes LTEs across six themes (Box 1,) using a semantically rich and structured format.

Box 1: Metadata Themes Captured by the GLTEN Schema
1. Experiment objectives
2. Experiment design: experimental factors, factor levels and factor level combinations; plot layouts; replication; cropping system and crop rotations.
3. Administration: ownership; management; contacts; site access; sample access; data access; funding.
4. Environmental characterisation: geo-location; elevation, slope and aspect; climate; baseline and manipulated soil properties; landscape.
5. Routine data collection.
6. Research outputs: datasets; publications; supporting documentation.

An intentional property of the GLTEN schema is the Experiment Design Period which supports capturing narrative knowledge in a more structured way. Design Periods are temporally bounded and capture significant changes or transition points for an experiment, including changes to objectives, experiment factors, design, methods, management or cropping. Within a design period all properties including experiment factors and cropping can be temporally bounded to denote minor changes. In the earlier Hoosfield Barley example, the decision to withhold phosphorous to study residual effects would mark the start of a new design period.

Despite the detail provided in the GLTEN schema, it isn't comprehensive, so to plug these gaps the schema provides a structure for referencing outputs.

3.2 Sources of Variability

LTEs operate over extended periods of time and so are subject to changes. For example Glendining and Poulton discuss the problems associated with changes to sampling protocols and analysis methods for interpreting soil organic matter (Glendining & Poulton, 1996). For the Park Grass Hay experiment, in 1960 a change to the harvest method was introduced which resulted in increased dry matter yields caused by fewer yield losses during harvest. Consequently, reported yields before and after 1960 are not directly comparable. To address this a conversion factor for post 1960 yield data has been determined to allow comparisons.

The role of the Curators is to understand and manage these changes to maintain consistency and comparability of routine data over time. The experiment narrative aims to capture and explain these changes as an aid to using the data, however this typically only extends to methodological change and extreme events.

Environmental variability over time is more difficult to describe and may only be understood through analysis of the data and raises the importance of linking LTE data to complimentary covariate observations. For example, using the Rothamsted temperature record the mean air temperature is known to be 1.1 °C higher than the 1878–1987 mean with the 10 warmest years on record occurring in the last 17 years and increases greatest in the autumn and winter months, and in night time temperatures (Perryman et al., 2020). Recent work using the Broadbalk and Hoosfield LTEs has demonstrated the importance of including weather temporal variation for crop yields (Addy et al., 2020).

3.3 Adopting FAIR

The FAIR principles provide a benchmark for assessing LTE data stewardship. e-RA 2013 data provision, when measured against FAIR only partially satisfied some of the principles, and important areas could be identified where the principles are not met (Table 1).

e-RA 2021, provides access to a new class of curated dataset, referred to as LTE Standard datasets which are developed following FAIR data principles and feature data standards use, data packaging, DOI assignment and a simplified dataset registration process.

3.3.1 Standard Long-Term Experiment Datasets

The LTE Standard datasets are intended to provide comprehensive and usable datasets as an alternative to the summary Open Access datasets and e-RA Data. The Open Access datasets, while providing a useful overview, present data at a coarse scale with limited utility for research. e-RA Data by contrast provides data at the resolution of plot years, but while this resolution is clearly more useful for research, the architecture of e-RA Data means datasets are provided without an identifier or context making it uncitable and difficult to interpret. This is of course a problem for any dataset, but in the case of long-term experiments the problem is exacerbated by the often complex experiment narrative.

The LTE Standard datasets aim to address the limitations of coarse open access data and non-FAIR compliant e-RA Data downloads by re-packaging the data in the e-RA's SQL Server database (which Open Access datasets summarise and e-RA Data queries) both in line with FAIR data principles and by excluding certain data.

The aim of LTE Standard datasets is to provide data with metadata and supporting information to allow researchers to independently reuse the data with confidence.

Table 1 Comparison of the e-RA 2013 data provision methods against the FAIR principles

FAIR principle	DET	Open access
Findable	Downloaded datasets are not provided with a DOI (F1), therefore cannot meet F3 and are not provided with metadata (F2). Experiment descriptions are available from the website but are not explicitly related to downloaded datasets. Data provided by the e-RA Data Curators would be provided with appropriate supporting documentation. DET is not registered in a searchable resource (F4)	Datasets are provided with a DOI (F1) and include metadata (F2), the identifier describing the resource (F3) and are registered in a searchable and indexable resource (F4), namely the Rothamsted Data Repository and DataCite Search
Accessible	DET uses a query interface to parameterise a dataset for download. The query used cannot be saved nor is a snapshot of the data downloaded retained with an identifier assigned, therefore it is not possible to retrieve a DET derived dataset by an identifier (A1) and since DET derived datasets are not linked to experiment metadata, there is no explicitly identified metadata to link to (A2)	Datasets have a DOI and are therefore retrievable by their identifier over the internet (A1). The datasets are supported by a landing page which is accessible even if the data is not (A2)
Interoperable	Datasets do not use any formal knowledge representation or controlled vocabularies and do not have qualified references to other resources, therefore do not meet any of the interoperability principles	Datasets are described using the DataCite Schema (I1), but do not use relevant vocabularies (I2). The datasets may have relationships to other resources, formally defined in the DataCite Schema (I3)
Reusable	Users are required to agree to a data access policy stating the conditions of use (R1.1) before access is granted	Datasets are provided with accurate and relevant attributes including a Creative Commons licence (R1.1), and provenance (R1.2), however, they do not use community relevant standards (R1.3)

One expectation for the LTE Standard datasets is to reduce the time spent by the e-RA Curators servicing data requests and free them to spend more time mobilising additional legacy datasets and supporting other researchers to manage LTE data. Nevertheless, supporting researchers to use LTE data will likely remain a core activity.

Data Exclusion

The LTE Standard datasets aim to provide comprehensive and well described subsets of LTE data which are internally consistent over time and treatment factors. This means some data may be excluded based on a set of four criteria (Box 2). Excluded data can still be requested, but with the caveat that it must be used with caution and may need additional support to use.

Box 2: Exclusion Criteria for Standard Datasets

C1. There is insufficient documentation to support interpretation and re-use of the data.

C2. A plot does not have continuity of treatments over time.

C3. A plot deviates from planned treatment regimes

C4. The treatments do not have relevance or comparability to other treatments.

For example, in the Broadbalk experiment, between 1987 and 1990, plots were split for a comparison of the modern wheat cultivar Brimstone and the older Squareheads Master, grown until 1967. In 2015, the spring wheat Mulika was grown as a wet autumn and winter prevented sowing of winter wheat. Following these criteria an LTE Standard wheat yield dataset would exclude the Squareheads Masters plot data and Mulika data based on criteria C4 and C3 respectively.

Data excluded from one dataset might be included in another, for example, the Squareheads Master data could be included in a separate dataset comparing it with Brimstone for the periods 1987–1990.

Fair Data Adoption

Previously, when the e-RA Curators provided data for a request it would conventionally be provided as an annotated Excel file with additional supporting documentation such as cropping and treatment plans, but without reference to existing data standards or best practices. Adoption of FAIR data principles requires a new formalisation in how the e-RA Data Curators manage and present LTE data. Key steps in the adoption of FAIR data include organising tabular data as non-proprietary CSV files containing only column headings and observations, and with notes and style formatting removed; publishing with a DOI and making full use of the DataCite metadata schema to identify related documents and research outputs; providing conditions of use, licencing and recommended citation; providing a plurality of relevant metadata; dataset enrichment using semantic annotation. Supporting and training Curators to adopt FAIR as a best practice are also required.

Dataset Formats and Packaging

Within agricultural sciences, field experiment disciplines such as plant breeding and phenotyping tend to be used by specialist of researchers using established data

Table 2 Standard dataset package contents

README. md	Readme file in Markdown. Contains information extracted from the datapackage. json and DOI metadata reformatted into Markdown
README. html	The Markdown file reformatted as browser viewable HTML
datapackage. json	Metadata following the frictionless tabular data package specification. It describes the contents of the data package and table definitions for each CSV file
CSV data files	Tabular CSV files containing the data. The observation data tables use the Tidy data format
Excel file	A single Excel file where each worksheet is matches one of the CSV data files. For each worksheet, the first row is reserved for field names, subsequent rows must contain only data relating to the defining field. Additional text annotations and the use of text and cell formatting to convey information are disallowed
Other documents	Other supporting documents

standards and community best practices. Long-term experiments by contrast have broader appeal across a range of disciplines (Perryman et al., 2018) spanning soil and environmental sciences, metagenomics, agronomy, ecology, plant pathology and even social and economic sciences.

A challenge for the Curators is to provide data in accessible formats for a diverse range of potential users. In some cases, the data format is proscribed by existing communities of practice, for example soil metagenomics data would naturally fit with deposition to an existing genomics repository using data standards for sequence data. However, other users may have differing expectations and experiences for the same types of data. For example, an LTE time series of plant trait data, a plant phenotyping scientist might reasonably expect to access and use data in the ISA-TAB format (Sansone et al., 2012) using the MIAPPE standard, while an agricultural systems modeller might expect data conforming to the ICASA standard (White et al., 2013). Maintaining a plurality of different representations of the same data would be a costly and place an unnecessary curation burden on the e-RA Data Curators.

To address this, we have adopted the Frictionless Data Package Specification (https://specs.frictionlessdata.io/data-package/) and tabular data formatting following Tidy Data principles (Wickham, 2014) to provide a structured dataset accompanied by supporting information (Table 2). Frictionless provides a simple container format for describing CSV data using a standard schema. It has good tool support including R (https://github.com/frictionlessdata/datapackage-r) and Python (https://github.com/frictionlessdata/frictionless-py) libraries and so provides a directly usable format in two languages commonly used by research scientists for data manipulation and analysis. The Tidy Data format is a readily understandable structure commonly used for analysis where columns are variables and rows are observations. In the case of an LTE, the observation is a plot year. The data are also provided in an Excel representation as this remains a popular file format for data exchange.

Semantic Annotation

The Frictionless specification includes an rdfType property which supports annotation of fields with an RDF Class. This property allows us to enrich the CSV data by adding a meaningful definition to each field using ontology concepts. Ontologies and controlled vocabularies used include Agronomy Ontology, Agrovoc, Plant Experimental Conditions Ontology (PECO), Trait Ontology, Environment Ontology and ChEBI. For example, in a dataset where nitrogen application is a treatment factor, the field defining the factor levels can be annotated with the PECO term nitrogen fertilizer exposure (http://purl.obolibrary.org/obo/PECO_0007102). If the factor levels are different forms of nitrogen, for example ammonium nitrate vs sodium nitrate, then these categorical values are further mapped to the ChEBI terms http://purl.obolibrary.org/obo/CHEBI_63038 and http://purl.obolibrary.org/obo/CHEBI_63005.

Providing this level of semantic annotation on the data improves the potential interoperability of the data and is a useful step for moving to a linked data format.

Publishing with DataCite DOIs

All datasets are published with DataCite DOIs and make maximum use of the DataCite schema. Publishing with a DOI improves the findability of the datasets and means they can be formally cited, and citations measured for impact reporting to funders. The DataCite Schema's RelatedIdentifier property is used extensively to link a dataset to related outputs including publications, other datasets and supporting documents such as experiment plans. This allows us to publish datasets in context with a plurality of relevant documentation.

A further advantage of DOIs and using the RelatedIdentifier property is it will allow us to generate a PID Graph (Aryani, 2019) by describing and uncovering relationships between datasets, experiments, supporting materials and publications (Fig. 1). However, the success of DOIs for measuring impact and in the PIDGraph depends on their adoption by researchers, dataset citation by authors and enforcement by journal editors.

Curry, in this volume, has discussed duplication issues facing gene banks which in part arise from data management practices of the time, and similar issues face LTE data from previous practices for sharing unidentified and unversioned data. We know there are older and duplicate versions of LTE datasets in circulation, so while adopting DOIs cannot remove these, going forward, their use provides a centralises the discovery of datasets and the relationships between datasets (Fig. 2).

Reducing Barriers to Access

Access to the new LTE Standard datasets requires user registration and this allows us to continue collecting data use metrics for reporting to funders. Including a

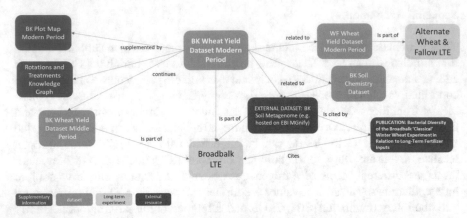

Fig. 2 PID graph showing relationships between experiments, datasets and supplementary material

registration wall is a recognised barrier to reuse (Sébastien Martin et al., 2013), and for e-RA, while there is anecdotal evidence that it has deterred requests and is viewed as archaic it is not possible to quantify this. To mitigate this potential risk the registration process has been simplified to provide a more streamlined experience compared to e-RA Data access. Users submit a one-time registration form which mandates entering a valid email address and optionally asks users to provide additional information on the intended use of the data. Users are sent a link to confirm their email and can then download the dataset.

To avoid deterring users with a lengthy form, most questions requesting information are optional. Completion of optional questions is encouraged by reminding potential users that continued funding and access to LTE data relies on our being able to demonstrate impact to funders and completing optional questions supports this.

The registration process will be monitored alongside other tracking methods such as Google Analytics and DOI metrics. If these are demonstrated to provide suitable reporting data, the need for a registration wall may be removed, however it is worth noting there are benefits from registration. By logging user emails and the datasets being accessed, subject to consent we can notify when new dataset versions and corrections are published.

3.4   Measuring Impact

Publishing datasets with a DOI should offer an attractive route to measuring the impact of LTE dataset reuse, however, there are currently cultural and technological limitations to this.

Dataset citation is not yet normalised across academic publishing; not all authors are in the habitat of citing datasets in reference sections and editors do not always enforce journal guidelines for dataset publishing. For example, a paper (Shtiliyanova et al., 2017) published in an Elsevier journal using Rothamsted Meteorological Station data did not cite the dataset despite it being published with a DOI and editorial guidelines encouraging authors to cite datasets in the reference list (https://www.elsevier.com/journals/computers-and-electronics-in-agriculture/0168-1699/guide-for-authors).

Nor is the infrastructure to support PID Graphs mature enough to provide reliable metrics. For example, the Broadbalk mean long-term winter wheat yields dataset (2017) has been formally cited in publications at least three times, but these citations are not reflected in results from querying the DataCite Commons (https://commons.datacite.org/) and DataCite GraphQL API, giving 1 and 0 results respectively.

For reporting, a further limitation of dataset citation is it only reports examples of data reuse in the public domain, it does not reveal unpublished works. Impact from unpublished work can only be reported to funders if users of the data volunteer this information.

3.5 Preventing New Data Loss/Supporting Best Practices for Externally Generated Data

As described previously, LTEs can be used as living laboratories by external researchers. These activities can result in new externally generated datasets, but there is no current mechanism to govern how these data are managed, it relies on the researchers generating the data to follow best practices. This is a serious data stewardship issue for the long-term experiments with consequence if datasets are not well stewarded:

- The National Capability is unaware of the full extent of data pertaining to an experiment
- The e-RA Data Curators are unaware of additional datasets relevant to a data request.
- Opportunities to create new time series from repeated, but irregular samplings or observations by different groups are missed
- Opportunities for collaboration or coordination of data collection activities may be missed.
- Opportunities to increase impact through re-use for both the National Capability and the external researchers generating the data are missed
- Costly data collection activities are repeated because the fate of data generated by previous researchers is unknown.

The Rothamsted Long-term experiments are maintained as a publicly funded resource for the benefit of all and this principle should be extended to externally generated datasets, regardless of their funding and as a condition of LTE access. In future, requests for access to use the Rothamsted LTEs and sample archive should be assessed not only on their scientific merit, but also on their sampling and analytical methods and data management plan. The LTE National Capability must have agency to prescribe actions to ensure externally generated data well managed, accessible and maintains continuity with other related LTE data. This means reviewing and agreeing DMPs to ensure there is an appropriate curation and publication pipeline, with a minimum expectation that metadata is published; recommending methodologies where these will facilitate comparability of new data with previously generated data, provided this does not conflict with the scientific question being addressed.

Enforcing these behaviours is clearly difficult, but a role of the e-RA Curators should be to liaise with the researcher to ensure the data curation terms agreed to are being followed. In cases where a researcher fails to follow the agreed DMP, the National Capability should reserve the right to deny future access to LTE resources.

In exchange for these stricter terms, the National Capability must be resourced to provide data management support and data hosting for external researchers or advise on appropriate alternative data hosting.

Adopting this as a best practice has the potential to create a virtuous circle for data linkage in which FAIRer datasets can be more easily identified and link, although it carries the risk of alienating researchers unwilling to comply with new access terms and burdening research with legal agreements.

3.6 Addressing Legacy Data

On the e-RA 2013 website the well-known Classical Experiments were extensively described with links to supporting publications and documents. For example the Broadbalk Experiment https://web.archive.org/web/20170210114304/http://www. era.rothamsted.ac.uk/Broadbalk, however, other experiments were more briefly described with only limited information, for example the Woburn Ley Arable Experiment https://web.archive.org/web/20170210235005/http://www.era. rothamsted.ac.uk/Other#SEC9.

A primary challenge for these less well-known experiments is finding information about them. Previously we stated in the 5 years following e-RA 2013's launch there were over 400 requests for data, but these headline figures hide significant variation between experiments. Over 85% of requests were for Broadbalk, Park Grass and Hoosfield, with Broadbalk accounting for nearly half of all requests. The remaining requests were spread over 27 other experiments listed on e-RA. This does not mean data from these experiments has less scientific value, rather it highlights their lower profile and lack of documentation and accessible data. For e-RA, experiment fame appears to be the determinant of the extensiveness of available data and documentation and drives a positive loop for maintaining the likelihood of re-use. Therefore, it is reasonable to assert less well-known experiments are caught in an opposite loop where a lack of data and documentation disincentivises requests and a lack of requests lowers the perceived research value of these experiments and priority for mobilising them and so they remain under-used. As can be seen in the case study (Box 3), prior lack of use is not an indicator of the research value of a legacy dataset.

To raise the profile of these experiments in e-RA 2021 we are describing them using the GLTEN Schema and publishing with the same visibility as the more famous experiments. This reduces the presentation bias between different LTEs and allows researchers to find and make a more informed decision on the potential usefulness of an experiment. Key metadata to characterise LTEs includes cropping system, climate and soil classifications, treatment factors, design and, available data. The site now indicates when data may be available but requires curation and at this point a request for data can be made. If the data are judged to be both useable (i.e. sufficient documentation exists to describe the data) and useful, the e-RA data curators will prepare a dataset for publication. Given the volume of legacy data available, this reactive strategy makes efficient use of staff time to support users and provide new, scientifically valuable datasets.

Box 3: The Long-Term Liming Experiment Case Study

In July 2016 the e-RA Data Curators received a request from the James Hutton Institute for long-term data on liming. After internal discussions the Rothamsted and Woburn Long-term Liming Experiments were suggested as potential datasets.

These experiments ran from 1962 to 1996 and e-RA provided limited information which was both difficult to find and insufficient for an external researcher to make an informed decision their usefulness. The potential value of the long-term liming experiments was only identified by a long serving staff member with deep knowledge of the long-term experiments.

The data was in poor state with the first 12 years data on paper and the remainder as early Genstat formats. Significant work was required to transcribe from and update data formats. Some data was organised by plot numbers while other data was organised by treatment, fortunately a paper key for the mapping between plots and treatments was found, otherwise a portion of the data would have been unusable.

Datasets for soil chemistry and yields were compiled and two papers investigating the effects of liming on yields and economic returns (Holland & Behrendt, 2020; Holland et al., 2019) published. The second paper on the economics of liming in arable crop rotations has since been reported on by Farmers Weekly (Clarke, 2021), demonstrating the applied agricultural interest in an experiment ended 25 years previously.

4   Conclusion

The new version of e-RA has made significant progress to make the Rothamsted Long-term Experiments a FAIRer data resource. This has required the e-RA team to look critically at how data has been managed and provided, understand where there are weaknesses and how they can be addressed. As with any change process this can be a difficult when existing conventions that work and, anxieties about giving researchers greater freedom to use the data are challenged.

Arguments against more open sharing of data are the fears of misrepresentation or misinterpretation. Earlier we gave an example of a paper not using a data citation (Shtiliyanova et al., 2017), in fact this paper also made a false assumption about the data available. Referencing the data would have highlighted this mistake. Data citation provides a degree of confidence in that data used to assert conclusions is available for verification. The challenge for the LTE Data Curators remains to ensure LTE data are presented with sufficient metadata to support independent re-use by researchers and internally consistent to avoid accidental misinterpretations.

The experience of e-RA demonstrates long-term data stewardship needs specialist data skills and continuing investment to maintain and develop both skills and the infrastructures to support the data. Importantly this support should be extended as a service to external researchers generating data and providing it should be viewed as a matter of self-interest as additional well curated accessible and interoperable data only enhances the overall value of the experiments. Neglecting data stewardship does a disservice to the work of every technician, field worker and lab assistant who created it and future generations who can benefit from it, and as can be seen from the example of the Long-term Liming Experiments, restoring neglected data into an accessible and useable product can require significant cost and effort.

To date, most of the effort on re-thinking e-RA has focused on providing a better experience for the users. The present state of the art for e-RA are adoption of the GLTEN schema to describe long-term experiments and the move to LTE Standard Datasets which provide published datasets following FAIR principles. This is a significant advancement towards linked data; DOI metadata can identify relationships between datasets, and semantic annotation supports data linkage between equivalently described datasets. With Frictionless data there are opportunities to develop schema profiles that can better meet the needs of different user communities. However, the fundamental unit for publishing and sharing data is the dataset and LTE Standard datasets are really a convenience for grouping related observations and variables as a coherent set. The next logical step for e-RA is to provide better linkage at the observational scale, however, this will bring a new set of challenges, notably how to cite and measure impact for dynamically accessed and linked data and how to capture the experiment narrative as a set of rules to prevent invalid combinations of observations.

Compared to many research institutes managing long-term agricultural experiments, Rothamsted is now relatively advanced in adopting the FAIR Data Principles and the data stewardship approaches being used can be a template for best practices within the long-term experiments community. From the experiences of colleagues working across the GLTEN to analyse data across multiple LTEs, there are remain significant blocks to integrating and re-using data from LTEs managed by different institutes. From the initial identification of appropriate experiments and available data, understanding the methods and experiment design, and wrangling data into a usable and interoperable form, the 80/20 rule, that 80% of time is spent finding, cleansing and organising data, still applies to LTE data.

Adopting the approaches outlined here by the wider LTE community, namely adopting the GLTEN schema for robust LTE descriptions and characterisation and implementing FAIR Data Principles using the Frictionless approach may provide more opportunities for impact and in turn demonstrate evidence for continued investment to maintain these unique resources. But for now, at least having access to data stewards with intimate knowledge of an LTE remains essential for successful re-use.

References

1862. *Memoranda of the plan and results of the Rothamsted field experiments.*
1928. *Rothamsted Experimental Station report for 1927–28 with the supplement to the guide to the experimental plots.* Rothamsted Experimental Station.
2017. Broadbalk mean long-term winter wheat grain yields. In R. Research (Ed.) (1 ed.). Electronic Rothamsted Archive.
2019. *Guide to the classical and other long-term experiments, datasets and sample archive.* Rothamsted Research.
Addy, J. W. G., Ellis, R. H., Macdonald, A. J., Semenov, M. A., & Mead, A. (2020). Investigating the effects of inter-annual weather variation (1968–2016) on the functional response of cereal grain yield to applied nitrogen, using data from the Rothamsted Long-Term Experiments. *Agricultural and Forest Meteorology, 284*, 107898.
Aryani, M. F. A. A. (2019). *Introducing the PID graph* [Online]. Available: https://blog.datacite.org/introducing-the-pid-graph/. Accessed 07 Feb 2021.
Brandon Whitehead, C. C., & Aubin, S. (2019). *39 hints to facilitate the use of semantics for data on agriculture and nutrition.* https://rd-alliance.org/
Clarke, A. (2021). *Why regular liming can raise profits by up to £436/ha/year* [Online]. Available: https://www.fwi.co.uk/arable/land-preparation/soils/why-regular-liming-can-raise-profits-by-up-to-436-ha-year. Accessed 20/02/2021.
Dyke, G. V. (1974). Chapter Seven – Long-term experiments. In G. V. Dyke (Ed.), *Comparative experiments with field crops.* Butterworth-Heinemann.
Edwards, C. A., & Lofty, J. R. (1982). Nitrogenous fertilizers and earthworm populations in agricultural soils. *Soil Biology and Biochemistry, 14*, 515–521.
Fan, M.-S., Zhao, F.-J., Poulton, P. R., & Mcgrath, S. P. (2008). Historical changes in the concentrations of selenium in soil and wheat grain from the Broadbalk experiment over the last 160 years. *Science of the Total Environment, 389*, 532–538.
Glendining, M. J., & Poulton, P. R. (1996). Interpretation difficulties with long-term experiments. In D. S. Powlson, P. Smith, & J. U. Smith (Eds.), *Evaluation of soil organic matter models, 1996* (pp. 99–109). Springer.
Group, D. M. W. (2019). *DataCite metadata schema documentation for the publication and citation of research data version 4.3.*
Hawkins, N. J., Cools, H. J., Sierotzki, H., Shaw, M. W., Knogge, W., Kelly, S. L., Kelly, D. E., & Fraaije, B. A. (2014). Paralog re-emergence: A novel, historically contingent mechanism in the evolution of antimicrobial resistance. *Molecular Biology and Evolution, 31*, 1793–1802.
Holland, J. E., & Behrendt, K. (2020). The economics of liming in arable crop rotations: Analysis of the 35-year Rothamsted and Woburn liming experiments. *Soil Use and Management*, n/a.

Holland, J. E., White, P. J., Glendining, M. J., Goulding, K. W. T., & Mcgrath, S. P. (2019). Yield responses of arable crops to liming – An evaluation of relationships between yields and soil pH from a long-term liming experiment. *European Journal of Agronomy, 105*, 176–188.

Johnston, A. E., & Poulton, P. R. (2018). The importance of long-term experiments in agriculture: Their management to ensure continued crop production and soil fertility; the Rothamsted experience. *European Journal of Soil Science, 69*, 113–125.

Lawes, J. B., & Gilbert, J. H. (1864). Report of experiments on the growth of wheat for twenty years in succession on the same land. *Journal of the Royal Agricultural Society of England, 25, 10*.

Macholdt, J., Piepho, H. P., Honermeier, B., Perryman, S., Macdonald, A., & Poulton, P. (2020). The effects of cropping sequence, fertilization and straw management on the yield stability of winter wheat (1986–2017) in the Broadbalk Wheat Experiment, Rothamsted, UK. *The Journal of Agricultural Science, 158*, 65–79.

Mariem, S. B., Gámez, A. L., Larraya, L., Fuertes-Mendizabal, T., Cañameras, N., Araus, J. L., Mcgrath, S. P., Hawkesford, M. J., Murua, C. G., Gaudeul, M., Medina, L., Paton, A., Cattivelli, L., Fangmeier, A., Bunce, J., Tausz-Posch, S., Macdonald, A. J., & Aranjuelo, I. (2020). Assessing the evolution of wheat grain traits during the last 166 years using archived samples. *Scientific Reports, 10*, 21828.

Morris, M. G. (1992). Responses of Auchenorhyncha (homoptera) to fertiliser and liming treatments at Park Grass, Rothamsted. *Agriculture, Ecosystems & Environment, 41*, 263–283.

Neal, A. L., Bacq-Labreuil, A., Zhang, X., Clark, I. M., Coleman, K., Mooney, S. J., Ritz, K., & Crawford, J. W. (2020). Soil as an extended composite phenotype of the microbial metagenome. *Scientific Reports, 10*, 10649.

Perryman, S. A. M., Castells-Brooke, N. I. D., Glendining, M. J., Goulding, K. W. T., Hawkesford, M. J., Macdonald, A. J., Ostler, R. J., Poulton, P. R., Rawlings, C. J., Scott, T., & Verrier, P. J. (2018). The electronic Rothamsted Archive (e-RA), an online resource for data from the Rothamsted long-term experiments. *Scientific Data, 5*, 180072.

Perryman, S. A. M., Scott, T., & Hall, C. (2020). *Annual mean air temperature at Rothamsted 1878–2019* (2nd ed.). Electronic Rothamsted Archive: Rothamsted Research.

Sansone, S.-A., Rocca-Serra, P., Field, D., Maguire, E., Taylor, C., Hofmann, O., Fang, H., Neumann, S., Tong, W., Amaral-Zettler, L., Begley, K., Booth, T., Bougueleret, L., Burns, G., Chapman, B., Clark, T., Coleman, L.-A., Copeland, J., Das, S., et al. (2012). Toward interoperable bioscience data. *Nature Genetics, 44*, 121–126.

Sébastien Martin, M. F., Turki, S., & Ihadjadene, M. (2013). Risk analysis to overcome barriers to open data. *European Journal of e-Government, 11*, 348–359.

Shtiliyanova, A., Bellocchi, G., Borras, D., Eza, U., Martin, R., & Carrère, P. (2017). Kriging-based approach to predict missing air temperature data. *Computers and Electronics in Agriculture, 142*, 440–449.

Stroud, J. L. (2018). Co-produced data: Open access tests trust. *Nature, 562*, 344.

White, J. W., Hunt, L. A., Boote, K. J., Jones, J. W., Koo, J., Kim, S., Porter, C. H., Wilkens, P. W., & Hoogenboom, G. (2013). Integrated description of agricultural field experiments and production: The ICASA Version 2.0 data standards. *Computers and Electronics in Agriculture, 96*, 1–12.

Wickham, H. (2014). Tidy data. *Journal of Statistical Software, 1*(10), 2014.

Wilkinson, M. D., Dumontier, M., Aalbersberg, I. J., Appleton, G., Axton, M., Baak, A., Blomberg, N., Boiten, J.-W., da Silva Santos, L. B., Bourne, P. E., Bouwman, J., Brookes, A. J., Clark, T., Crosas, M., Dillo, I., Dumon, O., Edmunds, S., Evelo, C. T., Finkers, R., et al. (2016). The FAIR Guiding Principles for scientific data management and stewardship. *Scientific Data, 3*, 160018.

Plant Science Data Integration, from Building Community Standards to Defining a Consistent Data Lifecycle

Cyril Pommier, Frederik Coppens, Hanna Ćwiek-Kupczyńska,
Daniel Faria, Sebastian Beier, Célia Miguel, Célia Michotey, Flora D'Anna,
Stuart Owen, and Kristina Gruden

Abstract FAIR (Findable, Accessible, Interoperable, Reusable) data principles for plant research build upon experience from other life science domains such as genomics. But plant specificities, e.g. plant-environment interactions or phenotypes, require tailored solutions. Major global players have joined forces to answer that challenge with the Minimal Information About a Plant Phenotyping Experiment

C. Pommier (✉) · C. Michotey
Université Paris-Saclay, INRAE, BioinfOmics, Plant Bioinformatics Facility, Versailles, France

Université Paris-Saclay, INRAE, URGI, Versailles, France
e-mail: cyril.pommier@inrae.fr

F. Coppens · F. D'Anna
Department of Plant Biotechnology and Bioinformatics, Ghent University, Ghent, Belgium

VIB Center for Plant Systems Biology, Ghent, Belgium

H. Ćwiek-Kupczyńska
Institute of Plant Genetics, Polish Academy of Sciences, Poznań, Poland

D. Faria
LASIGE, Faculdade de Ciências, Universidade de Lisboa, Lisbon, Portugal

BioData.pt, Instituto Gulbenkian de Ciência, Oeiras, Portugal

S. Beier
Leibniz Institute of Plant Genetics and Crop Plant Research (IPK) Gatersleben, Stadt Seeland, Germany

C. Miguel
Instituto de Biologia e Tecnologia Experimental, Oeiras, Portugal

Biosystems and Integrative Sciences Institute, Faculdade de Ciências, Universidade de Lisboa, Lisbon, Portugal

S. Owen
Department of Computer Science, University of Manchester, Manchester, UK

K. Gruden
Department of Biotechnology and Systems Biology, National Institute of Biology, Ljubljana, Slovenia

H. F. Williamson, S. Leonelli (eds.), *Towards Responsible Plant Data Linkage: Data Challenges for Agricultural Research and Development*,
https://doi.org/10.1007/978-3-031-13276-6_8

(MIAPPE, www.miappe.org) that handles general metadata organization and its companion web service API, the Breeding API (www.brapi.org). Both rely on two established data standards, the MultiCrop Passport Descriptors (MCPD) for identification of plant genetic resources and the Crop Ontology (www.cropontology.org) for trait documentation. Researcher communities' coordination and collaborative approaches have enabled the success and adoption of MIAPPE and led to a general data lifecycle description by ELIXIR Plant Sciences Community to identify gaps and needed developments. A priority has been placed on addressing the "first mile" of data publishing, i.e. the gathering and documentation of data by the researcher, which enables relevant data findability and reusability. Here we describe the existing ecosystem of tools and standards for plant scientists as well as their history, including their convergence through the use of MIAPPE for describing genotyping datasets.

1   Introduction

Plant research communities have invested a lot of effort not only in increasing biological knowledge of the plant realm but also in enabling greater sustainability of plant production. Indeed, climate change and the increasing world population have led to agricultural considerations being identified in 12 of the 17 United Nation Sustainable Development Goals (https://sdgs.un.org/goals). As a consequence, more and more data is being produced and we are now facing a range of Big Data challenges, as described in the 4Vs of Big Data (Velocity, Variability, Volume, Veracity) (De Mauro et al., 2016). Volume and Velocity may be less of an issue for plant sciences in comparison to other scientific fields such as astronomy for instance, but Variability is especially challenging both because of the genomic complexity of plants (polyploidy, for example) and because of the heterogeneous nature of plant phenomics. The latter encompasses all the observations and measures that can be made on a precisely identified plant material in a characterized environment. This very general definition of phenomics (Watt et al., 2020) includes diverse types of properties and variables measured at different physical (Tardieu et al., 2017) and temporal scales, ranging from field observation of plant populations to molecular cell characterizations, and for some research communities includes metabolomics or gene expression. The acquisition of these data is conducted in various experimental facilities like greenhouses, fields, phenotyping networks, or natural sites. It can be done using many different devices from manual measurements to high throughput means. The resulting complex and heterogeneous datasets include all the environment and phenotypic variable values at each relevant scale (plant, micro plot, and so on) and very importantly the identification of the phenotyped germplasm, *i.e.*, the plant material being experimented upon. In addition, there are often relationships between levels (*i.e.* physical scales such as microplot or plant individual and organs) inside datasets and between different datasets. The resulting rich wealth of data is usually formatted in a very heterogeneous manner and is difficult to integrate manually or automatically.

It is therefore a necessity not only to produce scientific data – from genetic and genomic to phenomic and environmental and up to systems biology – but also to develop the means for managing it, integrating it and therefore analysing it at high throughput dimensions, not only for model species such as Arabidopsis but also for crops and trees. This management is therefore a direct application of the FAIR (Findable, Accessible, Interoperable, Reusable) data principles (Wilkinson et al., 2016), especially crucial in regard to such a complex data life cycle. In the present paper we will describe the plant research data lifecycle and data management challenges as well as the solutions developed by different communities over the past years. This is followed by a focus on the 'first mile challenge' and some considerations on findability of data across distributed data repositories.

2   The Plant Data Life Cycle

The plant sciences community of the European Infrastructure for life sciences ELIXIR (Harrow et al., 2021) has been structured in recent years by funded activities such as the ELIXIR EXCELERATE Horizon 2020 project as well as its collaboration with the EMPHASIS European Infrastructure for Plant Phenotyping. With a switch from big structuring projects to a funding model that relies on the coordination of many smaller projects, including national projects such as ELIXIR implementation studies or European Open Science Cloud (EOSC) demonstrations, it has been necessary to create a roadmap to organize and coordinate the necessary activities. This roadmap (Pommier et al., 2021) needed first a general description of its objectives, through the definition of a data life cycle underlying the needs of plant science.

The building of this roadmap has been possible thanks to many years of collaborations within and between different groups with tangible results, including the groups responsible for the Minimum Information About a Plant Phenotyping Experiment (MIAPPE) standard (Papoutsoglou et al., 2020); BrAPI, the plant breeding API (Selby et al., 2019); the International Wheat Information System (WheatIS); and the transPLANT genomic infrastructure, among others. This facilitated a community of ideas that drafted the life cycle and the roadmap. We ensured the openness of the community both by welcoming new members and institutions and by setting up formal collaborations between ELIXIR and other groups such as EMPHASIS in particular. The structure of this community, *i.e.* a European infrastructure that relies on a network of national nodes supported and encouraged at each national level, has further pushed forward the activities. These activities are both bottom-up, with concrete use cases and demonstration datasets used as the basis for discussions by the persons in charge of actually running data related activities, and top-down, with strategic decisions made by principal investigators and node representatives to increase collaboration between communities and infrastructures. The existence of

Fig. 1 Schema of the data lifecycle, that begins with data gathering, then preparation for analysis through processing and integration, finally publication and sharing for knowledge extraction. Note that the integration encompasses both statistical data integration and normalization as well as data linking and mapping

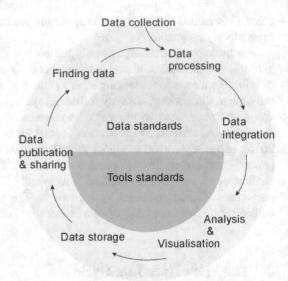

this bottom-up approach, mobilising concrete elements and utilising real datasets to demonstrate the validity of the data standards and the data life cycle elements, has been instrumental to trigger interest and collaboration within the ELIXIR plant community. Last but not least, the fact that the elements of the roadmap and our objectives have been included as deliverables of many projects ensures that they will be realised.

The activities described within the Roadmap are designed with the goal of enabling successful data handling across the complete data life cycle (Fig. 1). As a community, we will focus on tailoring data findability both at the level of describing the generated datasets in order to be discoverable as well as via developing tools for data retrieval. We will also define pipelines for efficient data pre-processing, integration, analysis and visualisation to enable successful biological interpretation of results. For the last part of the data life cycle we will work on data storage in accordance with different standards and describe it using appropriate vocabularies and ontologies. This will enable publication of plant data and scientific papers in accordance with FAIR principles.

3 Plant Data Management Challenges

Plant research communities handle different types of data, some of which are shared with other realms, like genomics, genetics and systems biology, while others are very specific, like phenotyping and plant-environment interactions. Existing data standards and management practices have offered practical solutions for genomics and genetics, but phenotyping needed a whole new framework. Several communities have built (separately or jointly) their solutions through three types of data standards: Semantic, Structural and Technical.

Semantic standards provide means for data description. They encompass controlled vocabularies, with term names and definitions, possibly organised as ontologies through the addition of semantic linking. For plant phenotyping, 10 years ago the Crop Ontology (Shrestha et al., 2012) formalized the Trait-Method-Scale model that has subsequently been embedded in MIAPPE. This model is aligned with the practices and approaches commonly applied by agronomists and phenotyping researchers, especially in terms of the terminology, organisation and range of descriptors for documenting observed and measured variables. Such semantic standards make researchers' life easier by providing a description framework and common vocabularies. They are therefore mostly driven by biologists. However, they don't answer the problem of data organisation.

Structure standards allow the organisation of datasets through schemas of metadata descriptors, *i.e.*, sets of fields, possibly grouped hierarchically, including mandatory and recommended information. They allow the description of data and defining of the interrelations between the different data files that gather all the measures and analyses done (something that is especially challenging in multilocal and/or multiannual experimental networks). The MIAPPE standard is a good example of such a standard. It takes advantage of the Investigation, Study, Assay (ISA) (Sansone et al., 2012) approach and encompasses elements taken from the Crop Ontology and the MultiCrop Passport Descriptor (MCPD) format, which is the reference for identifying and describing plant genetic resources and varieties within international genebanks (Alercia et al., 2015). In light of current research technologies and the increasing amounts of data being produced, those standards should allow one to organise data and metadata in a machine actionable way. At the same time, they must also be usable and still explicit enough to be effectively adopted by plant researchers. Therefore, they are built through close collaboration between computer scientists, biologists and agronomists who organise the data and provide their semantic description to ensure long term understandability and reusability.

Technical standards address interoperability challenges and data exchange between databases, tools and analysis environments. They include, for instance, web service APIs such as the Breeding API, a web service specification implementing MIAPPE. These standards are thus mostly computer scientist driven.

4   Plant Data Standards

The history of the Minimum Information About a Plant Phenotyping Experiment standard, MIAPPE, shows how a solution designed for a focused use case, plant phenotyping experiments, can be extended to other data types and in particular to genetic variation. The goal of MIAPPE (Papoutsoglou et al., 2020) is to support researchers in explaining plant phenotypes, *i.e.* the observable results of the growth of plants in specific environmental conditions. To disentangle genotype-environment interactions and identify the biological mechanisms leading to specific phenotypes, a good description of the biological material, environmental conditions and observed variables is needed. These constitute the cornerstones of plant phenotyping experiments, and thus also comprise the pivots of phenotyping experiment description in MIAPPE.

The description of biological material in a plant experiment encompasses characteristics of the studied genotypes, such as their origin (the source of the seeds/plant parts, their pedigree) or taxonomic classification. The environment description comprises geographical locations, environment type and growth conditions, and additional treatments applied. In addition, the design of the experiment needs to be included (e.g., spatial and hierarchical arrangements of observation units, temporal arrangement of actions and events). Such a description of a plant experiment, if sufficiently detailed, allows us to understand the genetics and environmental factors that interact and produce particular phenotypes.

Observations are carried out for individual observation units at desired time points, and in phenotyping assays they typically involve the measurements of macroscopic plant traits (anatomy, yield, physiology, etc.) and environmental variables (actual environmental conditions). However, these are not the only measurement types that can be observed. More and more frequently, notably in systems biology approaches, the same plant experiments are the source of samples for other assays, such as microscopic measurements and multi-omics studies. Development of new technologies and gradually decreasing costs make this type of analyses more affordable and allow plant researchers to repeat the assays in multiple environments and time points. These omics data need to be properly placed in the context of the whole plant experiment. Thus, MIAPPE constitutes a solid foundation not only to integrate different types of data but also to build bridges between communities.

This notion of integration of different data types is broad and encompasses multiple things. Indeed, for a biostatistician or a data scientist, integrating means normalizing the data, reducing it to make its analysis and understanding possible. But from a data management point of view, integrating datasets is about finding links and common keys among several datasets. Since plant science relies heavily on the integration between phenomic, environmental and omics data, it is necessary to link them using common pivot objects (Pommier et al., 2019), *i.e.* common keys. The work of the ELIXIR implementation study FONDUE sets up such common keys between databases and datasets. The concepts and descriptions from MIAPPE are used to describe the environment and the plant material in genotyping experiments published in EMBL-EBI data repositories, especially Biosamples, the European Nucleotide Archive (ENA) and the European Variation Archive (EVA). This allows both findability and interoperability with other data repositories. A BioSamples checklist, based on a reduced set of MIAPPE metadata, has been developed to describe the samples more precisely and uniformly. In addition, the FONDUE project is developing further recommendations for new metadata information in the header of genotyping files (Danecek et al., 2011), such as the BioSamples sample identifiers.

5 Plant Data Standards History, Use and Adoption

The community management and gathering around the data standards described above has been highly collaborative. The history of the use by ELIXIR of the Crop Ontology, initiated by Bioversity International of the CGIAR, is a good example of how we decided to join forces to avoid creating new data standards when possible. Giving up existing institution-specific standards in favor of international ones, *e.g.* Crop Ontology, allowed to avoid unnecessary competition. Furthermore, through common workshops, like the PhenoHarmonIS series of conferences, and projects, such as ELIXIR EXCELERATE, new communities have been invited to actively contribute to the Crop Ontology, either through dedicated sub-ontologies or at the level of the ontology's formal concepts.

The building of MIAPPE has been even more community intensive. It has been built through close collaboration among European research groups, initially between European Union FP7 projects transPLANT and the European Plant Phenotyping Network (EPPN), together with the CGIAR, later followed by European infrastructures ELIXIR, EMPHASIS. Eventually, MIAPPE became an open project, and multiple plant researchers were invited to contribute to its development and to adopt and introduce the standard in their communities. To collect as broad a range of feedback as possible, open collaboration and requests for comments were organised and advertised in plant-focused events (conferences, webinars, mailing lists). In parallel, MIAPPE has been kept up-to-date with other external activities (BrAPI and Crop Ontology, among others) and projects. Prioritizing the directions for development and drafting new versions were led by dedicated working groups in connection with the MIAPPE steering committee. Currently, MIAPPE governance is minimal and pragmatic, with the steering committee in charge of discussing and organizing decisions around the evolution of MIAPPE specifications, an issue tracker on github to follow evolution requests, and a website and mailing lists for announcements. An important part of the current life of MIAPPE is about outreach, promotion and further adoption of the standard. This is done through webinars, training or workshops and is handled by any member of the MIAPPE community. Coordination of decision making is done by the six members of the steering committee. Any group can propose an addition to MIAPPE, formalize it, and bring it to the committee, which will organize the adequate consultation. Therefore, the evolution of MIAPPE relies on the willingness of self-elected workgroups that will hammer down all the details of their propositions through meetings and workshops. It is noteworthy that some of the most important evolutions of MIAPPE were made thanks to several EU projects that funded a group of people to work on a given problem, and were required to deliver a working solution. The available time of members of the focus groups has been critical here. As a consequence, the decision-making process takes a lot of time but ensures both the quality of the evolutions and improvements, which are tested with real datasets, as well as the fairness of the decisions, to ensure any stakeholder constraints will be taken into account.

The building of BrAPI took a similar path. Indeed, in the BrAPI community, initiated by the CGIAR and Cornell University, several groups from the European infrastructures and the CGIAR brought forward their use cases with the set of necessary specifications and web service calls. Here again, like in MIAPPE, those theoretical models were both tested with real data and gathered in a consensus specification during dedicated hackathons. These one week events, occurring twice a year, were organised both with the dedication of the BrAPI community members, who took turns organising them, and thanks to the Gates Foundation, who initiated BrAPI by funding those events for 3 years. Here again, like with MIAPPE, a need for coordinated governance arose quickly. The proposition was made to have a dedicated full-time coordinator, located at Cornell University. This structure proved highly successful for two important reasons. First, the coordinator is dedicated full time to the technical aspects, hence ensuring the consistency of the versions and the release cycle. Second, he does very important community management by organising the discussions on the mailing lists and github, by offering support, by organising further hackathons and everything needed to keep BrAPI members involved. One of the most important things in his activity, like in MIAPPE, is to ensure equity by being neutral, hence ensuring that the needs of all stakeholders – *e.g.* the CGIAR, ELIXIR, EMPHASIS, and national institutes – are taken into account, and that BrAPI doesn't get driven mainly by the needs of its most active members. For that purpose, a review board has been created.

The dissemination of data standards toward their end users, biologists and computer scientists, can be achieved through training, publications, workshop standard registries and open science policies and projects. Training, oral communication and valorisation have already been discussed and are actively used within ELIXIR and EMPHASIS communities. The WheatIS of the Wheat Initiative of the G20 has built standard recommendations to guide wheat researchers (Yeumo et al., 2017). It is indeed important to offer researchers the possibility to share recommended standards lists, carefully tailored for their domain, within the plethora of available standards and good practices. There are two types of recommendations that can be used. The first type are general recommendations targeting mainly computer scientists but which can also be used by researchers and principal investigators. The main example is FAIRsharing (The FAIRsharing Community et al., 2019) which can also be used to build collections dedicated to some communities, such as the WheatIS Data Interoperability Guidelines (https://fairsharing.org/collection/WheatDataInteroperabilityGuidelines). Those can furthermore be hosted and promoted in dedicated community registries (http://wheatis.org/DataStandards.php) in order to foster good practices within one-stop community web portals.

There are also activities that promote the use of standards in the frame of general data management and stewardship practices, such as the ELIXIR Research Data Management (RDM) toolkit developed within ELIXIR CONVERGE (https://rdmkit.elixir-europe.org). It contains, among other things, guides for using specialised tool assemblies like the Plant Genomics Assembly (https://rdmkit.elixir-europe.org/plant_genomics_assembly.html) in order to promote the use of community standards, and supplements this with prominent data management tools and best practices. The RDM Toolkit is an interesting example of how data sharing practices can be brought to researchers, at a minimal and therefore affordable cost of invested time, and by putting its content directly in the hands of the researchers that will ultimately use it.

The example of the INRAE (National Research Institute for Agriculture, Food and Environment, France) forest tree community is interesting because it shows how the data lifecycle can be updated to cope with the constraints of data acquisition and data sharing through the use of the plant data standards described here. An automated data flow was set up to synchronize data shared by multiple information systems. Data produced by research teams are managed in local experimental information systems, each platform having its own. The system facilitates the daily management of those data (raw, analysed or inventory data). Publication and sharing is not done at this local level, however, but through a global information system using an automated workflow. This makes it possible to improve the datasets' visibility and interoperability in order to share this knowledge and enhance data quality, reuse and enrichment.

6 The First Mile Challenge

The first step of the data life cycle is to gather and organize the data needed to answer a given scientific question. This can be achieved by documenting the data during experimentation or by adding and organizing the necessary metadata to existing datasets. For phenomics, this documentation process relies on dedicated tools and laboratory information management systems (LIMS) such as PIPPA (Coppens et al., 2017), Breedbase (Fernandez-Pozo et al., 2015; Agbona et al., this volume) or PHIS (Neveu et al., 2018). For omics data, equivalent LIMS systems exist to run an experiment. But while those systems are commonly available with high throughput experimental platforms, there is a need for another tool both for classical experimentation management and for managing the data obtained from integration and reduction of the experimental datasets. The solution came from a joint activity between ELIXIR, EOSC and EMPHASIS to add MIAPPE to the FAIRDOM-SEEK (Wolstencroft et al., 2017) data management system.

From a community point of view, this collaboration went smoothly thanks to the quality of the existing software, the willingness of all partners and the existence of an accepted and published standard, MIAPPE. Indeed, there hasn't been any extended discussion on the selection of metadata and fields, something that commonly occurs in those types of development projects; MIAPPE was simply selected and implemented in FAIRDOM-SEEK. The fact that both MIAPPE and FAIRDOM-SEEK share the common backbone of the ISA tools helped a lot.

7 The Findability Challenge for Dispersed Community Data

Data discovery, i.e. the ability for researchers to find any dataset suitable for their scientific questions, is a very active domain nowadays, with many solutions. Indeed, to find data, one can either (i) query all relevant data repositories one after the other, (ii) use one or many data discovery web portals, or (iii) use general search engines such as google data search. The first solution isn't appropriate in our era of big data. The third approach might be too general, hence lacking specificity without the help of dedicated markups such as bioschemas.org. The second approach leads to building community global portals to cross data repository boundaries. The plant community has in particular built two of them: the WheatIS (http://wheatis.org/Search.php) and FAIDARE (https://urgi.versailles.inrae.fr/faidare/).

The WheatIS (Sen et al., 2020) is an interesting example from a community management point of view. Indeed, it showed that the success of this portal relies on several key points: (i) keeping data distributed in a global federation rather than gathering all of it in a single global data repository, (ii) sharing at least one critical need, (iii) having clear leadership that ensures mutual benefit and relies on engaged people from several institutions, and (iv) gathering experts and making it easy to join the data federation through technical simplicity.

FAIDARE goes one step beyond this; first, by extending the WheatIS species range to more crops and, second, by enabling MIAPPE and BrAPI data standards. This brings two main benefits: Including in the FAIDARE federation new BrAPI data repositories at no additional cost and providing refined search based on the MIAPPE/BrAPI metadata. The key to success of FAIDARE relies on its ability to extend the data federation it indexes by merging the BrAPI network with the WheatIS network, and later with the addition of Bioschemas.org sources.

8 Conclusion

Sharing data to ensure its useful reuse is complex and poses major technical, social and scientific challenges. The present paper has shown how some international communities, through their social interactions, managed to build technical solutions to enable FAIR data management throughout the data lifecycle. We have seen in particular that the most time-consuming challenge is the community management, not only to formalize the standards, but also to build adoption and train users in the long term. This can be done in a sustainable way if first adopters become trainers too, through 'train the trainers' initiatives, and ensure adoption by other research networks. The solutions presented in this paper are already helping a lot, but more work is needed to allow easy data management to be realised. This will occur through technical improvements and their sharing, adaptation and adoption, and a lot of activities are already ongoing in this regard. But some social and scientific aspects have not really been discussed yet, in particular the criterion to select data that needs to be shared for the future. Indeed, from raw data to computed, reduced data, there are huge volumes of data to be stored, much more than the expected storage capabilities. We know that to enable reproducibility we should aim at sharing raw data, but that is often where the highest volume lies. The research community is therefore waiting for debates and guidance to make the right choices on such future issues.

References

Alercia, A. et al. (2015). *FAO/Bioversity multi-crop passport descriptors V.2.1 [MCPD V.2.1]*. *DataCite*. https://doi.org/10.13140/rg.2.1.4280.2001

Coppens, F., et al. (2017). Unlocking the potential of plant phenotyping data through integration and data-driven approaches. *Current Opinion in Systems Biology, 4*, 58–63. https://doi.org/10.1016/j.coisb.2017.07.002

Danecek, P., et al. (2011). The variant call format and VCFtools. *Bioinformatics, 27*(15), 2156–2158. https://doi.org/10.1093/bioinformatics/btr330

De Mauro, A., et al. (2016). A formal definition of big data based on its essential features. *Library Review, 65*(3), 122–135. https://doi.org/10.1108/LR-06-2015-0061

Fernandez-Pozo, N., et al. (2015). The sol genomics network (SGN) – From genotype to phenotype to breeding. *Nucleic Acids Research, 43*(D1), D1036–D1041. https://doi.org/10.1093/nar/gku1195

Harrow, J., et al. (2021). ELIXIR-EXCELERATE: Establishing Europe's data infrastructure for the life science research of the future. *The EMBO Journal*. https://doi.org/10.15252/embj.2020107409

Neveu, P., et al. (2018). Dealing with multi-source and multi-scale information in plant phenomics: The ontology-driven phenotyping hybrid information system. *New Phytologist*. https://doi.org/10.1111/nph.15385

Papoutsoglou, E. A., et al. (2020). Enabling reusability of plant phenomic datasets with MIAPPE 1.1. *New Phytologist, 227*(1), 260–273. https://doi.org/10.1111/nph.16544

Pommier, C., et al. (2019). Applying FAIR principles to plant phenotypic data management in GnpIS. *Plant Phenomics, 2019*, 1–15. https://doi.org/10.34133/2019/1671403

Pommier, C., et al. (2021). *ELIXIR plant sciences 2020–2023 roadmap. F1000 Research Limited*. https://doi.org/10.7490/F1000RESEARCH.1118482.1

Sansone, S.-A., et al. (2012). Toward interoperable bioscience data. *Nature Genetics, 44*(2), 121–126. https://doi.org/10.1038/ng.1054

Selby, P., Abbeloos, R., Backlund. J. E., Basterrechea Salido, M., Bauchet, G., Benites-Alfaro, O. E., Birkett, C., Calaminos, V. C., Carceller, P., Cornut, G., Vasques Costa, B., Edwards, J.D., Finkers, R., Yanxin Gao, S., Ghaffar, M., Glaser, P., Guignon, V., Hok, P., Kilian, A., König, P., Lagare, J. E. B., Lange, M., Laporte, M. A., Larmande, P., LeBauer, D. S., Lyon, D. A., Marshall, D. S., Matthews, D., Milne, I., Mistry, N., Morales, N., Mueller, L.A., Neveu, P., Papoutsoglou, E., Pearce, B., Perez-Masias, I., Pommier, C., Ramírez-González, R. H., Rathore, A., Raquel, A. M., Raubach, S., Rife, T., Robbins, K., Rouard, M., Sarma, C., Scholz, U., Sempéré, G., Shaw, P. D., Simon, R., Soldevilla, N., Stephen, G., Sun, Q., Tovar, C., Uszynski, G., & Verouden, M. (2019). BrAPI consortium: BrAPI-an application programming interface for plant breeding applications. *Bioinformatics, 35*(20):4147–4155. https://doi.org/10.1093/bioinformatics/btz190. PMID: 30903186; PMCID: PMC6792114.

Sen, T. Z., et al. (2020). Building a successful international research community through data
    sharing: The case of the wheat information system (WheatIS). *F1000Research, 9*, 536. https://
    doi.org/10.12688/f1000research.23525.1

Shrestha, R., et al. (2012). Bridging the phenotypic and genetic data useful for integrated breeding
    through a data annotation using the crop ontology developed by the crop communities of
    practice. *Frontiers in Physiology, 3*. https://doi.org/10.3389/fphys.2012.00326

Tardieu, F., et al. (2017). Plant phenomics, from sensors to knowledge. *Current Biology, 27*(15),
    R770–R783. https://doi.org/10.1016/j.cub.2017.05.055

Watt, M., et al. (2020). Phenotyping: New windows into the plant for breeders. *Annual Review of
    Plant Biology, 71*(1), 689–712. https://doi.org/10.1146/annurev-arplant-042916-041124

Wilkinson, M. D., et al. (2016). The FAIR guiding principles for scientific data management and
    stewardship. *Scientific Data, 3*, 160018. https://doi.org/10.1038/sdata.2016.18

Wolstencroft, K., et al. (2017). FAIRDOMHub: A repository and collaboration environment for
    sharing systems biology research. *Nucleic Acids Research, 45*(D1), D404–D407. https://doi.org/
    10.1093/nar/gkw1032

Yeumo, E. D., et al. (2017). Developing data interoperability using standards: A wheat community
    use case. *F1000Research, 6*, 1843. https://doi.org/10.12688/f1000research.12234.1

The FAIRsharing Community, et al. (2019). FAIRsharing as a community approach to standards,
    repositories and policies. *Nature Biotechnology, 37*(4), 358–367. https://doi.org/10.1038/
    s41587-019-0080-8

Part III
Challenges from/for Institutions: Data Linkage, Governance and Regulation Across Borders

Preface

The third part of the book considers the institutions responsible for devising and implementing data governance strategies. Responsible practice here includes not only the design of rules and regulations that may support – rather than hinder – data work, but also regular monitoring of the extent to which these systems are being implemented, and most importantly, of their impact on plant research as well as agricultural and food systems. Responsibility means taking ownership of both the positive and negative social consequences of specific plant data practices, and taking action whenever a given governance method fails to support agricultural and social development. This in turn requires ongoing consideration of what constitutes desirable development, and for whom. The contributors to this part examine specific organizational, legal and policy structures that may help address this challenge. These range from the strategies for transnational coordination and international governance implemented by the CGIAR and the Food and Agriculture Organization, including efforts to manage germplasm collections and link them with digital data infrastructures, to novel mechanisms for trading data to facilitate agricultural innovation, such as blockchain and related transformations to existing intellectual property regimes. Central to such efforts is the recognition of the enormous inequities among agricultural research locations and stakeholders around the world, and the need to lend visibility and capacity to researchers, breeders, farmers and indigenous communities whose data practices are being unfairly exploited.

Data, Duplication, and Decentralisation: Gene Bank Management in the 1980s and 1990s

Helen Anne Curry

Abstract In the 1970s, the number of accessions held in national and international seed and gene banks increased steadily. This growth, initially a source of pride, was recognised as a liability by the 1980s. Too many accessions lacked the basic information necessary for researchers to access and use samples knowledgably. Many gene banks came under scrutiny for poor management practices and several found themselves accused of mishandling a 'global patrimony' entrusted to their care. In this paper, I explore one response to these concerns that attracted attention from many in the germplasm conservation community: creating linked, standardised databases of collections. Calls for more and better data about accessions often emphasised that these data would make collections easier to use and therefore more valued. Here I take a close look at the early history of data collation and standardisation as a means of 'rationalising' collections, a motivation that was not advertised as prominently. This historical example shows the infrastructures developed to facilitate data exchange in the context of seed and gene banking to have been tied up with both mundane imperatives to cut costs and lofty goals of building political bridges—in addition to the often-repeated ambition of making plant breeding more efficient.

1 Introduction

In 2010, the United Nations Food and Agriculture Organisation (FAO) released its *Second Report on the State of the World's Plant Genetic Resources for Food and Agriculture*, a status update on seed and gene bank collections and other activities aimed at conserving crop diversity worldwide. Based on accounts received from national, regional, and international institutions, the *Second Report* estimated that, collectively, the world's seed and gene banks maintained some 7.4 million

H. A. Curry (✉)
Georgia Institute of Technology, Atlanta, Georgia, USA
e-mail: hcurry3@gatech.edu

© The Author(s) 2023   163
H. F. Williamson, S. Leonelli (eds.), *Towards Responsible Plant Data Linkage: Data Challenges for Agricultural Research and Development*,
https://doi.org/10.1007/978-3-031-13276-6_9

accessions (that is, individual samples registered in collections), a 20% increase over the preceding 20 years (FAO, 2010). From the perspective of preserving examples of farmers' varieties and crop wild relatives—conceived since the 1970s as vanishing 'genetic resources', but recognised much earlier as potentially endangered by breeders' creations—the continued growth of collections would seem cause for celebration (Fenzi & Bonneuil, 2016; Bonneuil, 2019). Yet the compilers of the *Second Report* cautioned against the easy interpretation of growth as victory. As they noted, most of this expansion had *not* come from acquiring new materials in the field, despite the fact that many minor crops and wild relatives desperately needed such attention. The impressive increase in collections was instead 'the result of exchange and unplanned duplication' of existing accessions. Scientists' and administrators' concerns about the implications of undirected growth can be read into the conclusion to their opening summary statement: 'There is still a need for greater rationalization among collections globally' (FAO, 2010: xix).

This was not a new demand. As I describe in this paper, calls to 'rationalise' gene banks, especially though not exclusively through the elimination of duplicate materials, date to the late 1970s. That decade saw a steady increase in the numbers of accessions held in national and international collections, expansion that was driven by concern about rapid 'genetic erosion' in the wake of the Green Revolution and perceived widespread agricultural industrialisation. The quick expansion of collections, initially a source of pride, was by the end of the decade recognised as a liability. Too many accessions lacked the basic information necessary for researchers to make requests of gene bank managers, let alone put samples to work knowledgably in breeding programmes. In the early 1980s, many gene banks came under scrutiny for poor management practices, and several prominent banks found themselves accused of mishandling a 'global patrimony' entrusted to them by the international community. One response to these failings, real and perceived, attracted attention from many in the germplasm conservation community: creating linked, standardised databases of collections. Calls for more thorough and consistent data about accessions often emphasised, and still emphasise today, that these data will make collections easier to navigate and therefore more valued and more used (e.g. Weise et al., 2020).

In this chapter I take a close look at the early history of data collation and standardisation as a means of 'rationalising' gene bank collections, a motivation that was not advertised as prominently. For some researchers and collection managers, the identification of duplicates was thought to allow the channelling of limited time and money to only the most unique accessions, even creating the possibility of de-accessioning items known to be held elsewhere. My analysis calls attention to three elements of this history in particular. First, I note the diverging functions of evaluation data and other identifying information within seed and gene banks in the late 1970s and 1980s, when these were sought both to encourage greater use of collections (e.g. in breeding programmes) and also to better manage collections (e.g. to eliminate duplicates). Second, I examine the political motivations that lay behind some calls for rationalisation. Where rationalisation was to include the elimination of duplicates across gene banks, it promised to save precious time and money and also to forge trust and interdependence among politically divided scientists, institutions, and states. Third, I explore how the ability of rationalisation initiatives to meet either economic or political objectives was frustrated by technical hurdles in data management, limited personnel, financial constraints, and political obstacles.

Ultimately this historical example shows the infrastructures developed to facilitate data exchange in the context of seed and gene banking to have been tied up with mundane imperatives to cut costs and lofty goals of building political bridges—in addition to the often-repeated ambition of making plant breeding more efficient and effective. By following imperatives issued from above out into the 'field' where curators wrestled with the chaos of actual collections, I show that the technical magic bullet of database development demanded, rather than generated, economic and political resources.

The political and economic imperatives behind data sharing are often neglected in historians' and sociologists' assessments of data practices associated with gene banks, which tend to focus on actors' interest in deriving value from collections (e.g. Parry, 2004; Van Dooren, 2010; Fullilove, 2018; for an exception see Chacko, 2019). Over time, calls for more and better data (and better data curation, too) in and across gene banks have become entwined with even more ambitious data enterprises that seek to unify a vast array of information about crop germplasm (for examples see, in this volume, chapters by Harrison and Caccamo; Arnaud et al. and Devare, Arnaud and King). Unpicking the many competing factors—social, political, technical—that informed and impeded earlier efforts to build comprehensive data infrastructures may not only provide a richer historical picture, but also help today's data developers recognise and navigate the complexities of their own present and future work.

2 Seed Surfeits, Data Shortfalls and the Call for Rationalisation

Much of today's international infrastructure for the conservation of crop diversity— the breeders' and farmers' varieties and crop wild relatives collectively designated as 'plant genetic resources for food and agriculture'—was forged in the 1970s and 1980s. Over the preceding century, plant breeders and other agricultural experts had called with increasing urgency for cross-border coordination of efforts to conserve crop genetic diversity (Lehmann, 1981; Bonneuil, 2019). As a number of historians have described, it was the real and perceived effects of international agricultural aid programmes of the late 1950s and 1960s (e.g., the 'Green Revolution') that finally galvanised international initiatives. These centred on collecting and arranging for long-term storage of breeders' varieties, landraces, and crop wild relatives thought to be endangered (Pistorius, 1997; Fenzi & Bonneuil, 2016; Curry, 2017, 2022; Bonneuil, 2019). They also entailed further coordination efforts focused on data generation, systematisation, and exchange. Better data and more thorough data linkage were considered essential to making collections useful, manageable, accessible and cost effective (Curry & Leonelli, Forthcoming). However, as I describe here, it proved far easier to acquire samples than data about these, circumstances that drove and, paradoxically, frustrated calls for collection 'rationalisation'.

From 1974, an International Board for Plant Genetic Resources (IBPGR), organised under the Consultative Group on International Agricultural Research (CGIAR), attempted to coordinate the conservation efforts of national institutions and international agricultural research centres and to encourage further programmes through strategic sponsorship of collecting missions and conservation facilities (Curry, 2017). It also aspired to establish a network of 'base collections', which would link national and international gene banks with especially good infrastructure and management (Hanson et al., 1984; Thormann et al., 2019). This network would disperse the responsibility for administering an international gene bank—something that had been sought by many individuals and institutions in the preceding decades—across multiple sites. With administrators focused especially on technical capacities such as reliable temperature and humidity control, those sites ended up being the comparatively well-resourced genetic resources programmes of industrialised countries and the internationally financed agricultural research centres of the CGIAR (Peres, 2019).

By the mid-1980s, these international coordinating efforts had fostered hundreds of collecting missions and many new conservation programmes. A 1984 tally estimated that the IBPGR had been 'instrumental in fielding over 300 collecting missions' in 88 countries and in its first decade had placed over 100,000 samples in gene banks (Williams, 1984: 7). It had also come under intense scrutiny, in part because of its perceived effectiveness in securing seeds. In his influential 1979 book *Seeds of the Earth*, the Canadian activist Pat Mooney linked IBPGR sponsorship of collecting missions and its ferrying of seeds to well-resourced facilities in the United States and Europe (or to CGIAR institutions located in the Global South but managed largely from the North) to a long history of imperial exploitation. In Mooney's assessment, 'The emerging network of gene banks takes national genetic treasures from the Third World to be stored abroad. In effect, these national resources cross a technological frontier, robbing the world's original plant breeders—subsistence farmers—of their rightful heritage, and leaving Third World governments dependent upon the First World for their own germplasm' (Mooney, 1979: 102). Thanks in part to Mooney and a growing number of seed activists, the 1980s saw a powerful surge in critiques of, and resistance to, the international network of seed and gene bank facilities sought by IBPGR and its funders. These critiques eventually forced the reimagining of this global network (Aoki, 2008; Fenzi & Bonneuil, 2016).

Among other outcomes, the fight over control of seed fostered by activists like Mooney and pursued by the nonaligned states at FAO from the 1980s onward brought new scrutiny to seed banks (Fenzi, Forthcoming). Scientists and administrators associated with IBPGR needed to provide evidence that their work had been in the global interest and that its network of base collections was indeed keeping seeds safe and accessible to all potential users. Critics, meanwhile, needed proof of the opposite. Subsequent studies compiled many shortcomings of national and international conservation efforts: broken refrigeration systems, lost samples, restrictions on access (e.g., US Comptroller General, 1981; Goodman, 1984; Mooney, 1983). Even champions of the existing structures had to acknowledge that the putative success of gathering seed samples had created a significant influx of materials to conserve, and that this multiplied the labour needed in processing, monitoring, and evaluating samples (e.g., Frankel, 1984; Peeters & Williams, 1984). What's more, the burgeoning size of collections had not been accompanied by increasing demand. A 1984 study of seed and gene bank use conducted by IBPGR, and co-authored by its executive secretary, described 'a consensus of opinion that genebanks are not being used very extensively by breeders' (Peeters & Williams, 1984: 22).

The acknowledgment that seed and gene banks often struggled to stay abreast of maintenance and almost always failed to provide meaningful services to breeders prompted calls for new strategies in conservation. A demand for more and better information about gene bank accessions—that is, for good data—featured centrally in many of these calls. In a clear signal of change, Otto Frankel, an early and effective champion of urgent collecting missions and institution building in the 1960s and 1970s, now advised a slowed pace for these. Frankel thought the use of collections was hampered by the lack of information about individual accessions, especially their agronomic traits and how they might be expected to perform in different environments (Frankel, 1984). This was not idle speculation. In 1984, IBPGR estimated that 95% of samples in gene banks had no such evaluation data attached (Peeters & Williams, 1984: 24). This was despite the fact that IBPGR had, since its founding, emphasised the creation, standardisation and computerisation of such data (Curry & Leonelli, Forthcoming).

By the mid-1980s, the imbalance between the cascade of collections and the dribble of data to accompany these—and the mounting critiques of its work—led the IBPGR to articulate a change in policy, a 'period of consolidation' in which 'characterisation, documentation and the ready exchange of information' would predominate (Williams, 1984: 14). One of the chief obstacles to this vision was that few if any people had capacity to systematically generate data. Initially, the IBPGR had assumed that national and international agricultural institutes would create this essential information, for example by carrying out evaluation programmes to generate data for individual accessions on agronomic qualities and environmental adaptations. However, although 'it was thought [at IBPGR] that characterisation and preliminary evaluation would not be costly' this initial view was quickly revised. Delays in generating these data were intensified by funding and staff shortfalls at many national programmes. In addition, the assumption that breeders would contribute to gene bank work by submitting any data they produced about requested accessions had to be scrapped. As the executive secretary of IBPGR bluntly summarised, '[B]reeders have not been very forthcoming in offering their services in this respect' (Williams, 1984: 11). This was not necessarily a product of intransigence on the part of breeders. On the contrary, there simply were not many rewards to their spending time and energy returning information to gene banks about samples they studied. Even if they had done so, the data would likely have arrived in heterogenous forms, requiring further labour from curators, who along with breeders would also have been navigating the changing international standards for the crop descriptors scientists were exhorted to use (Curry & Leonelli, Forthcoming).

The labour- and resource-intensive nature of evaluation was complicated by another issue increasingly recognised as characteristic of the international conservation system: duplication. A 1984 study of seed and gene bank conservation based on a survey of some 760 scientists determined that '[a]t least 50% of the combined collections of most crop species are duplicate accessions' (Lyman, 1984: 5). In some ways, this was a definite advantage. Having extra copies meant that disruptions at one seed or gene bank need not cause undue alarm. It also presented a problem, particularly given the finite nature of resources. '[I]ndiscriminate duplication of entire collections at numerous genebanks is costly and unnecessary', the scientist preparing the report insisted, noting that '[r]edundant duplicates within the same bank are undesirable' (Lyman, 1984: 5). If one were to add the labour of evaluating accessions to that of maintaining them, the unintended costs of accession duplication would only intensify. A conundrum followed, however: Unnecessary duplication increased the costs of maintaining collections, including the costs of evaluating these, but evaluation was also needed to identify duplicates if collections were to be rationalised (Lyman, 1984: 17).

Otto Frankel thought it was absurd to expect that seed and gene banks, with their limited resources, would be able to produce evaluation data for the thousands of samples they now maintained. His proposal was instead the 'rationalization of *evaluation*' (emphasis mine) through the selection of a 'core collection' of samples that were thought to represent most of the genetic diversity in the collection. This would entail 'a drastic reduction in redundancy', at least in terms of genetic variation within the identified core, and therefore also reduce the energy devoted to evaluation (Frankel, 1984: 161; see also Brown, 1989, 1995). The 'rationalization of evaluation' through the use of core collections found influential champions in the 1980s and 1990s (e.g. Brown & Spillane, 1999). A chief selling point of core collections was not that they would reduce the overall number of samples, but instead that they would ensure that the widest possible range of genetic diversity would be maintained and used even in circumstances of constrained resources. In fact, the core collection concept appealed precisely because it meant that streamlined gene bank management could occur without a costly investment in eliminating duplicate samples via field evaluations of an entire collection or even by newly available biochemical and molecular techniques. In the 1980s and 1990s, these were, for the most part, prohibitively expensive to run for entire seed bank collections. As a committee of experts assembled under the aegis of the US National Research Council acknowledged in 1991, 'Elimination of redundancy in existing collections is not cost-effective' (NRC, 1993: 172).

3   The Techno-political Project of Collection Decentralization

The merits of eliminating duplicates were, as the foregoing discussion suggests, often debated by gene bank managers and other experts in technical and economic terms. Scientists advocated approaches that they felt would produce the best conservation outcomes at the least expense. Yet the drive for rationalisation through data generation and de-duplication was at times a product of political considerations as much as technical ones—even beyond the unquestionably political project of showing that the 'patrimony of humanity' was well and affordably cared for in seed banks. A European Cooperative Programme on the Conservation and Exchange of Crop Genetic Resources (ECP/GR), which first took shape in the late 1970s, imagined the elimination of duplicates across collections as a crucial step in the creation of a decentralised European gene bank. As I discuss here, realising such a bank would depend not only on better data, but also on strong ties and mutual confidence among institutions as well as among the governments that sponsored those institutions. In this context, rationalisation through the elimination of duplicates was a project that depended on existing geopolitical relationships—and attempted to forge new ones.

The initial conversations that led to the ECP/GR took place in 1975. Although this timeframe—in sync with the founding of the International Board for Plant Genetic Resources—points to the influence of international mobilisations on this European project, the origins of the ECP/GR lay in regional, not global, concerns. As early planning documents described, the initiative was imagined within the UNDP's European Office as contributing to that organisation's 'endeavour to establish cooperation between East and West European countries' (FAO, 1979: 1). Thanks to early imperial infrastructures for acquiring and maintaining plant materials from around the world (Brockway, 1979; Drayton, 2000) and increasing state emphases on strategic collections of crop diversity from the 1920s onward (Pistorius & van Wijk, 1997; Flitner, 2003; Saraiva, 2013; Bonneuil, 2019), European institutions collectively possessed an estimated two-thirds of the world's crop gene bank accessions (FAO, 1979: 15). The European Association for Research on Plant Breeding (EUCARPIA) had begun to link the activities of these institutions through its gene bank committee in 1966, focusing especially on coordinating collecting missions and agronomic characterisation of accessions. The planned ECP/GR would expand and deepen this coordination effort, with the aim of 'permitting direct access on the part of every plant breeder to the germplasm of the entire continent... thus making possible a previously unattainable level of plant breeding efficiency' (FAO, 1979: 17). Communication and harmonisation across European agricultural research organisations would benefit all breeders, and all nations, that participated. This could not be achieved simply through professional researchers and breeders acting independently out of their 'somewhat limited goodwill': it demanded the formal commitment of governments (FAO, 1980: 11).

This bridging of East and West to the benefit of all Europeans—and the 'Third World', too, as many planning documents insisted that shoring up the foundations of

European gene banks would ramify well beyond the continent—would depend especially on generating data about accessions and ensuring that both data and the systems used to record these were in reasonable harmony. In the most general terms, 'a major effort to describe and document all existing genetic resources collections in Europe' would be accompanied by an 'all-European genetic data exchange', the latter produced by finding various means of making diverse existing gene bank data management systems interoperable (FAO, 1979: 18–19).

Achieving this level of data exchange would not just make accessions more readily accessible to researchers. It was also imagined as a route to reducing duplication: duplication of collecting missions, duplication of evaluation and characterisation programs, duplication of accessions themselves, and—of course—duplication of the expenditures needed to conduct any of these activities (FAO, 1979: 20; FAO, 1980: 3). For example, 'The burden of collating comprehensive information about the genetic resources of crop plants could be shared between genebanks by each one accepting responsibility for the in-depth study of a particular crop (or crops)' and making the results available to all other collections (ECP/GR, 1981a: 25). Having converged on these objectives, more than 20 European countries agreed to launch the ECP/GR in 1980, with start-up funding from UNDP and administrative support from the United Nations Food and Agriculture Organisation (FAO).[1]

The programme's governing body met for the first time in December 1980. Observers to the initial meeting and other early convenings of this body included nations unwilling to become full participants but interested in the proceedings (most notably, the Soviet Union) as well as organisations with relevant expertise, resources or both. The International Board for Plant Genetic Resources was an obvious collaborator, and its international mandate was seen as the route for delivering the promised payoffs of harmonisation within Europe to the wider world, especially agricultural research programmes in developing countries. The IBPGR's still nascent understanding of network-building at a global scale was complemented by the input of several organisations with network-building expertise at a sub-regional level: the Nordic Gene Bank, the IBPGR's recently established Mediterranean Programme, and the genetic resources networks of both the European Economic Community and the Council for Mutual Economic Assistance (also known as COMECON) (FAO, 1979: 2; see also participant lists in meeting reports, e.g., ECP/GR, 1981a, b). These sub-regional groups had been established to facilitate precisely the type of coordination and exchange now imagined as a pan-European project. In a way, the European Cooperative Programme sought to knit together existing but geopolitically divided networks of researchers and institutions.

---

[1] By the second meeting of the ECP/GR governing body in 1981, 20 countries had formally agreed to the program: Bulgaria, Cyprus, Denmark, Finland, the German Democratic Republic, Greece, Hungary, Iceland, Israel, Italy, the Netherlands, Norway, Poland, Portugal, Romania, Spain, Sweden, Switzerland, the United Kingdom, and Yugoslavia. In addition, Belgium, France, and the Federal Republic of Germany had indicated their intention to participate (ECP/GR, 1981b: 23).

One way to forge lasting links was to create interdependence, unifying institutions by dividing labour. Although national representatives and other participants in the ECP/GR programme wanted ideally to establish 'one or a few centralized genebanks in Europe' they knew that the resources and will for creating new transnational institutions was in short supply. They therefore initially imagined existing institutions becoming 'lead centres' for a certain crop or several crops, taking on the responsibility for maintaining and providing access to all the accessions of that crop on behalf of participating countries. For example, an early list suggested the Plant Breeding and Acclimatization Institute in Radizkow, Poland would take responsibility for *Secale* (ryes), the National Vegetable Research Station in Wellesbourne, UK as the lead centre for *Allium* species (onions, garlic, leeks, etc.), and so on (ECP/GR, 1981a: 30–33; see also ECP/GR, 1981b: 38–44). This arrangement was thought to potentially economise on time and labour and—perhaps just as importantly—'create mutual interest and build up confidence by making countries and sub-regions mutually dependent' (ECP/GR, 1981a: 30).

Ultimately this vision of a decentralised European gene bank, to be created by networking specialised crop centres, gave way to a still more decentralised vision in which even the European crop collections would be generated by networking among different national collections rather than transferring responsibility to a single lead centre. With crop species as the definitive means of organising the network—rather than, say, ecological zones, regional boundaries, or working languages—the ECP/GR established 'crop committees' (later, 'crop working groups') for the crops its scientific advisors deemed most important to collect and conserve An initial selection of 12 crops was based on criteria that included evolutionary history (European indigeneity), biocultural factors (significant genetic diversity in European landraces, unique national appreciation), economic and agronomic importance, and technical considerations (state of existing collections, quality of existing data) (ECP/GR, 1982: 25–26). This list was then narrowed to just six: barley, forages, *Prunus* (plums, cherries, peaches, almonds, etc.), *Allium*, oat, and sunflower. A handful of experts on the selected crops, representing institutions with significant existing collections of these, constituted the working groups. Their main objective was to find means of actualising the overarching ECP/GR goal of enhancing cooperation and reducing duplication in specific projects (UNDP-IBPGR, 1984: 4).

As the ECP/GR moved toward implementation, after multiple years of negotiation and planning, data generation and data management took centre-stage. Ensuring cross-institution 'interoperability' of data about collections had been seen from the outset as a crucial mechanism for cross-country coordination. But emphasis on this aspect was likely heightened by the decision to fold ECP/GR into the work of IBPGR in 1983. In 1981, while planning was still in progress, the executive secretary of IBPGR, J. Trevor Williams, had exhorted ECP/GR participants not to delay on what he saw as the most crucial element of more effective gene bank coordination: generating data. He insisted that 'immediate action' was needed to 'put into order most of the collections by incorporating basic information into data bases'. These could then be used 'to sort out redundant duplicates thereby leading to the maintenance of perhaps smaller, but well documented and more useful, collections'. He was particularly concerned that the group didn't have a grasp of the true number of accessions it needed to manage, given the amount of 'redundant duplication' within and across institutions (ECP/GR, 1981b: 9–10).

Williams' concerns about the problems of European collections reflected issues that IBPGR was grappling with more generally in the early 1980s. As discussed above, these included especially the significant uptick in collecting of the 1970s arising from increased attention to genetic erosion as a conservation issue, a subsequent expansion in collections, and the vulnerabilities in collections management generated as a result. Given the extent to which the world's collections were in European hands, the IBPGR problem of seed surfeits and missing data could be understood as a largely European problem.

The emphasis on data and databases as essential and often neglected instruments for gene bank management contributed to the positioning of crop databases— information infrastructures containing 'all information on the germplasm of each crop kept in European genebanks' and maintained on a computer at a single leading institute—as a top priority for the crop working groups (ECP/GR, 1981b: Appendix VIII; UNDP-IBPGR, 1984: 9–15). These databases would be the chief tools by which the working groups (and by extension the ECP/GR) would coordinate conservation activities across Europe. They would be useful for not only rationalising collections by removing unwanted duplicates and moving towards decentralisation of collections, but also identifying gaps in European holdings that could be resolved through collecting missions and enabling strategic planning for evaluation and characterisation (UNDP-IBPGR, 1984). (See Fig. 1).

The general steps outlined for the crop working groups included, first, the compilation of accession lists for all of the samples of the relevant crop (and in some cases its wild relatives) in the gene banks of participating institutions. A second step was agreeing and implementing descriptors, that is, deciding on a consistent set of information to be associated with each accession and a consistent way of expressing that information. With complete accession lists in hand providing 'basic passport descriptors' (as opposed to more detailed evaluation or characterisation descriptors), working groups would be in a position to complete the third initial task, the identification of duplications. Coordination of further characterisation activities, with an eye to populating the database with still more useful agronomic data, would follow—or so the idealised workflow suggested (UNDP-IBPGR, 1984; Perret, 1985). In sum, European crop databases, created by asking experts from different institutions and nations to create, harmonise and pool data about existing crop gene bank collections, was the technical tool through which the political project of uniting European genetic resources—and by extension European scientists and governments—would be achieved.

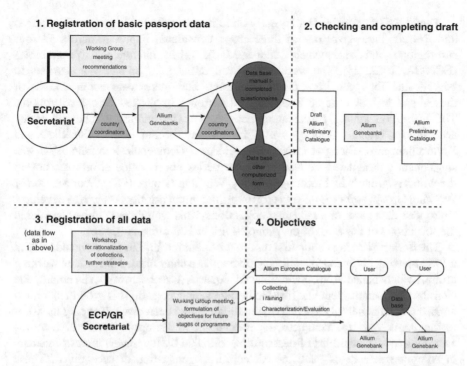

Fig. 1 In 1984, the ECP/GR imagined the implementation of crop databases using the example of *Allium*. 'Country coordinators' would ensure that basic passport data existed for all national collections. These data would be fed into a European catalogue (i.e. the crop database) via manual or computerised questionnaires. Once all data were registered and the catalogue was complete, the latter would be the basis for rationalisation of collections and further activities including collecting, training, and characterisation/evaluation. (From UNDP-IBPGR, 1984, pp. 10–11). Reprinted by permission of Bioversity International

4 Databases and De-duplication

The ECP/GR's Barley Working Group made impressive progress towards the goals of developing a database and deploying this to reduce duplication within and across collections in the programme's first two decades. Its impacts nonetheless fell far short of those projected at the outset, consisting mostly in investigating the tools necessary for accomplishing decentralisation through de-duplication. A close look at the working group's efforts in this period reveals the significant technical hurdles that database creation and decentralisation entailed. It also reveals the strategies that scientists and gene bank managers adopted in attempting to navigate the paradoxical situation outlined above—namely, that preventing costly duplication of data-generating evaluation programmes (along with other expenditures related to the maintenance of accessions) nonetheless depended on undertaking potentially costly data production and management exercises.

The ECP/GR prioritized barley early on in its proceedings. Together with forages and *Prunus*, barley was one of three crops considered important across all four sub-regions and inadequately addressed in other international programmes (ECP/GR, 1982: 25). The working group's initial tally of barley accessions in 'significant' European gene banks suggested that there were about 85,000 of these—and that at least 60% represented samples duplicated across collections. The institution among participating nations with the largest number of barley accessions (about 9400) was the Zentralinstitut für Genetik und Kulturpflanzenforschung at Gatersleben, German Democratic Republic. This was subsequently designated the lead centre for barley conservation efforts and hosted the working group's first meetings (Barley Working Group, 1983).[2] For the Barley Working Group, as for the other crop groups, the initial priority task was to develop a 'European data base' of all existing collections. This would come to be considered the 'backbone of the work of the group' (Dirk & Knüpffer, 2001: 50).

The Barley Working Group first outlined its plan for fulfilling its assigned tasks at a 1983 meeting in Gatersleben. Six members plus a chair formed the official working group, which hailed from institutions in Austria, Czechoslovakia, Denmark, the German Democratic Republic, the Netherlands, and Poland. This group elaborated a set of aims that hewed closely to the mandate it had been given. Topping its list of action items was 'the complete documentation of European barley collections' according to a standard list of descriptors, followed by 'the registration of this data in computer data bases', and the 'detection of replication of accessions'. These would make possible 'the rationalization of collections by agreement between participating gene banks with consequent elimination of potential waste of resources in the storage, multiplication, characterization and evaluation of redundant accessions' (Barley Working Group, 1983: 1). In other words, immediate improvements in data creation, management and exchange would reduce the costs of collection management—including further costly data generation (i.e. evaluation of duplicate accessions). Documentation, database development and de-duplication would also result in a decentralised European barley gene bank managed not by any one institution but by all.

The first version of the European Barley Database, assembled between 1984 and 1987, brought together the passport data associated with over 55,000 barley accessions from more than 30 European collections in 26 countries. The database was maintained in Gatersleben at the Zentralinstitut für Genetik und Kulturpflanzenforschung on an 8-bit microcomputer (Knüpffer, 1988a, b; Dirk & Knüpffer, 2001: 50). Participants in the initial 1983 meeting of the Barley Working Group had been asked to bring with them information—'if possible, computer printouts'—reporting on the contents of collections in their home country and

---

[2] Documents prepared for the 1983 meeting of the Barley Working Group suggested that more than 17,000 barley accessions were held by the VIR in Leningrad, which (assuming the numbers were even reasonably accurate) would have made this the largest collection not only in Europe but also worldwide.

neighbouring countries. Only four institutions provided the requested printouts, suggesting the extent to which most information regarding collections remained in records maintained chiefly by hand (Barley Working Group, 1983: 2) or otherwise difficult to share. The development of the database at Gatersleben was therefore a staged process requiring the acquisition of collection data from across institutions, the 'preprocessing' of these by a colleague in Sweden, the standardisation of the descriptors used in different data sets, and their eventual merger into a complete list of accessions maintained at Gatersleben (Knüpffer, 1988a, b).

With a first iteration of the database in hand in 1987, it was time to put it to use in coordinating across institutions, as imagined. As the database's chief developer noted, it was already possible to send inquires and requests to Gatersleben such as 'Print a list of all two-rowed winter barleys originating from the Far East' or 'Where could I ask for living seeds of the cultivars and strains listed below?' and receive a reply from the scientist in charge of the database (Knüpffer, 1988b: 19) But this service provision was hardly the reason the database had been developed. Duplication remained a key concern of the Barley Working Group; by 1988 members had 'repeatedly stressed' the need to eliminate as many 'redundant duplicates in European collections' as possible by systematic comparison of collections (Knüpffer, 1988a: 144). A second phase of the project was therefore to implement the database in precisely this way. The database would be used not only to identify where duplicates occurred but also to assign and track responsibility for maintaining the accessions remaining after redundant duplicates were eliminated.

The working group hoped that removing duplicate samples would be a step in the rationalisation of collection management and therefore also a step towards decentralisation. However, the identification of duplicates in the emerging database was itself 'a time-consuming procedure requiring much knowledge about the breeding and collecting history of a particular crop' (Knüpffer, 1988a: 150). De-duplication demanded data and new forms of data analysis. For a few individuals involved in the creation and curation of the European Barley Database, weeding out duplicates efficiently and effectively in the name of rationalisation became its own area of research. Duplicates were not a single class but comprised different types of genetic duplication, depending on their origin. 'Identical duplication' happened when a well-mixed sample was split in two, as happened for example in the creation of safety duplicates for off-site storage. 'Common duplication' referred to accessions arising from the same sample, for example when a new generation was grown out in order to renew a dwindling stock or multiply the seed to share beyond the bank. There were also 'partial' duplicates, 'compound' duplicates, and other known circumstances in which the genetic identify of samples overlapped significantly (van Hintum & Knüpffer, 1995: 128–129). (See Fig. 2).

Passport data—which, to reiterate, formed the foundation of the first generation of electronic databases of gene bank accessions—could not be used to discover many of these kinds of duplication. At the most basic level, the genetic makeup of a sample might have shifted during regeneration. The environmental conditions of grow-out, the size of the original population, mismanagement of seed lots—all these conditions and more could lead to divergence from one generation to the next. As a

Fig. 2 Eliminating
duplicates in the seed bank
first required a knowledge of
how they typically
originated and their genetic
relationship to the original
sample. (From van Hintum
& Knüpffer, 1995). Used
with permission of Kluwer
Academic Publishers,
permission conveyed
through Copyright
Clearance Center, Inc.

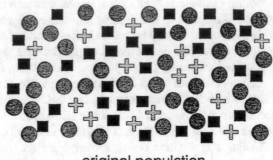

original population

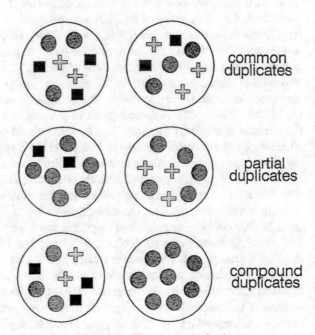

common
duplicates

partial
duplicates

compound
duplicates

genebank accessions sampled
from the original population

result, even 'common' duplicates (created for example by splitting a sample and sharing among two institutions) might actually become genetically distinct despite the fact that their identifying information remained exactly the same. Duplicates identified through passport data were therefore only 'probable' duplicates, genetically speaking, and not known duplicates (van Hintum & Knüpffer, 1995: 128). This limitation was compounded by fact that passport data were notoriously unreliable. Two scientists working on the database noted, as common occurrences, the 'omission of (parts of) the collection number or other collection data, errors in interpretation... typing errors, probable translation, transcription or transliteration errors or inconsistencies' (van Hintum & Visser, 1995: 137). The records, in other words, were too messy to be trusted. Samples with the same label might in fact be genetically different, while samples with different labels might be identical.

Grow-outs and evaluations could potentially resolve the accession-identity issues plaguing the database. However, the whole promise of using the database for collection rationalisation was that it would avoid, among other things, these often-costly activities. The European Barley Database developers therefore sought to engineer around poor data, and trialled different means of identifying duplicates that took into consideration probable errors and inconsistencies. The 'Soundex' method of locating accessions carrying phonetically similar labels, for example 'Closess IV', 'Colcess', 'Colcess IV', 'Colchicum', 'Colses', and 'Colsess'. Meanwhile the 'Keyword in Context' approach sussed out accessions with similar identifying information even if this hadn't been standardized to the same database fields (see discussion in van Hintum & Knüpffer, 1995). But even where these approaches could be considered successful in that they guided database managers to probable duplicates, understanding whether identified items could be considered genetically identical (or nearly so) still required biochemical intervention (van Hintum & Visser, 1995: 143–144). There were limits, then, to the promise of streamlined de-duplication via the database.

Decentralisation nonetheless remained a key goal, and the European Barley Database was seen as the central means of achieving it. In 1997, a statement generated by the Barley Working Group outlined the imperatives for creating a 'decentralized European Barley Collection'. It noted overall reductions in funding that negatively affected genetic resources work and emphasised that this 'strained economic situation is further aggravated by the duplication of both efforts and germplasm'. These 'economic constraints' in turn required not only priority setting but also 'the sharing of responsibilities' and, more generally, recognition that 'no single country in Europe can, on its own, conserve all barley genetic resources' (Maggioni et al., 1997: 111). That same year, a new version of the European Barley Database was released, now including more than 90,000 accessions; this represented the first update since the 1987 iteration and was made possible, after efforts to secure external funding failed, by local provision of a staff person to conduct the update over a six-month period (Knüpffer et al., 2001: 50). As in the case of its predecessor, the 'wide coverage and completeness of data' was touted as 'essential' to its full use, including 'a screening of the collections in Europe, made to identify unique samples or to locate duplicates' (Maggioni et al., 1997: 3). Ten years on, the identified problems and solutions remained much the same. But realisation of rationalisation remained elusive.

5   Conclusions. Out of Many, One?

At a 2000 meeting of the Barley Working Group, membership of which had by that time reached more than 30 scientists from institutions across Europe and beyond, participants discussed a proposal that had been floated at higher levels of the

European Cooperative Programme for Crop Genetic Resources. This was the aspiration 'to build virtually a decentralized European Genebank'—a new formulation of what had really been the aspiration of the programme all along. The recorded discussion of this proposal among the Barley Working Group reveals members' scepticism about its feasibility, but curiously not because of the challenges of data creation and management that limited the horizons of their own decentralisation efforts up to that point. Problems were envisioned with gene banks whose assigned accessions were not particularly useful to its core users, who would then be forced to look abroad for items of interest. Recommendations for enrolling national collections in the broader European Cooperative Initiative centred on identifying the accessions to be pooled and assigning specific accessions to the care of specific gene banks, rather than pointing out the significant investments that would be required to make this cost-cutting measure feasible (Knüpffer et al., 2001: 9–10).

Meanwhile, an expansion of the European Barley Database to link with other international collections brought the total number of accessions to more than 135,000 and made it an increasingly international, as opposed to European, enterprise. Although locating duplicates was listed as a key outcome of this further database development, the payoff of this identification was not in creating opportunities for eliminating redundancy, but simply better data sharing, 'allow[ing] links to be established between accessions and their evaluation data accessible in the respective databases' (Dirk & Knüpffer, 2001: 52). It is this vision of databases—as tools for sharing information, linking communities, pooling knowledge—that predominates in both celebratory and critical accounts of seed and gene banks' database development projects. However, as I have shown here, these projects have also been driven by desire for greater economy in the expenditure of scarce resources and ironically forestalled for lack of funds. They have gained traction as political initiatives, without consistent appreciation for the technical and political challenges of realising data linkage. This brings to the fore a different narrative about the history and politics of seed and gene banks and of the data infrastructures associated with these.

Data creation, harmonisation and centralisation remain key objectives across the agricultural sciences, perhaps even more so in an era of 'Big Data' than in the period covered by this chapter (see, e.g., Harper et al., 2018; Arnaud et al., 2020). In the intervening years, the transformation of technical capacities has made it easier to implement forms of data linkage that could only be aspirational in the mid-to-late twentieth century. Political and economic constraints nonetheless remain a significant concern in database development, as several contributions to this volume highlight (see also Leonelli, Forthcoming). Meanwhile many crucial domains of technical skill, such as data curatorship, remain undervalued (Leonelli, 2014; Strasser, 2019; Leonelli & Tempini, 2020). If recent history points to constancy in the vision of data linkage as a solution to the imperatives of international agricultural research and development, thereby affirming contemporary calls to resolve—finally—the technical obstacles to it, this history also points to the extent to which these technical projects were and are much more than that. They have served as means of deflecting criticism, vehicles for fostering geopolitical ties, cost-cutting measures, and more. Recognising these aims, which sometimes converge but may also be in contradiction, is crucial to forging effective and equitable programmes for gene bank data management in the future.

Acknowledgements This research was funded by the Wellcome Trust [Grant number 217968/Z/19/Z]. For the purpose of open access, the author has applied a CC BY public copyright licence to any Author Accepted Manuscript version arising from this submission.

The author is grateful to Jessica J. Lee for her help researching this article, to workshop participants and members of the 'From Collection to Cultivation' project team for helpful feedback on an early draft, and to Sabina Leonelli and Hugh Williamson for their editorial direction.

References

Aoki, K. (2008). *Seed wars: Controversies and cases on plant genetic resources and intellectual property*. Carolina University Press.

Arnaud, E., et al. (2020). The ontologies community of practice: A CGIAR initiative for big data in agrifood systems. *Patterns, 1*(7), 100105.

Barley Working Group. (1983). *Report of a Working Group held at the Zentralinstitut für Genetik und Kulturpflanzenforschung der Akademie der Wissenschaften der DDR Gatersleben* (AGP: IBPGR/83/68). UNDP–IBPGR.

Bonneuil, C. (2019). Seeing nature as a 'Universal store of genes': How biological diversity became 'Genetic resources'. *Studies in the History and Philosophy of Science Part C, 75*, 1–14.

Brockway, L. (1979). *Science and colonial expansion: The role of the British Royal Botanic Gardens* (Reprint, 2002). Yale University Press.

Brown, A. H. D. (1989). Core collections: A practical approach to genetic resources management. *Genome, 31*, 818–824.

Brown, A. H. D. (1995). The core collection at the crossroads. In T. Hodgkin et al. (Eds.), *Core collections of plant genetic resources* (pp. 3–19). IPGRI.

Brown, A. H. D., & Spillane, C. (1999). Implementing core collections: Principles, procedures, progress, problems and promise. In R. C. Johnson & T. Hodgkin (Eds.), *Core collections for today and tomorrow*. IPGRI.

Chacko, X. (2019). Creative practices of care: The subjectivity, agency, and affective labor of preparing seeds for long-term banking. *Culture, Agriculture, Food and Environment, 41*(2), 97–106.

Curry, H. A. (2017). From working collections to the World Germplasm Project: Agricultural modernization and genetic conservation at the Rockefeller Foundation. *History and Philosophy of the Life Sciences, 39*, Article 5.

Curry, H. A. (2022). *Endangered maize: Industrial agriculture and the crisis of extinction*. University of California Press.

Curry, H. A., & Leonelli, S. (Forthcoming). Describing crops in the CGIAR era. In H. Curry & T. Lorek (Eds.), *Research as development: Historical perspectives on agricultural science and international aid in the CGIAR era*.

Dirk, E., & Knüpffer, H. (2001). Appendix I. The European Barley Database. In H. Knüpffer et al. (Eds.), *Report of a Working Group on Barley. Sixth meeting, 3 December 2000* (pp. 50–63). IPGRI.

Drayton, R. (2000). *Nature's government: Science, Imperial Britain, and the 'improvement' of the world*. Yale University Press.

ECP/GR [FAO/UNDP European Cooperative Program for Conservation and Exchange of Crop Genetic Resources]. (1981a). *Report of the first governing board meeting, 15–18 December 1980*. UNDP and FAO.

ECP/GR [FAO/UNDP European Cooperative Program for Conservation and Exchange of Crop Genetic Resources]. (1981b). *Report of the second governing board meeting, 14–18 December 1981*. UNDP and FAO.

ECP/GR [FAO/UNDP Cooperative Programme for Conservation and Exchange of Plant Genetic Resources]. (1982). *Report of extra-ordinary meeting of the governing board, Geneva, 14–16 June 1982*. UNDP and FAO.

FAO [UN Food and Agriculture Organization]. (1979). *Report of the FAO/UNDP government consultation on the European Cooperative Programme for the Conservation and Exchange of Genetic Resources for Plant Breeding, Rome, 8–9 March 1979*. FAO.

FAO [UN Food and Agriculture Organization]. (1980). *Report of the FAO/UNDP governments consultation on the European Cooperative Programme for the Conservation and Exchange of Genetic Resources for Plant Breeding, Geneva, 17–19 December 1979*. FAO.

FAO [UN Food and Agriculture Organization]. (2010). *The second report on the state of the world's plant genetic resources for food and agriculture*. FAO.

Fenzi, M. (Forthcoming). Crop diversity under CGIAR's lens. In H. Curry & T. Lorek (Eds.), *Research as development: historical perspectives on agricultural science and international aid in the CGIAR era*.

Fenzi, M., & Bonneuil, C. (2016). From 'genetic resources' to 'ecosystems services': A century of science and global policies for crop diversity conservation. *Culture, Agriculture, Food and Environment, 38*(2), 72–83.

Flitner, M. (2003). Genetic geographies: A historical comparison of agrarian modernization and eugenic thought in Germany, the Soviet Union, and the United States. *Geoforum, 34*(2), 175–185.

Frankel, O. H. (1984). Genetic perspectives of germplasm conservation. In W. Arber, K. Illemensee, W. J. Peacock, & P. Starlinger (Eds.), *Genetic manipulation: Impact on man and society* (pp. 161–170). Cambridge University Press.

Fullilove, C. (2018). Microbiology and the imperatives of capital in international agro-biodiversity preservation. *Osiris, 33*(1), 294–318.

Goodman, M. M. (1984). An evaluation and critique of current germplasm programs. In *Report of the 1983 Plant Breeding Research Forum* (pp. 195–249). Pioneer Hi-Bred International.

Hanson, J., Williams, J. T., & Freund, R. (1984). *Institutes conserving crop germplasm: The IBPGR global network of genebanks*. IPBGR.

Harper, L., et al. (2018). AgBioData Consortium recommendations for sustainable genomics and genetics databases for agriculture. *Database, 2018*, bay088.

Knüpffer, H. (1988a). The European Barley Database of the ECP/GR: An introduction. *Kulturpflanze, 36*, 135–162.

Knüpffer, H. (1988b). The European Barley Data Base of the European Cooperative Programme for the Conservation and Exchange of Crop Genetic Resources (ECP/GR). *Plant Genetic Resources Newsletter, 75*(76), 17–20.

Knüpffer, H., et al. (2001). *Report of a Working Group on Barley, sixth meeting, 3 December 2000*. IPGRI.

Lehmann, C. O. (1981). Collecting European land-races and development of European gene banks-historical remarks. *Die Kulturpflanze, 29*(1), 29–40.

Leonelli, S. (2014). *Data-centric biology: A philosophical study*. University of Chicago Press.

Leonelli, S. (Forthcoming). How data cross borders: Globalising plant knowledge through transnational data management and its epistemic economy. In J. Krige (Ed.), *Transnational transactions: Negotiating the movement of knowledge across borders*. University of Chicago Press.

Leonelli, S., & Tempini, N. (2020). *Data journeys in the sciences*. Springer.

Lyman, J. M. (1984). Progress and planning for germplasm conservation of major food crops. *Plant Genetic Resources Newsletter, 60*, 3–21.

Maggioni, L., et al. (1997). *Report of a Working Group on Barley, fifth meeting, 10–12 July 1997.* IPGRI.

Mooney, P. R. (1979). *Seeds of the earth: A private or public resource?* International Coalition for Development Action.

Mooney, P. R. (1983). The law of the seed. *Development Dialogue, 1–2,* 1–172.

National Research Council (NRC) [United States]. (1993). *Managing global genetic resources: Agricultural crop issues and policies.* The National Academies Press.

Parry, B. (2004). *Trading the genome: Investigating the commodification of bio-information.* Columbia University Press.

Peeters, J. P., & Williams, J. T. (1984). Towards better use of genebanks with special reference to information. *Plant Genetic Resources Newsletter, 60,* 22–32.

Peres, S. (2019). Seed banking as cryopower: A cryopolitical account of the work of the International Board of Plant Genetic Resources, 1973–1984. *Culture, Agriculture, Food & Environment, 41*(2), 76–86.

Perret, P. M. (1985). European Programme for Conservation and Exchange of Crop Genetic Resources. *Plant Genetic Resources Newsletter, 61,* 5–12.

Pistorius, R. (1997). *Scientists, plants and politics: A history of the plant genetic resources movement.* IPGRI.

Pistorius, R., & van Wijk, J. (1997). *The exploitation of plant genetic information: Political strategies in crop development.* CABI.

Saraiva, T. (2013). Breeding Europe: Crop diversity, gene banks, and commoners. In N. Disco & E. Kranakis (Eds.), *Cosmopolitan commons: Sharing resources and risks across borders* (pp. 185–212). MIT Press.

Strasser, B. (2019). *Collecting experiments: Making big data biology.* University of Chicago Press.

Thormann, I., Engels, J. M. M., & Halewood, M. (2019). Are the old International Board for Plant Genetic Resources (IBPGR) base collections available through the Plant Treaty's multilateral system of access and benefit sharing? A review. *Genetic Resoures and Crop Evolution, 66,* 291–310.

UNDP-IBPGR [United Nations Development Programme-International Board for Plant Genetic Resources]. (1984). *The ECP/GR: An introduction to the European Cooperative Programme for the Conservation and Exchange of Plant Genetic Resources.* IBPGR.

US Comptroller General. (1981). *Report to the Congress of the United States: The Department of Agriculture can minimize the risk of potential crop failures* (CED-81-75, 10 April). US General Accounting Office.

Van Dooren, T. (2010). Banking seed: Use and value in the conservation of agricultural diversity. *Science as Culture, 18*(4), 373–395.

van Hintum, T. J. L., & Knüpffer, H. (1995). Duplication within and between germplasm collections. I. Identifying duplication on the basis of passport data. *Genetic Resources and Crop Evolution, 42,* 127–133.

van Hintum, T. J. L., & Visser, D. L. (1995). Duplication within and between germplasm collections. II. Duplication in four European barley collections. *Genetic Resources and Crop Evolution, 42,* 135–145.

Weise, S., Lohwasser, U., & Oppermann, M. (2020). Document or lose it – On the importance of information management for genetic resources conservation in Genebanks. *Plants, 9*(8), 1050. https://doi.org/10.3390/plants9081050

Williams, J. T. (1984). A decade of crop genetic resources research. In J. H. W. Holden & J. T. Williams (Eds.), *Crop genetic resources: Conservation & evaluation* (pp. 1–17). George Allen & Unwin.

Digital Sequence Information and Plant Genetic Resources: Global Policy Meets Interoperability

Daniele Manzella, Marco Marsella, Pankaj Jaiswal, Elizabeth Arnaud, and Brian King

Abstract Plant genetic resources are source genetic material for conducting research and breeding. The use of this material is subject to international and national regulations on access and benefit-sharing (ABS). With modern genetic technologies generating desired trait and gene function improvement by replicating genetic signatures, ABS must adapt to the new technological reality. As the constituencies of international ABS conventions discuss if and how to extend the application of the conventions to digital sequence information (DSI) derived from source material, the genomics science community resists any incumbrance to continued free and unrestricted access to such information. Based on current ABS discussions and the likely future co-existence of diverse policy regimes, this paper proposes interoperability among data systems as an essential tool to implement legal solutions for benefit-sharing as well as advance science and innovation objectives. Two information technology tools are suggested for associating DSI to plant genetic resources and reciprocal citations with data exchange, namely digital object identifiers and digital genetic objects. This paper concludes that interoperability should be experimented with in both its technical and social dimensions, in order to support long-term alliances between policy and science through data archives, knowledge bases and live specimen collection resources.

D. Manzella (✉) · M. Marsella
Food and Agriculture Organization of the United Nations, Rome, Italy
e-mail: Daniele.Manzella@fao.org; Marco.Marsella@fao.org

P. Jaiswal
Oregon State University, Corvallis, OR, USA
e-mail: Pankaj.Jaiswal@oregonstate.edu

E. Arnaud · B. King
The Alliance of Bioversity International and CIAT, Rome, Italy
e-mail: E.Arnaud@cgiar.org; B.King@cgiar.org

1   Introduction

In plant biology research and crop breeding programs, the value of plant genetic resources is determined by the seed and the propagation material, called source genetic material, that are important for conducting genomics, genetics, phenotype and trait evaluation, in-vivo and in-vitro experiments. Much of the experimental information, data, and the knowledge gained become important for the researchers when they are properly associated with the source genetic material, thus enabling further scientific discovery and future replication of the studies. Often, the physical plant material and the derivatives (including isolated protein, DNA and RNA) are however difficult to access due to various national and international regulations and exchange permits. On the one hand, this limited access restricts the use of existing genetic material; whereas on the other hand it can require more tracking of use and citing the source material for various purposes, including publication. Associating experimental information with its source genetic material begins to bring aspects of plant science within the purview of global agreements that establish rules for accessing the source genetic material for research and development and sharing the benefits of its utilization. Under one such agreement, namely the International Treaty on Plant Genetic Resources for Food and Agriculture (ITPGRFA), a Global Information System (GLIS) was established to facilitate the exchange of information on crop genetic material.[1]

Innovation at the intersection of digital technologies and life sciences is quickly changing the context of the global agreements on genetic resources. Modern bioscience relies on the extraction and processing of large volumes of "omics" data in digital form, and this has precipitated a re-examination of the founding principles of such global agreements, as they relate to matters such as identification of the resource, monitoring of its use and attribution of the benefits of such use (Aubry, 2019; Welch et al., 2017). While whole-genome sequences are increasingly available as a result of new-generation technologies, the collective capacity to actually analyze and benefit from the data is lagging behind (Halewood et al., 2018).

The interaction between policymakers driving the global agreements on genetic resources and the genomics science community can be problematic. In the governance frameworks of the global agreements, this interaction is viewed through the lens of digital sequence information (DSI), a term of uncertain meaning that functions as a placeholder in the discussions as to whether the informational component of genetic resources should be regulated under the same rules of access and benefit-sharing (ABS) that govern source genetic material.

The respective value propositions seem to radically differ. ABS is equity-driven and relies on normative standards (legislation, contracts) to implement controlled access regimes (Ruiz, 2015). The genomics science community prioritizes research efficiency and is guided by community standards and protocols, e.g. the Fort Lauderdale agreement, the Toronto agreement, FAIR data principles on data sharing

---

[1] http://www.fao.org/3/a-i0510e.pdf, see Article 17.

with international archives and publications (Toronto International Data Release Workshop Authors, 2003; Wellcome Trust, 2003; Wilkinson et al., 2016). According to these standards, all genomics data including derived DNA, RNA and protein sequences must remain public and accessible without restrictions in order to enable biologists to discover and realize the benefits of the material in research and application.[2] The access to derived DNA, RNA and protein sequences from the physical genetic material has opened up a new possibility in the research and innovation community enabled by genetic engineering technologies like CRISPR (Chen et al., 2019). Now, researchers have the ability to update genomes of a germplasm by replicating genetic signatures of a wild relative with sequences associated with desired trait and/or gene function improvements, without actually accessing the original seed material considered a global and national heritage. Thus, new proposals and insights are under discussion to revisit the mandates of ABS international agreements for protecting the community interests that take into account the related compensatory gains derived from the genetic signatures and sequences to achieve genetic gains and trait improvements.

In our paper, we introduce ABS policy discussions around DSI, and argue that interoperability among data systems will be essential to implement future legal solutions for benefit-sharing. With a view to pursuing such interoperability, we suggest possible mechanisms that may be well-aligned with the spirit of the international agreements, to develop optimal and timely recommendations for associating DSI to the plant genetic resources and reciprocal citations with data exchange. One mechanism is based on the integration between the federated system of databases of the International Nucleotide Sequence Database Collaboration (INSDC) and the current tools that are available to the plant science community through the GLIS. Another mechanism revolves around the proposed concept of Digital Genetic Object (DGO) as a way to introduce a precise definition of DSI that is functional to interoperability among biological data systems. In conclusion, we flag the need to continue approaching data interoperability with a dual focus on global policy and information technology.

2 Global Policy on Access and Benefit-Sharing and the Nexus with Interoperability

ABS is a construct of international agreements on genetic resources. In Article 2 of the Convention on Biological Diversity, genetic resources are defined as any material of plant, animal, microbial or other origin containing functional units of heredity, of actual or potential value.[3] ABS is aimed at exploiting, through controlled access,

---

[2] A great example is the open data sharing on COVID-19 viral genome sequences that is instrumental to developing vaccines.

[3] https://www.cbd.int/convention/text/, see Article 2.

the potential of those resources for various public policy objectives, e.g. nature conservation, food security, sustainable development, and at rewarding, through the fair and equitable sharing of the benefits of utilization, those who maintain the diverse genetic base.

In various ABS international fora, including the ITPGRFA at the Food and Agriculture Organization of the United Nations, discussions are taking place as to whether to regulate DSI within the remit of the agreements. The motivation to subsume DSI into the domain of ABS, is to realize the provisions of the agreements in the light of scientific and technological advancements. Thanks to such advancements, innovation increasingly relies on the intangible component of genetic resources, i.e. information and data. Although at present, the use of both tangible and intangible components of genetic resources co-exists, it is postulated that in the near future, additional detachment of the informational component from the physical organisms will occur (Morgera et al., 2020; Smyth et al., 2020).

The priority focus of the ABS community is on three issues. The first is the scope of DSI, that is, the data sets that DSI encompasses. The scope of DSI is still under consideration and options range from only the base sequence of genomic DNA to all information associated with genetic resources. Being cognizant of such a broad range of options, our examination considers categories of data which may fit into a functional definition of DSI, namely: DNA, RNA, protein, genetic markers (with or without sequences), non-coding features and other data categories (Houssen et al., 2020; Brink et al., 2021) and specifically suggests ways to link these data to other ontologized knowledge to accommodate expansive views of DSI. As best practice, existing ontology may be used where each concept bears a unique and resolvable identifier, called a Uniform Resource Identifier, for which the definition, context of use and semantic relationships are validated by a large community (Arnaud et al., 2020). For example, the Sequence Ontology, the Protein Ontology, and the Gene Ontology, which include concepts and definitions of Genomic Objects along with other relevant ontologies, such as the NCBI taxonomy for species, and metadata standards, such as the Biosample record, may provide a useful point of departure.[4,5] As discussed in another chapter of this book, cross-domain ontologies have the potential to reduce concept proliferation (see Devare et al).The second issue is terminology, that is, a scientifically accurate term that can be applied in the governance of the international legal agreements. Terms that are under consideration include genetic sequence data, genomics information, natural information (Convention on Biological Diversity, 2020). The third issue is traceability of DSI in databases and in research and development activities that utilize DSI (Convention on Biological Diversity *cit.*). Traceability relates data to a particular genetic resource or to any source that implies the utilization of a genetic resource.

---

[4] http://www.sequenceontology.org/

[5] https://fairsharing.org/biodbcore-000008/.BioSamples records include mandatory fields linked to data standards for genomic data and additional fields for particular standards.

While scientists continue to rely on open access to sequence data, ABS policy demands benefit-sharing and brings this open system into question (Rohden et al., 2021). In the course of such discussions, policy options for utilization of DSI and benefit-sharing have begun emerging. The spectrum of such options is ample. It ranges from free and unrestricted access to genomics data coupled with the financing of benefit-sharing through a multilateral fund, to controlled access to databases and a transactional approach to benefit-sharing, with "club-approach" solutions, such as membership or cloud-based fees, commons licenses, also being proposed (Hartman-Sholz et al., 2020).[6]

Once the ABS policy discussions are complete, the expectation is that a functional definition of DSI will be agreed upon and solutions will be put in place in the framework of the ABS agreements to address the utilization of DSI and benefit-sharing. The current fragmentation of the global ABS framework illustrates an example of a regime complex, with overlapping institutions that interact among themselves on patterns of hierarchy and differentiation (Randall Henning et al., 2020). In the light of such institutional complexity and given the plurality of policy options that are being discussed for DSI, it is likely that different solutions will go through an initial phase of experimentation and thus co exist, e.g. for different categories of genetic resources and derived DSI. By way of example, some genetic resources and the derived DSI may be reserved to national sovereignty and the ensuing control of access and use, and others may be grouped into one or multiple global pools and administered in accordance with open access standards, coupled with multilateral benefit-sharing mechanisms, including pursuant to Article 10 of the Nagoya Protocol.[7]

As diverse policy options are likely to co-exist, it is foreseeable that data aggregation and interoperability will play a key role in implementing corresponding solutions for benefit-sharing. Identifying data sets as DSI and associating DSI to defined genetic resources will be necessary to impute individual or aggregate benefits to the use of identified data and resources.

A number of data sets that are under consideration as DSI are stored and accessed in a variety of databases including the INSDC (Rohden et al., 2021). For the international policy decisions on the scope of DSI to be channeled to actual producers and users of sequence data annotated with DSI, harmonization with the database system and the underlying technology and standards will be highly

---

[6] In this paper, the authors do not express any preference for any of the options.

[7] Article 10 of the Nagoya Protocol on Access to Genetic Resources and the Fair and Equitable Sharing of Benefits Arising from their Utilization (ABS) to the Convention on Biological Diversity provides that "Parties shall consider the need for and modalities of a global multilateral benefit-sharing mechanism to address the fair and equitable sharing of benefits derived from the utilization of genetic resources and traditional knowledge associated with genetic resources that occur in transboundary situations or for which it is not possible to grant or obtain prior informed consent. The benefits shared by users of genetic resources and traditional knowledge associated with genetic resources through this mechanism shall be used to support the conservation of biological diversity and the sustainable use of its components globally." https://www.cbd.int/abs/text/

desirable, if not indispensable. In practice, sequence databases will need to be interoperable with each other in order to identify DSI for legal purposes. The digital nature of sequence information renders it mandatory to propose, in parallel to the current legal discussion, solutions for the interoperability of the data to complement decisions about ABS. Clear, precise definitions of types of genetic material and data must be put into practice through improved data aggregation and interoperability, and increased integration among information systems.

As improved data curation, standardization, identification of provenance, aggregation, exchange and interoperability may support the unfolding institutional processes related to DSI, the outcomes of such processes are likely to elicit varied responses by the science community, based on different assumptions about the degree of choice, awareness, and self-interest. Academic scientists' responses to new regulatory controls on biological material inputs to research show a degree of variation that is shaped by both micro-level, cognitive and macro-level, institutional factors (Oliver, 1991; Welch et al., 2019). Within such a spectrum of responses, some researchers may not be inclined at all to support ABS policy processes in relation to DSI. Nevertheless, others may be willing to pursue anticipatory action with respect to DSI policy development, for instance to increase legitimacy and social qualification that are instrumental to resource mobilization. Such anticipatory action may offer other considerable benefits, such as exerting influence on implementation and co-opting technical standards.

3 Interoperability in the Global Information System of the International Treaty: Possible Applications to Exchanges of DSI

The ITPGRFA is one of the international instruments that compose the global ABS architecture. GLIS is founded on the principle of integration with existing information systems. It implements data aggregation and interoperability by associating information and knowledge to plant genetic resources in *ex-situ*, *in-situ* and on farm conditions to facilitate research and breeding for food and agriculture, as shown in Fig. 1 below. This association is pursued through permanent unique identifiers. Among the different identifier technologies considered, Digital Object Identifiers (DOIs) emerged as a very powerful mechanism to establish linkages to all sorts of information. DOIs are a well-established standard originally developed for the publication sector that has recently expanded its reach to many other application fields.[8]

Among other desirable qualities, DOIs are well known in the research space, offer advanced services such as EventData and the PID Graph, are widely adopted by the publishing sector and dataset repositories, and support flexible metadata structures

---

[8] https://www.doi.org

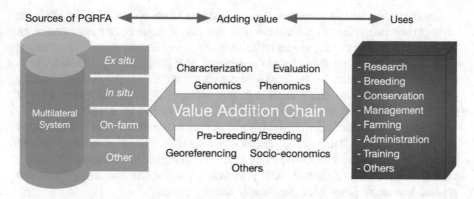

Fig. 1 Diagram of the Global Information System (GLIS). (Reproduced by permission of the International Treaty on Plant Genetic Resources for Food and Agriculture)

allowing representation of object types of very different nature. All these characteristics will come handy when we describe our proposed solution.[9]

Although the ITPGRFA community has not directly tackled the complex ABS legal issues through GLIS, it has acquired experience on the technical implementation of information-sharing in the context of global policy and multilateral cooperation, and has implemented a blended approach to interoperability combining technical standards with iterative learning and multilateral dialogue (Morgera et al., 2020).[10] Such features of GLIS make it suitable to explore possible further integration options between repositories of source genetic material and large repositories of genomics data.

As mentioned above, the diversified components of DSI are likely spread across multiple repositories and databases, each one designed and refined over time to meet the demands of its own user community. The multiplicity of repositories and user requirements undoubtedly poses a variety of challenges that cannot be reduced or solved through a uniform, standard solution. The approach of this paper is to initially tackle data aggregation and interoperability through identification. The identification and ability to link the array of component parts that are all suggested as being part of DSI, depending on the definition adopted, could enable creating the relationships between those component parts and ultimately improve the information discovery and insight about the plant genetic resources themselves.

---

[9]EventData is a joint initiative of Crossref and Datacite, the two leading DOI Registration Agencies (see https://www.crossref.org/services/event-data). PID Graph is a tool funded under the EU project FREYA that collects and makes available references of DOIs to other DOIs and other PIDs (such as ORCID or ROR). See https://www.project-freya.eu/en/pid-graph/the-pid-graph

[10]By "ITPGRFA community", we mean: State party delegates and the broad set of non-State actors who regularly participate in official meetings, including representatives of international and academic agricultural research, private sector, civil society, farmers.

Insofar as nucleotide sequence data as well as other components of DSI can express their full value in conjunction with passport data and other information on the source genetic material, the identification of such material emerges as an actual challenge that needs solutions for the attainment of interoperability. The identification of source material is an area where many genomics repositories currently provide little precision and traceability. For example, INSDC does not offer an accurate identification of the original material, as shown in Fig. 2 below.

The "source" block under "FEATURES" at the bottom of the page is a formatted text attribute that is not mandatory. The rice cultivar name "IR64" is indeed provided but this may not be sufficient to properly identify the original material nor would be a locally assigned identifier, such as a genebank accession number, as cultivars and genebank accessions are often genetically heterogeneous.

Such deficiency may be imputed to the fact that, until the deployment of DOIs by GLIS, there has been no practical solution to accurately and permanently reference a sample of crop germplasm across information systems. During the last 3 years, DOIs have addressed the issues arising with locally assigned identifiers that may cause

```
LOCUS       DQ884074                 257 bp    mRNA    linear   EST 24-FEB-2011
DEFINITION  DQ884074 Oryza sativa (indica cultivar-group) cv. IR64 cDNA-AFLP
            fragment Oryza sativa Indica Group cDNA clone 51_9b, mRNA sequence.
ACCESSION   DQ884074
VERSION     DQ884074.1
DBLINK      BioSample: SAMN00165158
KEYWORDS    EST.
SOURCE      Oryza sativa Indica Group (long-grained rice)
  ORGANISM  Oryza sativa Indica Group
            Eukaryota; Viridiplantae; Streptophyta; Embryophyta; Tracheophyta;
            Spermatophyta; Magnoliopsida; Liliopsida; Poales; Poaceae; BOP
            clade; Oryzoideae; Oryzeae; Oryzinae; Oryza; Oryza sativa.
REFERENCE   1  (bases 1 to 257)
  AUTHORS   Ventelon-Debout,M., Tranchant-Dubreuil,C., Nguyen,T.-T.-H.,
            Bangratz,M., Sire,C., Delseny,M. and Brugidou,C.
  TITLE     Rice Yellow Mottle Virus stress responsive genes from susceptible
            and tolerant rice genotypes
  JOURNAL   BMC Plant Biol. 8, 26 (2008)
  PUBMED    18315879
COMMENT     Contact: Ventelon-Debout M
            UMR 5096
            Institut de Recherche pour le Developpement
            911 avenue d'Agropolis, BP54501, Montpellier, 34394, France.
FEATURES             Location/Qualifiers
     source          1..257
                     /organism="Oryza sativa Indica Group"
                     /mol_type="mRNA"
                     /cultivar="IR64"
                     /db_xref="taxon:39946"
                     /clone="51_9b"
                     /clone_lib="SAMN00165158 Oryza sativa (indica
                     cultivar-group) cv. IR64 cDNA-AFLP fragment"
ORIGIN
        1 cgaacatggc cccaccagcc attggcttga aagtgttgtt gctctagcaa ccatcatgga
       61 agctagttga ccattagcaa ttggtactaa aagagccttg aaaagagcaa gcaagagagg
      121 gggaaacagc gatcggagta ctcggacagc tgattgctcg atctccaacg cactctttcc
      181 cttgaccatg gggcggttgg caacttcaat cagtccagtg tagccaggct tttcagaatc
      241 ccatccaacc tcctaca
//
```

Fig. 2 Example of an International Nucleotide Sequence Database Collaboration (INSDC) Accession record

collisions when taken out of the assigning institution's context. Through GLIS, assigning DOIs to plant genetic resources is a rapid process that can be performed in a variety of ways from a simple web form with a handful of mandatory attributes to powerful XML-based, system-to-system messaging.

Following the principle of requiring minimal changes to existing systems while maximizing advantages for users, one practical pathway of integration between repositories of samples of source germplasm and ISNDC genomics repositories would be to mention the original material's DOI in the "source" feature of the Accession record establishing a proper link between DSI data sets and the original material. As genomics researchers may sometimes have to multiply source genetic material that is provided in insufficient quantity, with the risk that the genetic identity of the resulting material may be altered, a new DOI could be assigned to the material that is actually sequenced. Such a new DOI would be related to the material received, thanks to specific GLIS features, and would be cited in the Accession Number. The potential of this DOI feature is clearly not limited to the INSDC data ecosystem. It could also be deployed to establish permanent relationships among multiple data sets that the definition of DSI may comprise, and between those aggregate data sets and source genetic material.

In the INSDC scenario, when displaying the Accession detail page, the system could detect the DOI in the "source" feature and transform it, through a trivial string manipulation, into a URL to the doi.org resolver leading to the landing page associated with it. This mechanism would work irrespective of the DOI being assigned by GLIS or by any other authority and irrespective of it being associated with a plant or other lifeform. This simple transformation would already significantly improve the user experience and add real value to the Accession record.

While this minimalist approach may benefit some INSDC users, it may need complementation for other user communities that GLIS serves. In this perspective, the link to the INSDC Accession could also be provided in the GLIS DOI detail page, as shown in Fig. 3 below.

Besides providing passport information, GLIS collects links to websites where additional information on the PGRFA can be found and maintains a graph showing how the material was obtained, as illustrated in Fig. 4 below where the nodes are the DOIs associated to the materials and the arcs are the relationships linking each node to its progenitor(s). It also lists publications and datasets citing the PGRFA's DOI. This feature is based on the EventData service, jointly developed by Crossref and DataCite,[11] and allows for automatic discovery of publications and datasets citing the current DOI.

Ideally, should INSDC opt to assign DOIs to its Accessions and properly cite the DOIs reported in the "source" feature, the link to the INSDC Accession would automatically appear in the GLIS landing page for that material thanks to Event Data. In turn, INSDC could directly benefit from Event Data services to discover publications and datasets citing the Accession's DOI.

---

[11] https://www.datacite.org

Fig. 3 Global Information System (GLIS) DOI landing page

4 Introducing Digital Genetic Objects for Precision of Definition and Interoperability of DSI

GLIS is an enabler of a global architecture for accessing and sharing germplasm and related information. The GLIS and associated DOIs provide an approach for accurately and permanently referencing crop germplasm across information systems and, as noted above, integration of DOI into the INDSC architecture may offer tangible benefits. Currently, GLIS DOIs are assigned primarily at the genebank accession level, which can contain significant – and in many cases undiscovered – genetic diversity. An incremental option to improve interoperability relies on a finer

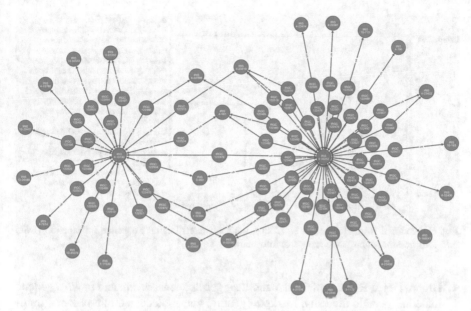

Fig. 4 Example graph displaying genetic lineages of a set of rice germplasm accessions from the International Rice Research Institute (IRRI)

definition and mechanisms for linking data so that GLIS DOIs can be assigned to associated DSI in a scalable and interoperable fashion. DGOs can provide this linkage.

As has been noted, there is a diversity of views of what comprises DSI. In 2018, the Ad-Hoc Technical Committee (AHTEG) of the Convention on Biological Diversity considered the following information (Convention on Biological Diversity, 2018):

1. The nucleic acid sequence reads and the associated data;
2. Information on the sequence assembly, its annotation and genetic mapping;
3. Information on gene expression;
4. Data on macromolecules and cellular metabolites;
5. Information on ecological relationships and abiotic factors of the environment;
6. Function, such as behavioral data;
7. Structure, including morphological data and phenotype;
8. Information related to taxonomy;
9. Modalities of use.

In preparation for a new meeting of the AHTEG, four possible cumulative groups of information were categorized (Houssen et al. *cit.*):

1. Narrow: DNA and RNA
2. Intermediate: DNA, RNA and proteins
3. Intermediate: DNA, RNA, proteins and metabolites

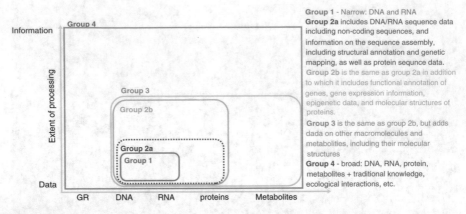

Fig. 5 Modified from Houssen et al. (2020), who clustered Digital Sequence Information (DSI) into four possible cumulative groups of information

4. Broad: DNA, RNA, protein, metabolites, germplasms, *in situ* and *in vitro* genetic material, genetic diversity, markers (genetic and molecular), microbiome, traditional knowledge, ecological interactions (Fig. 5).

In 2020, the AHTEG considered the first three groups as possibly constituting DSI and excluded the fourth group (Convention on Biological Diversity, 2020). We argue that a new identifier for DSI in information systems – DGOs – can help both the ABS and genomic science communities manage the complexity of DSI at the level of multiple data sets and association with specific PGRFA and be equipped with new approaches to facilitate implementation of policy decisions on the scope of DSI.

DGOs are knowledge objects created to precisely describe distinct types of DSI, objects that can be assigned GLIS DOIs and also annotated using community-driven reference vocabularies and ontologies to link to wider bodies of knowledge. Such an approach would accommodate narrow (e.g. just DNA or RNA) or broad (e.g. incorporating traditional knowledge or ecological interactions at organism, population and systems-level) definitions of DSI, and facilitate easier flow of data and knowledge across the spectrum of potential definitions.

For material for which there is an associated DOI, discrete DGOs may be created for each unique type of DSI, falling roughly within groups 1–3 of the scheme above. Each DGO can in turn then be assigned a DOI linked to the accession DOI, facilitating discovery of the associated data via GLIS and other international information systems such as INSDC. DGOs can link to a diversity of, and facilitate discovery between, bodies of knowledge related to even the broadest interpretations of what comprises DSI via data annotation leveraging reference ontologies and vocabularies. This approach points the way to describing these data in terms of their agronomic, environmental, phenotypic characteristics, and placing them more precisely in time and space to facilitate broad discovery and use of these data.

The precision of definition made possible by DGOs and the ability to link these knowledge objects to both the GLIS system through DOIs and to other bodies of knowledge through ontology-derived annotation can enable a cross-cutting 'interoperability layer' linking systems and existing data standards across operational domains. DGOs can represent an accelerator of scientific discovery and enhancement to public information systems through data interoperability, meeting needs of both the ABS and scientific communities.

5 Benefits and Possible Roadblocks

Despite the low investment required and the significant benefit for users of the technical options presented above, the experience with GLIS DOIs shows that there would be roadblocks to consider. First and foremost, user motivation to consistently adhere to the new workflow based on citing the original material's DOI, including assigning a new DOI to the original material if necessary, would be increased by the immediate advantage of being able to access at least passport data available through GLIS. However, awareness will have to be raised about this new approach and its advantages.

GLIS has also experienced some unexpected setbacks when dealing with publications and datasets, which would reverberate into the INSDC association. For technical reasons, most publisher systems have difficulties in properly handling *data citation*, i.e. referencing DOIs not associated with bibliographic references, such as GLIS DOIs. The current solution is to list GLIS DOIs among the bibliographic references but this encounters some resistance by editors because those "references" look odd, lacking traditional elements such as title, publisher and so on. Dataset repositories, on the other hand, implement heterogeneous practices: some support *data citation* properly while others do not.[12]

The DGO solution would have to resolve technical challenges. As the approach outlined in Fig. 6 could generate many thousands of DGOs, this will require not only precision of definition but also some operational decisions about when and how they are assigned, how to manage and store the associated data. Reference ontologies have not fully been used in this way, and would need to be fine-tuned. One initiative step will be to create a DGO ontology based on the diversity of discrete types of DGOs in groups 1–3 of DSI. Another will be to examine related reference ontologies and their suitability to linking to DGOs. In some cases, they will need to be fine-tuned to link to the material.

The pilot integration between INSDC and GLIS would pave the way for other DSI repositories towards a proper relationship between DSI and the original material. The cost/benefit ratio would be very small and would greatly improve science

---

[12] https://www.crossref.org/blog/data-citation-what-and-how-for-publishers and https://www.crossref.org/blog/why-data-citation-matters-to-publishers-and-data-repositories

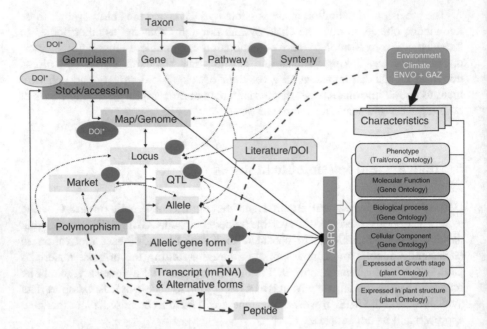

Fig. 6 Digital Genetic Objects (DGO; Dark blue colored bubbles) are knowledge objects that can be assigned to distinct types of DSI, allowing more precise definition for each, their semantic relationships and derivations. DGOs can be annotated and using relevant ontologies, as well as assigned DOIs, and both methods would serve to connect DSI to wider bodies of knowledge. Currently only literature-based DOI citations hold all the unstructured information in the natural language form in the published articles. A majority carry incomplete or insufficient information and metadata to build semantic relationships between various DGOs

and the life of users of both systems. Arguably, it would also motivate users to properly reference the original material. To date, the "source" block is not much populated, likely because there is little added value in referencing the original material in a non-actionable, potentially inaccurate way.

Once interoperability with INSDC is achieved, it would be a potent success story for future extensions to any other type of database or repository of information associated to plant genetic resources registered in GLIS, leading to a coordinated constellation of systems on, for example, phenomics, traditional knowledge and technologies.

Data interoperability resulting from successful application of DGOs could provide new linkages of DSI to information systems for genetic materials, and form the basis of interoperability across research and operational domains to help build more integrated research insights and analytic infrastructures for accelerating discovery and use. Some potential high-value use-cases supported by increased data interoperability include:

- accelerating understanding of genetic diversity within genebanks, through a cross-cutting data standard for describing results from diversity studies;
- increased accretion of knowledge related to the material from other domains such as breeding or on-farm research;
- revealing duplication in collections and informing the "right" level of duplication in light of long-term commitments for preservation of the genetic material;
- precision of definition supporting data integration, in turn helping to bridge research and operational domains;
- eased ability to compare data from multiple sequences, a key way to enhance their value (Laird & Wynberg, 2018);
- easier linkage of data on genetic discovery generated with newer forms of measurement (e.g. multispectral imagery) and linkages of associated databases;
- linking DSI to wider bodies of ontologized knowledge;
- improved access to data on the complex interactions between genomics, environment, and management practices—critical for predictive modeling.

The solutions discussed in this paper in relation to plant genetic resources could apply to other biological domains (e.g. microbes, fungi, land and aquatic animals and other eukaryotes) in the INSDC collection and beyond. Establishing the connection between INSDC Accessions and the corresponding biological materials as well as applying DGOs to link DSI across taxonomic groups could be of increasing importance for synthetic biology (Rohden et al. *cit.*) and facilitate study of horizontal gene transfer.

In addition to these potential benefits supporting use of DSI and materials by the scientific community, DGOs appear to provide key capabilities in support of issues for the ABS community noted earlier: greater precision of definition can help with fine-tuning the *terminology* associated with DSI. The ability to annotate DSI (via DGOs) with diverse bodies of knowledge makes it more possible to accommodate very narrow or broad views on the *scope* of DSI. The ability to link precise defined and well-described data is a necessary precondition for improving overall *traceability* of data and the associated materials. Data standards, however, are only as good as their use in information systems, by stakeholder communities, and complex institutional contexts. Concrete pilots will be needed to test the viability of DGOs at the intersection of these dimensions.

6   Conclusion: A Common Pathway Between Global Policy on Genetic Resources and Information Technology and Data Science

The consideration of DSI by ABS policymakers requires a harmonious relationship with the genomics science community. We postulate that data aggregation and interoperability are fields where the much-needed reciprocal adjustment in processes and the blending of different rules may occur (Leonelli, 2019). Given the value of associating DSI with source genetic material, interoperability solutions should be tested based on existing genetic resource information systems. In this paper, we have suggested interoperability solutions between GLIS and INSDC as well as the introduction of DGOs into biological data systems. The insertion of DOIs into the "source" feature of the INSDC Accession record would enable relationships with passport data and other information on plant genetic resources. DGOs would further improve interoperability through a finer definition of DSI component parts and mechanisms for linking data across research and operational domains. In conjunction with these technical features, the interoperability solutions proposed in this paper would enable the smooth association of genetic resources and data in multiple repositories with applicable legal regimes governing their use.

As far as the international genomics science community that routinely manages the genetic material and the data is open to learn, develop and adopt best practices, and the genetic resources policy community seeks dialogue and cooperation, the opportunity to test and refine the two suggestions made may exist.

At the practical level, if the proposals of this paper are broadly acceptable to the scientific community, engagement with the communities maintaining relevant ontologies, metadata standards for genetic and genomic data, and annotation tools would be advisable in order to study a functional definition of DSI out of the existing ontologies, and identify gaps in both metadata and semantics in order to support interoperability of the annotated data as well as facilitate the alignment with multiple ABS policy options. Governance and oversight of this experimental system would require careful consideration in order to pursue implementation of interoperability not only as syntactic or semantic levels through data formats and communication protocols, but also as cross-domain, so to include social, policy and organizational aspects that impact on the performance of the information technology systems. This proposition resonates with the emphasis made in other chapters of this book on the key role of governance in structuring transdisciplinary collaborations across academic and non-academic communities (Louafi et al. this volume; Devare et al. this volume).

The proposals of this paper may just be one small step towards building new global standards for access and exchange of plant genetic resources and plant sequence data. Mindful of both technical opportunities and governance challenges, our hope is that this paper will be conducive to experimenting interoperability in both its technical and social dimensions, and thus represent a factual contribution in the direction of long-term alliances between policy and science through data archives, knowledge bases and live specimen collection resources.

Disclaimer This publication reflects the technical opinions of its authors, which are not necessarily those of the respective organizations of affiliation.

References

Arnaud, E., Laporte, M. A., Kim, S., Aubert, C., Leonelli, S., Cooper, L., Jaiswal, P., Kruseman, G., Shrestha, R., Buttigieg, P. L., Mungall, C., Pietragalla, J., Agbona, A., Muliro, J., Detras, J., Hualla, V., Rathore, A., Das, R., Dieng, I., & King, B. (2020). The ontologies community of practice: An initiative by the CGIAR platform for big data in agriculture. *Patterns, 1*, 100–105.

Aubry, S. (2019). The future of digital sequence information for plant genetic resources for food and agriculture. *Frontiers in Plant Science*. https://doi.org/10.3389/fpls.2019.01046

Convention on Biological Diversity. (2020). *Report of the Ad Hoc technical expert group on digital sequence information on genetic resources*. https://www.cbd.int/doc/c/911e/cc8b/de7d7fba3 a8374ba4a2fbf53/dsi-ahteg-2020-01-07-en.docx. Accessed 28 Sep 2021.

Brink, M., & van Hintum, T. (2021). Practical consequences of digital sequence information (DSI) definitions and access and benefit-sharing scenarios from a plant genebank's perspective. *Plants, People, Planet*. https://doi.org/10.1002/ppp3.10201

Chen, K., Wang, Y., Zhang, R., Zhang, H., & Gao, C. (2019). CRISPR/Cas genome editing and precision plant breeding in agriculture. *Annual Review of Plant Biology, 70*(1), 667–697.

Convention on Biological Diversity. (2018) *Report of the Ad Hoc technical expert group on digital sequence information on genetic resources*. https://www.cbd.int/doc/c/7ea1/36b3/7ccf84 9897a4c7abe49502b2/sbstta-22-inf-04-en.pdf. Accessed 28 Sep 2021.

Halewood, M., Lopez Noriega, I., Ellis, D., Roa, C., Rouard, M., & Hamilton, R. S. (2018). Using genomic sequence information to increase conservation and sustainable use of crop diversity and benefit-sharing. *Biopreservation and Biobanking*. https://doi.org/10.1089/bio.2018.0043

Hartman-Scholz, A., Hillebrand, U., Freitag, J., Cancio, I., dos S. Ribeiro, C., Haringhuizen, G., Oldham, P., Saxena, D., Seitz, C., Thiele, T., & van Zimmeren, E. (2020). *Finding compromise on ABS & DSI in the CBD: Requirements & policy ideas from a scientific perspective*. https://www.dsmz.de/fileadmin/user_upload/Collection_allg/Final_WiLDSI_White_Paper_Oct7_2020.pdf. Accessed 28 Sep 2021.

Houssen, W., Sara, R., & Jaspars, M. (2020). Digital sequence information: Concept, scope and current use. In *Digital sequence information: Concept, scope and current use*. https://www.cbd. int/doc/c/fef9/2f90/70f037ccc5da885dfb293e88/dsi-ahteg-2020-01-03-en.pdf.   Accessed   on 28 Sep 2021.

Laird, S., & Wynberg, R. (2018). A fact-finding and scoping study on digital sequence information on genetic resources in the context of the convention on biological diversity and the Nagoya protocol. In *Fact-finding and scoping study on digital sequence information on genetic resources in the context of the convention on biological diversity and the Nagoya protocol*. https://www.cbd.int/doc/c/079f/2dc5/2d20217d1cdacac787524d8e/dsi-ahteg-2018-01-03-en. pdf. Accessed 28 Sep 2021.

Leonelli, S. (2019). Data – From objects to assets. *Nature, 574*, 317–320.

Morgera, E., Switer, S., & Geelhoed, M. (2020). *Study for the European Commission on 'Possible Ways to Address Digital Sequence Information – Legal and Policy Aspects'*. https://ec.europa. eu/environment/nature/biodiversity/international/abs/pdf/Final_study_legal_and_policy_aspects.pdf. Accessed 28 Sep 2021.

Oliver, C. (1991). Strategic responses to institutional processes. *The Academy of Management Review, 16*, 145–179.

Randall Henning, C., & Pratt, T. (2020). *Hierarchy and differentiation in international regime complexes: A theoretical framework for comparative research*. https://www.peio.me/wp-content/uploads/2020/01/PEIO13_paper_66.pdf. Accessed 28 Sep 2021.

Rohden, F., Huang, S., Dröge, G., & Hartman-Sholz, A. (2020). Combined study on Digital Sequence Information (DSI) in public and private databases and traceability. In *Combined study on digital sequence information in public and private databases and traceability*. https://www.cbd.int/doc/c/1f8f/d793/57cb114ca40cb6468f479584/dsi-ahteg-2020-01-04-en. pdf. Accessed 28 Sep 2021.

Rohden, F., & Hartman-Sholz, A. (2021). The international political process around Digital Sequence Information under the Convention on Biological Diversity and the 2018–2020 intersessional period. *Plants, People, Planet*. https://doi.org/10.1002/ppp3.10198

Ruiz, M. (2015). *Genetic resources as natural information: Implications for the convention on biological diversity and Nagoya protocol*. Routledge.

Smyth, S. J., Macall, D. M., Phillips, P. W. B., & de Beer, J. (2020). Implications of biological information digitization: Access and benefit sharing of plant genetic resources. *J World Intellect Prop, 23*, 267–287. https://doi.org/10.1111/jwip.12151

Toronto International Data Release Workshop Authors. (2009). Prepublication data sharing. *Nature, 461*, 168–170.

Welch, E., Bagley, M., Kuiken, T., & Louafi, S. (2017). *Potential implications of new synthetic biology and genomics research trajectories on the international treaty on plant genetic resources for food and agriculture*. https://doi.org/10.2139/ssrn.3173781

Welch, E., Taggart, G., Feeney, M. K., & Siciliano, M. (2019). Navigating the labyrinth: Academic scientists' responses to new regulatory controls on biological material inputs to research. *Environmental Science & Policy, 110*, 136–146. https://doi.org/10.1016/j.envsci.2019.08.001

Wellcome Trust. (2003). *Sharing data from large-scale biological research projects: A system of tripartite responsibility*. http://www.genome.gov/Pages/Research/WellcomeReport0303.pdf. Accessed 28 Sep 2021.

Wilkinson, M., Dumontier, M., Aalbersberg, I., et al. (2016). The FAIR Guiding Principles for scientific data management and stewardship. *Sci Data, 3*(160018). https://doi.org/10.1038/sdata.2016.18

Governing Agricultural Data: Challenges and Recommendations

Medha Devare, Elizabeth Arnaud, Erick Antezana, and Brian King

Abstract The biomedical domain has shown that *in silico* analyses over vast data pools enhances the speed and scale of scientific innovation. This can hold true in agricultural research and guide similar multi-stakeholder action in service of global food security as well (Streich et al. Curr Opin Biotechnol 61:217–225. Retrieved from https://doi.org/10.1016/j.copbio.2020.01.010, 2020). However, entrenched research culture and data and standards governance issues to enable data interoperability and ease of reuse continue to be roadblocks in the agricultural research for development sector. Effective operationalization of the FAIR Data Principles towards Findable, Accessible, Interoperable, and Reusable data requires that agricultural researchers accept that their responsibilities in a digital age include the stewardship of data assets to assure long-term preservation, access and reuse. The development and adoption of common agricultural data standards are key to assuring good stewardship, but face several challenges, including limited awareness about standards compliance; lagging data science capacity; emphasis on data collection rather than reuse; and limited fund allocation for data and standards management. Community-based hurdles around the development and governance of standards and fostering their adoption also abound. This chapter discusses challenges and possible solutions to making FAIR agricultural data assets the norm rather than the exception to catalyze a much-needed revolution towards "translational agriculture".

M. Devare (✉)
International Food Policy Research Institute (IFPRI), Washington, DC, USA
e-mail: m.devare@cgiar.org

E. Arnaud · B. King
Alliance of Bioversity International and CIAT, Rome, Italy
e-mail: e.arnaud@cgiar.org; b.king@cgiar.org

E. Antezana
Department of Biology, Norwegian University of Science and Technology (NTNU), Trondheim, Norway

Bayer CropScience SA-NV, Diegem, Belgium
e-mail: erick.antezana@gmail.com

1 Background

The COVID-19 pandemic has caused great human loss and economic suffering worldwide, but it may prove to be a ground-breaking model for agile collaborative science. This is exemplified by rapid and powerful approaches to data sharing, including the COVID-19 Open Research Dataset (Semantic Scholar, 2020) and mature biomedical ontologies (Bodenreider, 2005; Robinson & Haendel, 2020). The COVID-19 Open Research Dataset by the Allen Institute for Artificial Intelligence in collaboration with several research institutes gave researchers free and open tools and data to develop new insights about the novel coronavirus. The shared standards and data coupled with the collaborative application of massive computing power enabled research efforts worldwide to model and identify over 70 promising compounds for treatment in just under 2 days—a result that would otherwise have likely taken years (Quitzau, 2020). This is a shining example of *in silico* analyses over vast data pools enhancing the speed and scale of scientific innovation that may also be applied towards agricultural research and guide similar multi-stakeholder action in service of global food security (Streich et al., 2020).

Responding agilely and hyper-locally to challenges in the agricultural sector necessitates building on prior research. While much of the conversation in the agricultural research for development sector focuses on the need to appropriately scale promising solutions, these solutions must also be agile in responding to changing local conditions, be they weather, markets, or others. This, in turn, requires decision support tools that mine problem-relevant open pools of data, and data products that not only meet the Findability and Accessibility ("Open Access") criteria of the FAIR Data Principles but are also interpretable and reusable by humans as well as machines (Thessen & Patterson, 2011; Wilkinson et al., 2016). The biomedical sector began coalescing around the need for open, interoperable and machine-readable data by the mid-1980s to early 1990s with the creation of powerful open databases, standards and toolkits under the aegis of the National Center for Biotechnology Information (Smith, 2013). NCBI paved the way for rapid data-driven, transparent development of therapies and medical innovation.

In comparison, the agricultural sector has lagged in making data assets open and interoperable, with the possible exception of precision agriculture and work involving genetic and "omics", and technologies such as those related to developing plant germplasm or insect pest detection. Agriculture has moved in this direction only in the last few years (Smalley, 2018) partly because data assets still too often exist on individual laptops. Even when data is accessible on public repositories, it has traditionally been summary tables or metadata, rather than the raw and well-described data needed for analyses and further innovation. Further, where such data has gradually become available over the last 5–7 years, it tends to be opaquely annotated – if at all – and not interoperable or easily reusable as data variables are not described using standards, but typically by individual choice. Private sector has been increasingly amassing and mining location-specific agricultural data since the early to mid-2000s through Internet of Things (IoT), Big Data, AI, Blockchain and allied technologies in the service of precision agriculture and smart farming solutions (Rijmenam, 2013; Noyes, 2014; Pham & Stack, 2018). However, much of this data remains proprietary, and responsive only to – at best – company-specific standards and bespoke tools, making governance (including ownership) and linking of relevant but disparate data difficult (Rosenbaum, 2010).

It is only recently that agricultural public sector entities and researchers – and more importantly, their funders – are beginning to acknowledge the importance of data standards, and to specify open licenses and FAIR requirements (European Commission Expert Group on FAIR Data, 2018; Bill and Melinda Gates Foundation, 2021). CGIAR (https://www.cgiar.org/), the world's largest global agricultural innovation network launched the Gates Foundation-supported Open Access, Open Data Initiative in 2015 to facilitate culture change and technological support for open research outputs across the 15 globally-dispersed CGIAR agricultural research for development centers. The initiative built on the ratification of CGIAR's Open Access and Data Management Policy (CGIAR, 2013), and the momentum of this effort continued with greater emphasis on FAIR data through the Platform for Big Data in Agriculture (https://bigdata.cgiar.org/) which began in 2017. The Platform's work has resulted in a number of open tools and services, a revised Open and FAIR Data Assets Policy (CGIAR, 2021), and capacity enhancement to support FAIR research outputs.

There are several ongoing efforts to build knowledge bases and open data portals, including by CGIAR (the GARDIAN data ecosystem), the European Union (-European Data Portal), the United States Department of Agriculture (Ag Data Commons), and similar databases of compilations maintained by a number of research, academic, and funding entities in the agricultural space. These three exemplars explicitly pursue FAIRness, through alignment with established metadata schemas and semantic standards such as controlled vocabularies and ontologies to describe data variables. Such approaches enable mining and linking of data (e.g., as Linked Open Data), but adherence to standards remains challenging and is still elusive for a variety of reasons.

With the exception of bioinformaticians in fields like crop breeding or germplasm diversity studies, researchers encounter several hurdles to the adoption of data standards, and these are particularly entrenched in "non-digital natives". The challenges include limited awareness on how to mine and derive value from standards-compliant, interoperable data pools; limited data science capacity for *in silico* analyses, with a related emphasis on the collection, rather than reuse of existing data; and limited fund and time allocation towards data management and the collaborative development of standards. Other, more community-based issues relate to the collective development and governance of standards, and to coalescing "critical mass" around consistent adoption. Thus, while FAIR data assets are foundational needs for an evidence-driven, agile, and collaborative approach to enhancing the impact of research and development in the agricultural domain, the discipline is in its infancy in realizing the potential of consistent application of the FAIR Principles. Throughout an institution or set of entities in a disciplinary domain, the consistent adoption of the FAIR Principles and associated data standards and approaches relies on good governance (Koers et al., 2020).

But what exactly does "good governance" mean? It may be useful to first frame how we view this idea in the context of data, in line with Stedman and Vaughan's recent writing (2020), that defines data governance to be a cross-cutting concern to assure success across the data life cycle. Thus, the availability, usability, security, and trustworthiness of data are all dependent on its governance, which also includes development and oversight of data standards, policies, and compliance with these. This paper discusses governance challenges and possible solutions to enabling interoperability of agricultural data assets as a critical requirement in catalyzing a move from prescriptive, "one size fits all" recommendations, to more site-specific options that are agilely developed in response to local constraints and scenarios.

2 Challenges and Solutions

Effective operationalization of the FAIR Principles towards agricultural research data assets that support easier interpretation and linking requires as a foundational paradigm that researchers accept that their responsibilities do not end with data collection and manuscript publishing, as was the norm in a pre-digital age. As stated by Wilkinson et al. (2016), data-intensive science increasingly means "...assisting both humans and their computational agents in the discovery of, access to, and integration and analysis of task-appropriate scientific data and other scholarly digital objects." The reach of research therefore extends not just to data collection for personal analysis and publishing, but to stewarding or resourcing the stewardship of data assets to ensure long-term preservation, wide access and reuse. The development and adoption of common standards embodied by metadata schemas, ontologies and controlled vocabularies are critical to good data stewardship and reuse but developing and maintaining these efforts in agricultural research has been difficult. Despite data sharing and reuse being more accepted in other domains including the environmental and biomedical, the consistent use of standards is spotty even in these domains. For instance, in a survey of 100 ecological and evolutionary research datasets over half of the databases had issues including missing metadata; 64% were archived in a way that rendered reuse partially or entirely impossible due to poor or missing metadata, and/or non-machine-readable formatting (Roche et al., 2015).

2.1 Research Culture and How Researchers Understand Scientific Inquiry

The more traditional view of science is a prediction-based, hypothesis-driven approach as articulated by Karl Popper in 1963 (Brockman, 2015). Although this view is no longer central to some scientific domains, it remains quite relevant in agriculture. A consequence is that data is considered as more a by-product than a driver of research, and its governance, defined by Leonelli (2019) to be "...the strategies and tools employed to identify, manage, and disseminate data..." is typically not sufficiently valued or resourced. Leonelli challenges the traditional view of data as fixed and context-independent, and the notions of data quality and reliability as universal rather than influenced by context and purpose. The author's relational view of data (Fig. 1) argues instead that the presentation, selection, and use of data based on purpose and context is critical to knowledge creation. Thus, this relational view posits that data are often altered through production, dissemination,

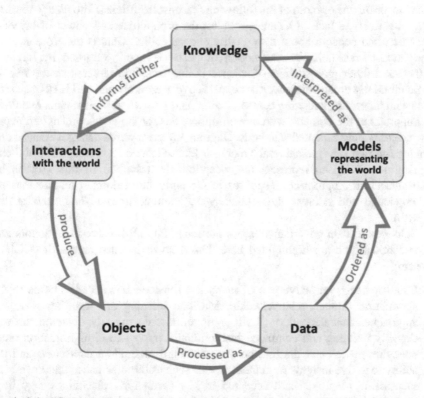

Fig. 1 Scientific inquiry according to the relational view of data (Leonelli, 2019). This mutually reinforcing view includes interactions of scientific subjects with the world, which produce objects that are documented as data. These data are managed and visualized to produce models that represent particular phenomena, leading to the creation of knowledge that can in turn inform future inquiry. (Reproduced without modification from Leonelli (2019), under CC-BY 4.0 licence)

and reuse for different purposes, imbuing their handling and management with more importance. Such a relational view is very relevant to the modern reality of digital technologies and capabilities, and particularly true for agricultural research – which necessitates context-based re-purposing of data.

Agricultural research culture is also influenced by the fact that it is traditionally field-based, involving time-consuming data gathering from experiments that typically run over several seasons/years and are generally conducted along the lines of Popper's falsification-based hypothesis testing. Among the few exceptions to this, though relatively recent, are climate science, precision technologies, and disciplines like genomics in, say, germplasm development. Until recently, agricultural research rewarded those with strong field know-how and ability to employ the Popperian method over more quantitative or digital smarts, resulting in a culture of "my research, my data". Our experience at CGIAR suggests that except for a few (e.g., geneticists, bioinformaticians, and the rare agronomist), the notion and use of *in silico* analysis involving secondary data is relatively new for agricultural scientists. In keeping with this, Denk (2017) suggests that researchers' reluctance to use open data hinges on one or more of the following reasons: Insufficient knowledge to mine data effectively, a lack of awareness about the capabilities and power of big data analytics, and concern about data quality and reliability. Data is therefore seen as peripheral to research, and the notion of "data-centrism" espoused by Leonelli (2019) and other philosophers of science is the exception rather than the rule in agricultural research. Data governance, particularly around open and FAIR research data with its goal of widening access, mining, and reuse therefore remains relatively unimportant in the domain, with direct implications for the development and maintenance of widely accepted standards. Ongoing efforts towards data governance and linking through the International Treaty on Plant Genetic Resources for Food and Agriculture (PGRFA) represent an exception, as described in this volume by Manzella et al. Appropriate responses in the agricultural domain require that we acknowledge and address these challenges, learning from efforts such as the PGRFA.

Solutions to data governance issues are manifold and involve many actors and approaches. Some are highlighted here, based on experience across the CGIAR system:

- Data science is an active part of many life sciences areas but has come to the agricultural domain relatively late. Machine learning and big data analytics approaches that depend on FAIR agricultural data must be fostered through capacity building and continued institutional support and hiring/retention practices that make clear the link between standards-compliant data pools and the ability to derive insights from them. Fields such as bioinformatics that have been successfully deployed and accepted in key agricultural disciplines may be a model to follow, and indeed, the notion of "ag informatics" now exists.
- The adoption of best practices throughout the data life cycle including the use of standards that enable data aggregation should be an expected part of agricultural research, with high value assigned to contributions toward strong data outcomes.

Clarity around open and FAIR, and associated data schemas and standards must be part of contractual language for new hires. KPIs that explicitly acknowledge FAIR data and data-driven science and innovation should form part of researcher annual evaluations. As efforts around standards development, maintenance, and use require funding, allocation of budget towards best practices in data management that includes these aspects should be required, not recommended practice (10–25% is suggested by many project funders, including the EU). Together with these, data stewards must be valued and empowered for success.

- Data sharing requirements that specify repository, data, and allied standards must be implemented, ideally via data sharing templates and checklists that facilitate consistency across research units and institutions – and their partners – easing governance considerations. Addressing ownership issues by democratizing data authorship and upload to standards-compliant repositories is likely to be a foundational aspect of buy-in to these.

- Robust institutional data policy and strategy frameworks are crucial to prioritizing open and FAIR data, and formalizing many of the above points, yet several academic and research entities in agriculture lack these, thereby missing the opportunity to effectively prioritize and leverage a strengthened open and FAIR data culture. A case in point is the 76 Land Grant Universities (LGUs) in the United States, set up in 1862 to focus on curricula in practical agriculture, life sciences, and other disciplines. Most of the LGUs have no explicit policy governing open data sharing, with recommendations urging exploration of the relative advantages of selective commercialization vs. fully open access approaches to advance science and support for sustained investment in research and development (Barham et al., 2017). Uncertain or missing policy/strategy means that researchers are not held to expectations relating to data stewardship. It makes governance related to linking across multi-disciplinary agricultural data challenging within any institution, let alone across the LGUs and beyond. Where data policies do exist, few explicitly require the consistent use of data standards.

- Research funders and publishers play a key role in changing institutional data culture towards openness and FAIRness. Funders who require open and FAIR data to be shared in specified time frames along with publications, and who hold grantees accountable for this are crucial catalysts of culture change regarding data sharing and reuse. Although data journals are cropping up rapidly and the sharing of data underlying publications is increasingly expected by scientific publishers, this is still not the norm even in the biomedical realm. A study by Vasilevsky et al. (2017) indicated that just under 40 of 318 biomedical journals explicitly required data sharing as a prerequisite for publication.

It is important to note that a key reinforcer of open and FAIR data sharing is the "re-examination" of data, either for quality and/or reuse in new analyses. Without such benefits, the carrots and sticks outlined above may only result in partial success. This idea and several of those above are summarized in Fig. 2, from a 2020 manuscript by Sielemann et al.

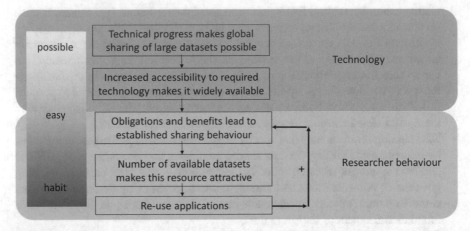

Fig. 2 The evolution of data sharing behavior. (Reproduced without modification from Sielemann et al. (2020), under CC-BY 4.0 licence. https://doi.org/10.7717/peerj.9954/fig-1)

2.2 Governance Issues and Repercussions Around Data and Data Standards

Technical challenges to governance towards greater openness and interoperability of data (which standards confer) are generally easier to address than those that are cultural, or subject to the legal frameworks of countries or rights of stakeholders (Sara & Devare, 2020). The latter may include intellectual property rights, confidentiality and/or privacy, farmers' rights, sensitivity (e.g., sensitive information relating to, say, harvesting forest species), farmers' rights and privacy (see also Leonelli and Williamson, this volume; Zampati, this volume).

Research data scenarios most likely to require robust governance frameworks include those that:

- Concern vulnerable peoples (including indigenous communities);
- Contain personally identifiable information that could be used to identify individuals or communities;
- Include anonymized data in which re-identification could result in significant harm;
- Concern genetic resources (including Digital Sequence Information) and any associated traditional knowledge;
- Include sensitive political data (including weather or health-related data, which in in some countries is subject to formal or informal reporting restrictions).

Governance arrangements in the above scenarios require due diligence in how the data is described and managed, acknowledging and addressing restrictions that may arise due to the need for:

Prior Informed Consent Human subject data is typically subject to ethical standards requiring approval of an oversight body (such as an Internal Review Board), and prior informed consent from research participants which is purpose-specific. Prior informed consent also features prominently in the context of restricted use of data, privacy protection, and ABS compliance (see below).

Restricted Use of Data Including Commercialization and/or Commercial Use of Data Use of data in a manner inconsistent with the informed consent or contractual obligations under which it was obtained can have legal as well as reputational repercussions. Accordingly, it must be proactively handled subject to appropriate data protection measures.

Proprietary, Commercially Sensitive and or Confidential Data Public disclosure of data that is proprietary, commercially sensitive or confidential in nature can have legal as well as reputational repercussions, and must be subject to robust data protection measures.

IP and Contractual Rights Over the Data and Results or Innovations Generated Using the Data Access and use of data may be subject to intellectual property and contractual rights governing the use of data as well as derivatives of the data (e.g., CC-BY-SA and other licenses requiring share-alike terms) and downstream products developed using the data.

Privacy Protection and Human Subject Rights Personal data (i.e., directly identifying data) or data that could potentially be used to identify an individual (i.e., indirectly identifying data such as GPS coordinates on their own or in combination with other data) can be subject to requirements complicated by a fragmented regulatory landscape governing data protection, privacy and the rights of data subjects (e.g., the EU's 2018 General Data Protection Directive).

Access and Benefit Sharing (ABS) Compliance Accessing biological resources and associated information (such as genomic information and traditional knowledge) can be subject to best-practice or regulatory requirements concerning prior informed consent and mutually agreed terms governing access to, and the sharing of benefits (monetary and non-monetary) resulting from, research and development concerning the biological resources or associated information (e.g., as addressed by the Nagoya Protocol on Access and Benefit Sharing).

Agricultural Data Codes of Conduct These are tools to facilitate better data governance frameworks, particularly as agricultural data are increasingly collected through digital sensors often embedding Artificial Intelligence-based analytics. Few countries have a code of conduct for farm data; an exception is the European Union Code of Conduct for Data Sharing by Contractual Agreement developed in 2018 (Wiseman et al., 2019) Such codes encompass many of the above points, in attempting to provide principles about rights and responsibilities supporting a transparent data governance that engages farmers in decision-making, guaranteeing their full access to data collected from them. To address the lack of global guidelines, GODAN has recently published a generic toolkit (https://www.godan.info/codes) to guide scientists or collectors of agricultural data to create a customized code that can be validated with the national authorities of the countries where data will be collected (Zampati, this volume). Implementation of such a code of conduct by public and private institutions may help plug the data governance gap particularly visible in the public sector.

As a case in point, agricultural research for development institutions in the CGIAR System have been attempting to tackle governance needs relating to the above while addressing cultural aspects in an *ad-hoc* way. With a new, more centralized CGIAR model envisioned, it is expected that governance frameworks will also be more uniformly applicable and backed by accountability. A nascent model is proposed for this new modality by a Data Assets Management Task Team operationalized in 2021 to address concerns around research data, with governance key among these (Fig. 3).

This data asset governance model recognizes that good governance goes beyond technical solutions, depending also on the ability of appropriately organized and empowered bodies with clear roles and responsibilities to create and assure a culture of best data practices, and compliance with legal structures. The proposed structure is briefly described here, and envisages 3 primary cascading areas of intervention, at the: (1) *strategic*, pan-CGIAR research portfolio plane; (2) *tactical*, research initiative level (with several initiatives forming the portfolio); and (3) *operational*, research team level within initiatives.

In this scenario, a strategic level Data Governance Committee (DGC) includes data scientists, domain experts (researchers), IT and legal personnel, and data stewards (with data asset management and standards expertise) and provides oversight across all three levels while primarily interfacing with tactical level teams. It is

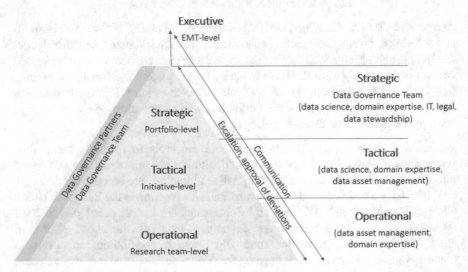

Fig. 3 Proposal for data governance under the One CGIAR model. (Modified from L. Mwanzia, pers. comm.)

solely responsible for strategic governance that determines organizational recommendations and decisions concerning all aspects of data governance, including policy implementation, repository management, standards governance and implementation, data asset management (e.g., concerning sensitive data, metadata), analytics needs etc.

Tactical level teams operate at the research initiative level and ensure that each initiative's data asset management and analysis approaches are aligned with the strategy and best practices suggested by the DGC. These teams are empowered to implement data governance principles, procedures and practices (including those around standards) as set out by the DGC, and involve data scientists, domain experts (researchers), and data asset managers (with standards expertise), with IT and legal expertise called on as needed. They provide oversight for operational level teams working on their research initiative's data asset management. Operational level teams include data asset managers and domain experts (researchers), working as part of or with research teams within an initiative to help them manage and share well-annotated, standardized data assets aligned with best practices as suggested by the DGT via tactical teams.

Considering effective use of standards in data management more specifically, governance relating to the creation, maintenance, and effective use of standards continues to be a hurdle in almost all scientific realms (McCourt et al., 2007; Zu & Wu, 2010), in no small part due to proliferation and overlap of the standards themselves. For example, there are a number of agricultural data standards, from metadata schemas aligned with industry standards such as Dublin Core (e.g. the CG Core Metadata Schema used by CGIAR Centers; https://github.com/AgriculturalSemantics/cg-core) to ontologies such as the Crop Ontology (CO; Shrestha et al., 2010; Cooper et al., 2018; Arnaud et al., 2020; https://www.cropontology.org/). The CGIAR Platform for Big Data in Agriculture also supports a beta version of the Agronomy Ontology (Devare et al., 2016; https://bigdata.cgiar.org/resources/agronomy-ontology/) and an early prototype socioeconomic ontology, with work just begun on small scale fisheries and aquaculture and livestock-related ontologies – with some overlap almost certain across these growing resources. As noted already, there are also several well-established standards used by researchers working in the crop genetics or genomics domains. The literature reflects the authors' experience across CGIAR and its stakeholders, that despite existing standards and growing awareness of their importance to enable linking across heterogeneous agricultural data, their development, maintenance, and adoption remains challenging (Wolfert et al., 2017; Bahlo et al., 2019; Drury et al., 2019).

Governance around standards in the public sector has been especially difficult, but efforts are ongoing even as new needs arise for interoperability standards around such concepts as Digital Sequence Information (DSI). As argued by Manzella et al. (this volume), interoperability across data systems is critical to enable legal solutions addressing access and benefit sharing (ABS) associated with the plant genetic material covered by the International Treaty on Plant Genetic Resources for Food and Agriculture (ITPGRFA). The authors cite the need for an ontology to model the ways DSI is defined by scientists and policy makers, as it will enable mediation across these two communities in arriving at common understanding of what DSI entails. While reaching consensus across these communities on DSI and a DOI-like standard applied to digital sequences (Digital Genomic Object identifier, or DGO) is not elementary, it will support traceability of data not only for scientific use, but also for ABS, by enabling provenance of genetic material. This work is likely to also spawn a new governance model for other use cases, that addresses interoperability for access and the needs of academic and non-academic communities. We envisage potentially similar ABS implications at the nexus of agricultural research (academic) and development (non-academic actors and beneficiaries) as machine learning, AI, and IoT applications are driven by multi-disciplinary and multi-instrument data streams.

A governance framework for the Crop Ontology project referenced above was solidified and could form a model for governance across the organization and the sector itself (Fig. 4). Governance around the ICASA variables developed by the Agricultural Model Intercomparison and Improvement Project (AgMIP) based at the University of Florida to help crop modelers harmonize crop simulation data is another noteworthy effort (White et al., 2013).

In sum, there are several reasons for such governance-related issues around data and data standards; here we have drawn on our experience across public and private sectors (i.e., CGIAR and Bayer CropScience) to illustrate a few, providing models and suggestions to overcome them where possible.

- "Invisible" data governance. While data governance may be embedded to various degrees in private sector organizations dealing with R&D data, it is not as common in the public sector, and in either case, is often not recognized as a critical function. Data governance efforts therefore tend to be ad-hoc, invisibly

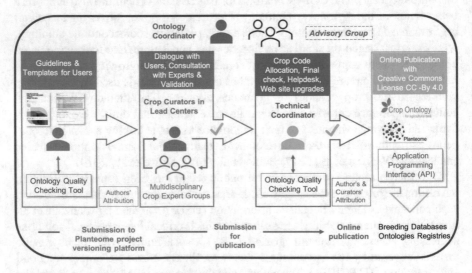

Fig. 4 Elements of a governance framework for the Crop Ontology. (Source: Crop Ontology Governance and Stewardship Framework (Arnaud et al., 2022))

keeping alive data systems, platforms and processes in spite of a lack of proper assignment and recognition of roles and responsibilities. Data stewardship, which usually ensures data accuracy, quality, and completion, is a role absorbed by data scientists or data enthusiasts. These persons must devote time to perform those data maintenance tasks on top of their core duties because a data governance strategy is absent. Another common problem when data governance strategies exist concerns their scoping of activities. Ideally all data should have a form of governance, ranging from a light setup with a few data stewards, for example, to a complex setup with several stewards, data owners and an overarching data council. In practical implementations only certain data areas are typically governed due to priorities, funding availability and staff resources. One model trialing at a couple of CGIAR Centers is the formation of institutional, multi-stakeholder governance teams, as suggested by Stedman and Vaughan (2020). Such a team might be composed of a leadership representative, research program leads and scientists, data scientists, IT professionals, data managers, and possibly, someone with IP or legal expertise. Data architects may be part of it, along with a Chief Data Officer or their equivalent. Growers are not part of these teams, but institutions including CGIAR are increasingly tweaking data consent statements towards dynamic consent models that empower farmers in voicing how data about their farms might be shared or used. This model, with some changes, is presented above (Fig. 3) and is gaining traction as one that could be widely implemented across the CGIAR System.

- Strategy. A data governance strategy should be driven by data practitioners' needs and not by IT tools or technical requirements (e.g., development of a data mapping tool) as is often the case. Governance bodies should devise a plan based on the relevant R&D data requirements, potentially resulting in IT tools or systems only if well-considered requirements dictate. In the agricultural research domain this is more often done the other way around, where platforms, tools, and systems dictate roles and responsibilities. An effective strategy must recognize that governance is primarily about people and not directly about tools or technologies. These latter are important but are far from the sole determinants of process and organizational efforts. Typically, R&D organizations and digitalization efforts start by implementing standards (e.g., controlled vocabularies, ontologies) across their data systems. They then move into the organizational aspects (that is, governance) as the data standards get used by more platforms and users, which demand better checkpoints and data maintenance for sustainable and reusable data and data products. Standardizing data is a very good initial step towards reusability and sustainability, but its success and the continuity of activities depends on a governance strategy and planning.

- Leadership support, governance teams and valuing data management. As already mentioned, the policy environment around the use and governance of data and standards is often poor. At CGIAR, the 2013 Open Access and Data Management Policy emphasized "open" but only tangentially referenced data standards and semantic interoperability, and accountability is missing. Recognizing the importance of data interoperability, CGIAR leadership supported a revision of this policy in 2021 to address FAIR data standards and their governance – including implementation, oversight, and compliance – without which loopholes bloom and adoption can wane. As noted above, many agricultural research and education entities lack policy frameworks, high-level support, and governance teams with formalized roles. This needs to change for governance around open and FAIR data to become less challenging.

- Governance plan deployment. A grassroots data governance initiative without high-level management support is condemned to fade or fail altogether. Another cause of failure is the relatively long-term deployment of a governance plan, for which it is important to define clear and concrete deliverables (e.g., appointment of data stewards, definition of data shareability policies, integration of data standards). The benefit of a governance setup is typically not only invisible to high-level management but also to end users who could further influence the investment of resources towards governance activities. Different tactics could be employed to mitigate these situations: proof of concept implementation on small data sets and limited to a few systems, including key players as part of decision bodies; implementing adequate data stewardship recognition mechanisms; a non-disruptive governance model, ensuring a balanced distribution of data standardization efforts; avoiding over-engineered governance plans that slow processes; and partitioning the data asset ecosystem into manageable but relevant pieces (e.g. governing data on traits and related assets as events).

- Developing and maintaining standards to enable linking data is time and effort-intensive, and funder support elusive. Ontologies can help standardize the heterogeneous data that the agricultural sector deals with, thereby enabling humans and machines to more easily mine and link such cross-disciplinary data. Best practices are typically followed in developing these ontologies, including technical considerations and the involvement of domain experts working with ontologists to build and validate content (Rudnicki et al., 2016; Garijo & Poveda-Villalón, 2020). However, such consultative processes often present difficult governance issues, in that they involve compromise on preferred individual approaches in favor of standard terminology that works more generally. Some of these issues can be mitigated by inherent properties of ontologies (as compared with controlled vocabularies), one of which allows for the addition of synonyms with their contexts and definitions. However, the process of arriving at a consensus choice of concepts that accurately and sufficiently cover a particular domain can be fraught and involve huge amounts of time and discussion. Lastly, funders and institution leadership typically balk at supporting what is often seen as the tedious underpinnings of data management, making such efforts difficult to sustain. Some of these challenges were articulated by respondents to a survey conducted by Geller et al. (2018) to determine why ontologies typically tend to be sparsely updated (Table 1). These situations can be improved if (1) a more progressive data culture and explicit policy and accountability environment as referenced above is in place; (2) data is routinely re-examined and reused to generate new value, in turn demonstrating the value of standardization; and critically, (3) there is wide-ranging support to allocate budget for these efforts.

Table 1 Results from a survey to assess primary reasons for sparse updates of ontologies. Curators of 83 ontologies were contacted, with a response rate of 48/83, or 58%

Main catagories of responses

Groups	Reason	Number of Ontologies	Ontologies
Group 1	Incorporated into other ontologies / systems	9	ICECI, DDO, PAE, TAO, CSEO, EHDA, MAT, AAO, EHDAA
Group 2	Lack of funding, time and man-power, interruption in funding	15	BIOMODELS, RH-MESH, TGMA, ONSTR, HFO, SDO, IDOMAL, CCONT, DERMO, APAONTO, ATOL, GALEN, GMO, MCCL, NPO
Group 3	Updating in large time gaps / slow development	7	GEXO, RETO, REXO, ICNP, IDOBRU, HINO, CHMO
Group 4	Organization / Project ended	6	OBI_BCGO, NIGO, GENE-CDS, TRAK, ONTOPNEUMO, DINTO
Group 5	Paused for publication / redesigning	3	ESSO, ROO, GMM
Group 6	Concepts valid / Serves the purpose	3	EHDAA2, ONTOLURGENCE, IMGT-ONTOLOGY
Group 7	Not updating on BioPortal	3	OMIT, PMA, COGAT
Unassigned		?	HUGO, NIFSUBCELL
Total		48	

Reproduced with permission from Geller et al. (2018), under CC licence

- Collaborative development and maintenance of standards to link agricultural data. Who decides what a standard should encompass, what standards to use for particular types of data, how to build critical mass around adoption? While a governance team may have a critical role to play in these concerns, the development and maintenance of data standards are thorny issues that require broader collaboration. An example may be where the boundaries are drawn around a particular domain standard; for instance, ontology concepts to be added to the Agronomy Ontology vs. the Environmental Ontology. Successful governance and maintenance of these ontologies involves working not just with ontology and subject experts within an organization like the CGIAR system, or even within any given domain, but forging strong relationships across domain ontologies to suggest new terms in the right domain ontology and reduce concept proliferation. For a multi-center entity like CGIAR, the governance of repository-level metadata also requires agreement from data and information managers around a common, widely responsive but industry-aligned standard, in this case, the Dublin Core-based CGIAR Core Metadata Schema (https://github.com/

AgriculturalSemantics/cg-core) which is broadly applicable to wide-ranging agricultural use cases. One model for cross-institutional domain-based governance is provided by the CGIAR Platform for Big Data in Agriculture (https://bigdata. cgiar.org/), launched in 2017 with the objective of increasing the impact of agricultural research and development by turning open and FAIR data into a powerful tool for discovery, while integrating principles of responsible and ethical data use (see box).

> The CGIAR Big Data Platform hinges on several Communities of Practice (CoPs) (Agronomy Data, Crop Modelling, Data and Information Management, Geospatial Data, Livestock Data, Ontologies, and Socio-Economic Data) which engage research domain experts and practitioners, data and information managers, and ethics and IP specialists from CGIAR and a variety of stakeholders. Such CoPs could play a key role in helping to provide data and standards-focused governance, in the form of interactions across entities and individuals developing standards and other data solutions and providing cross-learning opportunities and guidance (Arnaud et al., 2020). While this is not yet the case for all the Big Data Platform CoPs, some are deeply involved in such activities, including the Ontologies and Data, Geospatial, and Information and Data Management CoPs. All these CoPs are also instrumental in facilitating the use of data standards, helped by the Platform's GARDIAN data ecosystem (https://gardian.bigdata.cgiar.org/). GARDIAN enables the discovery of data assets produced by the CGIAR network and other key stakeholders in the public domain, and provides a data-to-analytics and visualization environment and model pipelines to realize the value of increasingly FAIRer data, bolstering the work of the CoPs.

- Standards make data easier to link and use – but what about data owners? Industry has already recognized that digital agriculture and an associated constellation of powerful technologies (e.g., IoT, remote sensing, AI) that can mine well-harmonized data offer huge potential for hyperlocal, tailored agricultural recommendations (e.g. for fertilizer). As addressed in more depth by Zampati et al. (this volume), such technologies raise critical legal and ethical questions. While farmers increasingly acknowledge the benefits of such standards-reliant technologies, they are also beginning to express concern about losing ownership over their data. Data ownership issues are often exacerbated by concerns about privacy as technology increasingly facilitates data triangulation to expose personally-identifiable information, yet the foregoing discussion has thus far omitted mention of the data owner. Data cooperatives are a recent model for governing farm data, with several examples in the US, such as the Ag Data Coalition (ADC) and the Grower Information Services Cooperative (GiSC). Some data cooperatives (like ADC) offer secure data repository solutions that enable farmers to store their data and decide with which platforms, agencies or research entities to share it. Others like GiSC, offer a repository and also perform analytics to give farmers greater insight into their production practices and can negotiate opportunities to monetize data on their behalf. There are subscription-based approaches like the Farmer Business Network (FBN) which offer data platforms and analytics over the data pool to offer the farmer farm-specific management and profitability insights such as yield by soil type or fertilizer, and input price comparisons. Similar approaches are also emerging in developing countries, for example, Digital Green is building FarmStack, a data exchange platform for farmers in India, with features similar to ADC (https://farmstack.digitalgreen.org/). Yara and IBM have also launched a joint effort to enable farmers to securely share data and retain determination around who uses it and how, benefiting monetarily in the process (Yara International, 2020). Central to these newer models is the placement and valuing of the data owner in the mix of stakeholders that determine how data gets managed and used.

3 Conclusions

Good data governance practices are the beating heart of innovation and impact, particularly through their impact on data interoperability and reusability by humans and machines. Data governance ensures that data assets remain widely available and interoperable, but are also secure, trustworthy and not misused (Stedman & Vaughan, 2020). As the digital landscape and data capabilities become more sophisticated, these latter concerns are especially important to assure that governance efforts address the gamut, from policy through standards, to ethics, to assure that sensitive data, rules for data use, data sharing agreements and allied efforts are considered in the light of managing both institutional and individual risk. We have attempted to address legal, technical and cultural challenges to data and standards governance, outlining some models for successful governance to enable data linking that cover a range of aspects, from administrative and financial enablers to the human and technical considerations. In doing so we have briefly addressed considerations such as policy environment and governance teams and data cooperatives – both within and across organizational structures; funding support; capacity and awareness of the human actors in the data ecosystem; and technical infrastructure that allow data owners to have a higher level of self-determination over their information.

All interventions aiming for impactful data governance must recognize the human experiences involved before improved practices can be recommended or required. Thus, we have touched upon the epistemology of scientific research in general, and agricultural research in particular as largely being hypothesis-driven rather than inductive and empirical as fields embracing data science and big data technologies tend to be. As might be expected, the business of how research is viewed and conducted is likely to be a key determinant of how the data it produces is handled, as appears to be borne out by our experiences with CGIAR researchers for whom the notion of *in silico* analysis involving secondary data is relatively new.

There are many unaddressed questions and gaps that remain regarding data governance as it relates to interoperability and enabling data linkages. Blockchain is being increasingly explored as a data provenance solution, a way to enable data security, traceability, and accountability (Liang et al., 2017; Ramachandran & Kantarcioglou, 2017; Devan, 2018; Shabani, 2019; Kochupillai, this volume). The Food Trust Blockchain has already been launched by IBM as a food traceability platform and adopted by large retailers, fruit and meat wholesalers, and multinationals in the food products sector (Stanley, 2018). Closer to the research world, Blockchain has been proposed as a solution to handling electronic medical records (EMRs) to give patients access to their medical records across providers and treatment sites via an immutable record. As envisioned by Azaria et al. (2016), the application of Blockchain to EMRs via a decentralized records management system called MedRec allows researchers and other medical stakeholders to mine aggregated, anonymized data. In return, these actors sustain and secure the network via a "Proof of Work" algorithm that is tamper-proof, involving individual nodes competing to solve computational "puzzles" before another block of content can be added to the chain. The work required of "miners" to append blocks assures that it is difficult to rewrite history on the Blockchain. Azaria et al. therefore propose empowering researchers through big data pools, while involving patients and care providers in choices around the release of their (meta)data.

Such models that include farm data and the farmer as the determinant of how her data is used, and by whom, are beginning to gain traction through the notion of data cooperatives. While Blockchain is still in its infancy as an enabler in these ecosystems, it is likely to gain prominence in the near-term, as data economies grow across sectors. How data standards mesh with and augment Blockchain capabilities is not clear but requires consideration in the near-term. What seems clearer, in fact, is the potential of Blockchain technology to provide accountability and traceability in the standards development process, even if privacy is not generally a concern. That data standards are critical for enabling interoperability is generally accepted; as this paper attempts to make clear, there remain some unexplored considerations around their governance, along with questions of what constellations of actors ought to be involved in standards development and maintenance.

References

Arnaud, E., Hazekamp, T., Laporte, M-A., Antezana, E., Andres Hernandez, L., Pot, D., Shrestha, R., Dreher, K., Castiblanco, V., Menda, N., Fabio Guerrero, A., Hualle, V., Salas, E., Mendes, T., Makunde, G., Chaves, I., Rathore, A., Das, R., Afolabi, A., Pietragalla, J., Pommier, C., Michotey, C., Detras, J., McNally, K., Borja, N., Winger, L., Cooper, L., Jaiswal, P., Mauleon, R., & Yu, J. (2022). *Crop ontology governance and stewardship framework*. Retrieved from https://hdl.handle.net/10568/118001

Arnaud, E., Laporte, M.-A., Kim, S., Aubert, C., Leonelli, S., Miro, B., Cooper, L., Jaiswal, P., Kruseman, G., Shrestha, R., Buttigieg, P. L., Mungall, C. J., Pietragalla, J., Agbona, A., Muliro, J., Detras, J., Hualla, V., Rathore, A., Das, R. R., Dieng, I., Bauchet, G., Menda, N., Pommier,

C., Shaw, F., Lyon, D., Mwanzia, L., Juarez, H., Bonaiuti, E., Chiputwa, B., Obileye, O., Auzoux, S., Dzalé Yeumo, E., Mueller, L. A., Silverstein, K., Lafargue, A., Antezana, E., Devare, M., & King, B. (2020). The ontologies community of practice: A CGIAR initiative for big data in agrifood systems. *Patterns, 1*(7). Retrieved from https://doi.org/10.1016/j.patter. 2020.100105

Azaria, A., Ekblaw, A., Vieira, T., & Lippman, A. (2016). *MedRec: Using Blockchain for medical data access and permission management.* 2nd International Conference on Open and Big Data. Retrieved from http://www.pitt.edu/~babay/courses/cs3551/papers/MedRec.pdf

Bahlo, C., Dahlhaus, P., Thompson, H., & Trotter, M. (2019). The role of interoperable data standards in precision livestock farming in extensive livestock systems: A review. *Computers and Electronics in Agriculture, 156*, 459–466. Retrieved from https://doi.org/10.1016/j.compag. 2018.12.007

Barham, B., Goldman, I., van Rijn, J., Foltz, J., & Agnes, M. I. (2017). Land-Grant University faculty attitudes in and engagement with open source scholarship and commercialization. *Agricultural and Environmental Letters, 2*(1). Retrieved from https://doi.org/10.2134/ael2017. 03.0008

Bill and Melinda Gates Foundation. (2021). *Bill and Melinda Gates Foundation Open Access Policy.* Retrieved from: https://www.gatesfoundation.org/How-We-Work/General-Information/ Open-Access-Policy

Bodenreider, O. J. (2005). Biomedical ontologies. *Pacific Symposium on Biocomputing, 76–78.* Retrieved from https://www.ncbi.nlm.nih.gov/pmc/articles/PMC4300097/

Brockman, J. (2015, July 5). *Popper vs. A Conversation with Peter Coveney.* Retrieved from Edge Conversations: https://www.edge.org/conversation/peter_coveney-popper versus-bacon

CGIAR. (2013, October 2). *CGIAR open access and data management policy.* Retrieved from https://cgspace.cgiar.org/bitstream/handle/10947/4488/Open%20Access%20Data%20Manage ment%20Policy.pdf?sequence=1&isAllowed=y

CGIAR. (2021, April 16). *CGIAR open and FAIR data assets policy.* Retrieved from https:// cgspace.cgiar.org/bitstream/handle/10568/113623/CGIAR_OFDA_Policy_Approved_1 6April2021.pdf?sequence=1&isAllowed=y

Cooper, L., Meier, A., Laporte, M. A., Elser, J. L., Mungall, C., Sinn, B. T., Cavaliere, D., Carbon, S., Dunn, N. A., Smith, B., Qu, B., Preece, J., Zhang, E., Todorovic, S., Gkoutos, G., Doonan, J. H., Stevenson, D. W., Arnaud, E., & Jaiswal, P. (2018). The Planteome database: An integrated resource for reference ontologies, plant genomics and Phenomics. *Nucleic Acids Research, 46*(D1), D1168–D11804. Retrieved from https://doi.org/10.1093/nar/gkx1152

Denk, F. (2017). Don't let useful data go to waste. *Nature, 543*(7643), 7. Retrieved from https:// www.nature.com/news/don-t-let-useful-data-go-to-waste-1.21555

Devan, G. (2018). *How Blockchain technology is revolutionizing data provenance.* Retrieved from medium.com: https://medium.com/blockpool/how-blockchain-technology-is-revolutionizing- data-provenance-e47610019390

Devare, M., Aubert, C., Laporte, M.-A., Valette, L., Arnaud, E., & Buttigieg, P. L. (2016). Data- driven agricultural research for development – A need for data harmonization via semantics. In P. Jaiswal & R. Hoehndorf (Eds.), *7th international conference on biomedical ontologies, ICBO 2016* (Vol. 1747:2). CEUR Workshop Proceedings.

Roche, D. G., Kruuk, L. E. B., Lanfear, R., & Binning, S. A. (2015). Public data archiving in ecology and evolution: How well are we doing? *PLOS Biology, 1–12.* Retrieved from https:// doi.org/10.1371/journal.pbio.1002295

Drury, B., Fernandes, R., Moura, M.-F., & de Andrade Lopes, A. (2019). A survey of semantic web technology for agriculture. *Information Processing in Agriculture, 6*(4), 487–501. Retrieved from https://doi.org/10.1016/j.inpa.2019.02.001

European Commission Expert Group on FAIR Data. (2018). *Final report and action plan: Turning FAIR into reality.* Retrieved from Publications Office of the EU: https://op.europa.eu/s/oHHB

Garijo, D., & Poveda-Villalón, M. (2020). *Best practices for implementing FAIR vocabularies and ontologies on the web. arXiv.org.* Computer Science: Digital Libraries. Retrieved from https:// arxiv.org/abs/2003.13084v1

Geller, J., Keloth, V. K., & Musen, M. A. (2018). *How sustainable are biomedical ontologies?* Proceedings, AMIA Annual Symposium. https://www.ncbi.nlm.nih.gov/pmc/articles/PMC63 71329/

Koers, H., D. Bangert, E. Hermans, R. van Horik, M. de Jong, and M. Mokrane. (2020). *Recommendations for services in a FAIR data ecosystem.*. https://dx.doi.org/10.1016%2Fj. patter.2020.100058

Liang, X., Shetty, S., Tosh, D., Kamhoua, C., Kwiat, K., & Njilla, L. (2017, May). ProvChain: A Blockchain-based data provenance architecture in cloud environment with enhanced privacy and availability *CCGrid '17: Proceedings of the 17th IEEE/ACM International Symposium on Cluster, Cloud and Grid Computing* (pp 468–477). Retrieved from https://doi.org/10.1109/ CCGRID.2017.8

Leonelli, S. (2019). Data governance is key to interpretation: Reconceptualizing data in data science. *Harvard Data Science Review*. Retrieved from https://doi.org/10.1162/99608f92. 17405bb6.

McCourt, B., R. A. Harrington, K. Fox, C. D. Hamilton, K. Booher, W. E. Hammond, A. Walden, M. Nahm. (2007). *Data standards: At the intersection of sites, clinical research networks, and standards development initiatives*. Retrieved from Therapeutic Innovation and Regulatory Science: https://doi.org/10.1177/009286150704100313

Noyes, K. (2014). *Cropping up on every farm: Big Data technology*. Retrieved from Fortune: https://fortune.com/2014/05/30/cropping-up-on-every-farm-big-data-technology/

Pham, X., & Stack, M. (2018). How data analytics is transforming agriculture. *Business Horizons, 61*(1), 125–133. Retrieved from https://doi.org/10.1016/j.bushor.2017.09.011

Quitzau, A. (2020). *IBM Supercomputer Summit Attacks Coronavirus...* Retrieved from IBM Digital Nordic: https://www.ibm.com/blogs/nordic-msp/ibm-supercomputer-summit-attacks-coronavirus/

Ramachandran, A., & Kantarcioglu, M. (2017). Using Blockchain and smart contracts for secure data provenance management. arXiv.org. *Computer Science: Cryptography and Security*. https://arxiv.org/abs/1709.10000

Rijmenam, M. V. (2013). *John Deere is revolutionizing farming with big data*. Retrieved from Datafloq: https://datafloq.com/read/john-deere-revolutionizing-farming-big-data/511

Robinson, P. N., & Haendel, M. A. (2020). *Ontologies, knowledge representation, and machine learning for translational research: Recent contributions*. Retrieved from https://doi.org/10. 1055/s-0040-1701991

Rosenbaum, S. (2010). *Data governance and stewardship: Designing data stewardship entities and advancing data access*. https://doi.org/10.1111/2Fj.1475-6773.2010.01140.x

Rudnicki, R., Smith, B., Malyuta, T., & Mandrick, W. (2016). *White paper: Best practices of ontology development*. CUBRC Advantage Through Technology. Retrieved from https://www. nist.gov/system/files/documents/2019/05/30/nist-ai-rfi-cubrc_inc_002.pdf

Sara, R., & Devare, M. (2020). *Excellence in agronomy (EiA) initiative: Best practice guidelines to support global access implementation* [Guidance note for CGIAR initiative].

Semantic Scholar. (2020). *CORD-19*. Retrieved from COVID-19 Open Research Dataset: https:// www.semanticscholar.org/cord19

Shabani, M. (2019). Blockchain-based platforms for genomic data sharing: A decentralized approach in response to governance problems? *Journal of the American Medical Informatics Association*. Retrieved from https://dx.doi.org/10.1093%2Fjamia%2Focy149

Shrestha, R., Arnaud, E., Mauleon, R., Senger, M., Davenport, G. F., Hancock, D., Morrison, N., Bruskiewich, R., & McLaren, G. (2010). Multifunctional crop trait ontology for breeders' data: Field book, annotation, data discovery and semantic enrichment of the literature. *AoB Plants*. Retrieved from https://doi.org/10.1093/aobpla/plq008

Sieleman, K., A. Hafner, B. Pucker. (2020). The reuse of public datasets in the life sciences: Potential risks and rewards. PeerJ 8:e9954 https://doi.org/10.7717/peerj.9954

Smalley, E. (2018). *In silico* farming drives next wave in agriculture. *Nature Biotechnology, 36*(9), 783–784.

Smith, K. (2013). *A brief history of NCBI's formation and growth*. Retrieved from The NCBI Handbook [Internet]. 2nd Edition: https://www.ncbi.nlm.nih.gov/books/NBK148949/

Stanley, A. (2018). Ready to rumble: IBM launches food trust Blockchain for commercial use. *Forbes*, October 8, 2018. Retrieved from https://www.forbes.com/sites/astanley/2018/10/08/ready-to-rumble-ibm-launches-food-trust-blockchain-for-commercial-use/?sh=13736cb97439

Stedman, C., & Vaughan, J. (2020). *What is data governance and why does it matter?* Tech Target. Retrieved from Tech Accelerator. https://searchdatamanagement.techtarget.com/definition/data-governance?_ga=2.159940984.476600454.1612269086-1594224955.1612269086&_gl=1*14r7hxp*_ga*MTU5NDIyNDk1NS4xNjEyMjY5MDg2*_ga_RRBYR9CGB9*MTYxMjI2OTA4OC4xLjEuMTYxMjI3MTAzMy4w

Streich, J., Romero, J., Gazolla, J. G., Kainer, D., Cliff, A., Prates, E. T., Brown, J. B., Khoury, S., Tuskan, G. A., Garvin, M., Jacobson, D., & Harfouche, A. L. (2020). Can Exascale computing and explainable artificial intelligence applied to plant biology deliver on the United Nations sustainable development goals? *Current Opinion in Biotechnology, 61*, 217–225. Retrieved from https://doi.org/10.1016/j.copbio.2020.01.010

Thessen, A. E., & Patterson, D. J. (2011). Data issues in the life sciences. *Zookeys, 150*, 15–51. Retrieved from https://zookeys.pensoft.net/articles.php?id=3041

Vasilevsky, N. A., Minnier, J., Haendel, M. A., & Champieux, R. E. (2017). Reproducible and reusable research: Are journal data sharing policies meeting the mark? *PeerJ, 5*, e3208. Retrieved from https://doi.org/10.7717/peerj.3208

White, J. W., Hunt, L. A., Boote, K. J., Jones, J. W., Koo, J., Kim, S., Porter, C. H., Wilkens, P. W., & Hoogenboom, G. (2013). Integrated description of agricultural field experiments and production: The ICASA version 2.0 data standards. *Computers and Electronics in Agriculture, 96*, 1–12.

Wilkinson, M. D., Dumontier, M., Aalbersberg, I. J. J., Appleton, G., Axton, M., Baak, A., Blomberg, N., Boiten, J.-W., da Silva Santos, L. B., Bourne, P. E., Bouwman, J., Brookes, A. J., Clark, T., Crosas, M., Dillo, I., Dumon, O., Edmunds, S., Evelo, C. T., Finkers, R., Gonzalez-Beltran, A., Gray, A. J. G., Groth, P., Goble, C., Grethe, J. S., Heringa, J., 't Hoen, P. A. C., Hooft, R., Kuhn, T., Kok, R., Kok, J., Lusher, S. J., Martone, M. E., Mons, A., Packer, A. L., Persson, B., Rocca-Serra, P., Roos, M., van Schaik, R., Sansone, S.-A., Schultes, E., Sengstag, T., Slater, T., Strawn, G., Swertz, M. A., Thompson, M., van der Lei, J., van Mulligen, E., Velterop, J., Waagmeester, A., Wittenburg, P., Wolstencroft, K., Zhao, J., & Mons, B. (2016). The FAIR Guiding Principles for scientific data management and stewardship. *Scientific Data, 3*(1), 160018. https://doi.org/10.1038/sdata.2016.18

Wiseman, L., Pesce, V., Zampati, F., Sullivan, S., Addison, C., & Drolet, J. (2019). *Review of codes of conduct, voluntary guidelines and principles relevant for farm data sharing* (CTA working paper 19/01). CTA. Retrieved from https://hdl.handle.net/10568/106587

Wolfert, S., Ge, L., Verdouw, C., & Bogaardt, M.-J. (2017). Big data in smart farming–a review. *Agricultural Systems, 153*, 69–80. Retrieved from https://www.sciencedirect.com/science/article/pii/S0308521X16303754?via%3Dihub

Yara International. (2020). *Yara and IBM launch an open collaboration for farm and field data to advance sustainable food production*. January 23, 2020. Retrieved from https://www.yara.com/corporate-releases/yara-and-ibm-launch-an-open-collaboration-for-farm-and-field-data-to-advance-sustainable-food-production/

Zhu, H., & Wu, H. (2010). Assessing quality of data standards: Framework and illustration using XBRL GAAP Taxonomy. In S. Sánchez-Alonso & I. N. Athanasiadis (Eds.), *Metadata and Semantic Research. MTSR 2010. Communications in Computer and Information Science* (Vol. 108). Springer. https://doi.org/10.1007/978-3-642-16552-8_26

Creating a Digital Marketplace for Agrobiodiversity and Plant Genetic Sequence Data: Legal and Ethical Considerations of an AI and Blockchain Based Solution

Mrinalini Kochupillai and Julia Köninger

Abstract The EU regulation on 'Organic Production and Labelling of Organic Products' opens the door for the creation of an EU-wide marketplace for agrobiodiversity contained in so-called "heterogeneous materials". However, the creation of such a marketplace presupposes the existence of optimal demand and supply of agrobiodiversity, linked plant genetic sequence data and local/traditional knowledge on how best to use agrobiodiversity. Farmers' tendency to prefer genetically uniform "high yielding" seeds and the adoption of chemical intensive farming have compromised the *supply* of agrobiodiversity. At the same time, regulatory regimes have disincentivized the use of agrobiodiversity in research and breeding programs, resulting in a lack of *demand* for agrobiodiversity. This chapter argues that these trends result from (inadvertent) inequities in existing regulatory frameworks that primarily support uni-directional data/knowledge flows from the formal sector (academia, industry) to the informal sector (farmers). We propose ways in which rapidly evolving technologies like blockchain/DLTs and AI/Machine Learning can (and should) diversify the direction of scientific research as well as of data/knowledge flows in the agricultural sector. The chapter thus provides food for thought for developing novel regulatory frameworks and ethical business models for robust digital marketplaces for agrobiodiversity for the benefit of farmers, researchers, and the environment.

M. Kochupillai (✉)
School of Engineering and Design, Technical University of Munich, Munich, Germany

SIRN (Sustainable Innovations Research and Education Network, sirn.eu), Munich, Germany

Munich Intellectual Property Law Center (MIPLC), Munich, Germany
e-mail: m.kochupillai@tum.de

J. Köninger
University of Vigo (Under the Collaborative Doctoral Partnership Programme with the Joint Research Centre of the European Commission), Vigo, Spain
e-mail: julia.koninger@uvigo.es

© The Author(s) 2023
H. F. Williamson, S. Leonelli (eds.), *Towards Responsible Plant Data Linkage: Data Challenges for Agricultural Research and Development*,
https://doi.org/10.1007/978-3-031-13276-6_12

1 Introduction[1]

1.1 Agrobiodiversity and Its Rapid Depletion

The EU Organic Regulation (EU 2018/848) on 'Organic Production and Labelling of Organic Products' opens the door for the creation of an EU wide marketplace for agrobiodiversity contained in so called "heterogeneous materials". However, the creation of such a marketplace presupposes the existence of optimal demand as well as supply of agrobiodiversity, linked plant genetic sequence data and local/traditional knowledge on how best to use this agrobiodiversity. According to estimates, about 75% of crop (on-soil) genetic diversity has been lost with farmers abandoning locally adapted heterogeneous seeds for genetically uniform "high yielding" ones. Associated adoption of chemical intensive farming has also led to loss of in-soil, beneficial microbial diversity. These, together, have a negative impact on *supply* of agrobiodiversity and its beneficial components. At the other end of the spectrum, regulatory regimes under well-intended laws create bureaucratic hurdles that disincentivize legal and transparent use of agrobiodiversity in research and breeding programs, creating a lack of *demand* for agrobiodiversity. Consequently, active and robust marketplaces for agrobiodiversity, and for derivatives thereof, have failed to evolve. This paper argues that these trends result from (inadvertent) inequities in existing regulatory frameworks that primarily support uni-directional data/knowledge flows from the formal sector (academia, industry) to the informal sector (farmers). The article argues that with the rapid evolution of technologies such as blockchain/DLTs and AI/Machine Learning, the direction of scientific research as well as of data/knowledge flows in the agricultural sector can and should be diversified. Such technologies and platforms based thereon can support: (i) secure and "controllable" data/knowledge sharing by the informal sector; (ii) accrual of fair, inclusive and equitable economic benefits for those sharing data, and (iii) traceability, for ensuring accurate economic benefit sharing on the one hand, and determining legal liability on the other, on a case by case basis. The article aims to provide food for thought for further multi-disciplinary and multi-stakeholder research, and for developing novel regulatory frameworks and ethical business models for robust digital marketplaces for agrobiodiversity for the benefit of farmers, researchers and the environment.

---

[1] This contribution builds on the position paper submitted to the Government of India in 2019, on means of promoting sustainable seed innovations in India (Kochupillai et al., 2019).

Author contributions: Kochupillai conceptualized the paper, wrote the first draft and edited and shortened the paper for final submission. Köninger contributed insights from the EU Green Deal and Common Agricultural Policy, edited and shortened the paper for final submission, and identified, managed and formatted all relevant references.

1.2 Agrobiodiversity and "Missing Markets"

Data is increasingly considered a tradeable commodity. In the context of agrobiodiversity and Plant Genetic Resources (PGRs), data as well as physical materials associated therewith have been internationally recognised as a valuable and conditionally tradeable commodity, at least since the adoption of the Convention on Biological Diversity (CBD), (CBD, 1992) and later, the International Treaty on Plant Genetic Resources for Food and Agriculture (Fraleigh & Davidson, 2003; FAO, 2011) ("the seed treaty"). Although the economic, social, as well as environmental value of such data and materials has been underscored by research from various disciplinary perspectives (Dulloo et al., 2010; Saatkamp et al., 2019), estimates emerging from the UN FAO state that more than 75% of crop genetic diversity has been lost since the widespread adoption of conventional agriculture based on a very few crop varieties (FAO, 1999).

To counteract that development, existing legal regimes, including under the CBD, the seed treaty, and various intellectual property (IP) protection laws, are well-intended to support *in* and *ex situ* conservation of (agro)biodiversity and to ensure:

(i) equitable access and benefit sharing (ABS) in the transfer and usage of agrobiodiversity (including agrobiodiversity conserved in situ)
(ii) optimal incentives for research and innovation with agrobiodiversity

Sui generis systems for the protection of plant varieties, such as the one adopted by India, also seek to incentivize the creation of new varieties along with the conservation of old/indigenous ones. Yet, extensive empirical research in earlier studies found that:

> the 'market failure' theory which is often used to justify the introduction of intellectual property rights for various fields of innovation can be better used as a justification to deny or limit intellectual property protection for plant varieties. This is because unlike in other fields of technology where the introduction of IPRs may address market failures, in the plant varieties sector, the introduction of such rights would worsen the existing negative externalities and produce new market failures. Regimes beyond those designed to protect intellectual property rights would therefore be necessary to promote sustainable innovation in plant varieties in general, and *in situ* conservation of agrobiodiversity in particular (Kochupillai, 2016; Kochupillai 2019a).

The new market failures emerging from current regulatory regimes governing the agricultural seeds sector, include the non-emergence of specific, desirable markets ("missing markets"), such as a market for agrobiodiversity.[2] How can *in situ* agrobiodiversity conservation (and downstream improvement, innovation and research with it) be incentivized, if IP protection is not the appropriate or adequate route? To answer this question, it is useful to take a quick look at why

---

[2] In this paper, we use the term "Agrobiodiversity" broadly to include indigenous/heterogenous seeds, but also traditional knowledge or know-how of best (farming) practices to enhance unique features of these seeds and resulting crop/produces.

agrobiodiversity is increasingly becoming a focus area for the EU and what are the barriers preventing the emergence of a robust, *equitable* marketplace for agrobiodiversity, and associated traditional knowledge and data.

1.2.1   Agrobiodiversity: What and Why

In 2018, the European Parliament adopted the new regulation EU 2018/848 on 'Organic Production and Labelling of Organic Products'. The recitals to this Regulation state, inter alia, that "Research in the Union on plant reproductive material that does not fulfil the "variety" definition as regards uniformity shows that there could be benefits of using such diverse material, in particular with regard to organic production, for example, to reduce the spread of diseases, to improve resilience and to increase biodiversity" EU 2018/848, recital 36". The Regulation refers to plant reproductive materials (e.g. seeds) that do not fulfil the "variety" definition as regards "uniformity", as "heterogeneous materials".

Heterogeneous materials include plant and seed agrobiodiversity, particularly landraces, indigenous seeds and farmers' varieties (collectively referred to hereinafter as "agrobiodiversity" or "heterogeneous materials"). Together, they host a wealth of PGRs that contribute significantly to global food security and sustainable agriculture, as also seen in case of the Baladi landraces in Palestine.[3] While agrobiodiversity can be conserved *ex situ*, problems of viability over prolonged storage and constantly changing biotic and abiotic conditions on field necessitate *in situ* agrobiodiversity conservation.

1.2.2   The "Supply Side" Story of Agrobiodiversity

Yet, agrobiodiversity conserved *in situ* is increasingly scarce and in "short supply" (Van de Wouw et al., 2010). Studies, both theoretical and empirical, have shown that farmers have little incentive to continue *in situ* conservation of agrobiodiversity (Kochupillai, 2016; Swanson & Goeschl, 2000). Most farmers are not able to recognise, capture and trade the 'value' inherent in their heterogeneous seeds. Unique nutritional or medicinal properties, colour and flavour of produce resulting from such seeds, often remain largely unknown to the wider farming community, to the research community, as well as to consumers. The few farmers cultivating unique varieties are often unable to obtain a fair price or a ready market for their produce. Some research has been undertaken seeking means of protecting heterogeneous materials, particularly in the form of "farmers' varieties" or land races (Borowiak, 2004). However, the tracing of the origin of material and the downstream access and usage of the material pose fundamental challenges. This results also from the characteristics of agrobiodiversity: being heterogeneous and variable, making it

---

[3] See the chapter by Fullilove in this volume.

not eligible for IP protection under regimes such as the plant breeders' rights system (Kochupillai et al., 2021).[4] The scarcity of agrobiodiversity is aggravated by national and regional crop procurement policies that are mostly focused on the procurement of crops produced using high yielding, uniform varieties that give "standard" produce.

1.2.3 The "Demand Side" Story of Agrobiodiversity

Incentives to increase supply are also hampered by the lack of demand: well-meaning regulations such as the CBD, the seed treaty, and their national counterparts are aimed at ensuring fair access and benefit sharing (i.e., preventing biopiracy). However, they create multiple bureaucratic hurdles, disincentivizing research with agrobiodiversity (Fusi et al., 2019; Halewood et al., 2018; Mekonnen & Spielman, 2018). Plant breeders increasingly rely on their own *ex situ* reserves that contain valuable germplasm, albeit for a very limited number of crop species: world nutrition is primarily based on ten crops, of which three, namely, rice, maize and wheat, contribute nearly 60% of the calories and proteins obtained by humans from plants (FAO, 1999).

Where agrobiodiversity and associated PGRs *are* accessed, existing bureaucratic and legal barriers often disincentivize honest access and use practices, leading to biopiracy, creating mistrust between suppliers and seekers of agrobiodiversity.

Sub-optimal research on locally relevant heterogeneous seeds also weakens consumer interest in buying (demand) produce emerging from such seeds, not least because of unknown or unconfirmed nutritional and other properties. Lack of legitimate and adequately compensated demand for research-related end uses, coupled with low consumer demand, further limit farmer incentives to cultivate and keep up the 'supply' of agrobiodiversity.

Further, the culture of sharing prevalent among farmers (Mcguire & Sperling, 2016) their inability to monitor the chain of transfer of ownership and the specific end use(s) to which their seeds are put (e.g. consumption or downstream research), together with the inability of IP protection regimes to grant meaningful protection to farmers' seed innovations (Kochupillai, 2019a) make any legal incentivisation schemes difficult to enforce.

This article argues that a novel understanding of the concept of 'value' linked to blockchain technology, together with its immutable, time-stamped record-keeping feature, can help overcome several of the above problems (and more, as described below). To make these arguments, in the following section (Sect. 2), we identify the current state of affairs vis-à-vis the creation of a marketplace for agrobiodiversity and its components. First, we describe the inequities created by existing agricultural regulations that promote uni-directional (top-down) flow of knowledge and

---

[4]The Plant Breeders' Rights (PBR) Regime was originally established under the UPOV Convention.

materials (from the formal sector, i.e. the seed industry/research institutes, to the informal sector, i.e. small and subsistence farmers). Second, we identify hurdles currently preventing the emergence of a marketplace for agrobiodiversity. Third, we identify the monetizable components of agrobiodiversity that currently remain un- or under-recognised as valuable resources.

In Sect. 3, we investigate how a blockchain/DLT based solution can help "mine" the "value" of agrobiodiversity conserved in situ, helping overcome the hurdles identified in Sect. 2. Section 4 looks into the legal and ethical considerations that need to be borne in mind while adopting a DLT/blockchain based solution as envisaged in Sect. 3.

2 Identifying Regulatory Inequities to Diversify Directions of Data, Knowledge and Value Flows

2.1 Agricultural Regulations Creating Inequities and Uni-directional Flow of Knowledge and Materials

Existing regulations (particularly in Europe) focus on creating, testing, certifying and regulating "uniform," homogenous seeds. While such seeds promise high yield in closely regulated farming environments, they often fail to perform in marginal environments (e.g. environments prevailing in small and subsistence farms).

Further, until recently, most European countries had outlawed the sale of heterogeneous (non-uniform), local seeds for agriculture, creating a marketplace dominated by top-down flows of knowledge and materials: seeds and knowledge on how to cultivate them (including with what types of inputs) are determined and dispersed by the formal sector (e.g. seed industry, research centres) to farmers, including to small and subsistence farmers.

This regulation-guided top-down flow of knowledge and materials has led to a focus on developing plant-data linkage solutions that primarily cater to managing data and information associated with the creation of uniform seeds for conventional farming systems. These solutions, therefore, are primarily developed for the formal seed sector and rely mostly on *ex situ* seed banks and private or proprietary germplasm reserves and plant genetic sequence data. This uni-directional flow of knowledge[5] and materials, supported by existing legal and regulatory thickets, leads to a plethora of inequitable (unintended/unforeseen) consequences:

(i) Inequitable exclusion of farmers from seed markets: As stated above, the sale of non-uniform, local, heterogeneous seeds for agriculture was outlawed by several EU countries until recently. Further, the sale of uniform seeds was pre-conditioned on fulfilment of registration and certification requirements,

---

[5] Fullilove in this volume.

which are both complex and expensive. As a result, regulations made it difficult for small and marginal farmers to become seed innovators and sellers of heterogeneous seeds, preventing the development of a marketplace for (local) agrobiodiversity, associated data, and farmers' know-how.

Excluding specific categories of persons from the competitive market for seeds or creating an uneven playing field for their participation, while being legal in most countries of Europe, is against the principle of fairness in ethics, and against equity under human rights law. With the evolution of scientific understanding, there is a growing legal and factual need to facilitate the development of a marketplace for heterogeneous seeds, including farmers' local seeds that are a rich storehouse of agrobiodiversity and PGRs.

The inequitable exclusion of farmers from the domain of seed innovations and seed sales, also leads to several socio-cultural and demographic problems: in several rural areas, farming is no more considered an honourable profession, younger generations prefer to move to cities seeking more promising careers, leading to increased rural-urban migration and alienation from land, local cultures and values. This situation compromises both social and environmental sustainability.

(ii) Inequitable incentive structures under IPR regimes: The structure of existing intellectual property rights (IPR) regimes and associated policies create inequities in incentive structures and in the innovation ecosystem. They are inappropriate or inadequate to recognise, reward and optimally incentivize farmer level (informal) *in situ* innovations on and with locally adapted agrobiodiversity. Indeed, empirical research has revealed that current IPR regimes, because of their design and based on practical matters of legal enforcement, are only able to incentivize formal (private/public sector) innovations (Kochupillai, 2016; Kochupillai 2019a; Kochupillai et al., 2021; Henry & Stiglitz, 2010).

Notably, farmers are the original custodians and generators of agrobiodiversity, i.e. of heterogencous seeds. They are also, traditionally, innovators, actively engaged in the art and science of seed selection, seed saving (storing) and seed improvement from one generation to the next. Yet, decades of policy and legal focus on certified, uniform, "improved seed" for agriculture emerging from the formal sector has reduced farmers to mere "users" and net 'takers' of know-how, leaving them with little negotiation power.[6]

As a result, farmers' art and science of improving local agrobiodiversity either remain unknown, unrecognised, undervalued, or merely labelled as "conservation", although it requires a great deal of innovation through (i) careful seed selection, (ii) locally suited innovations for seed storage, and (iii) indigenous, traditional and locally improvised means of improving soil

---

[6] See Zampati in this volume who also states that "data asymmetries and imbalances as well as monopolies are quite present/dominant in the agricultural sector".

biodiversity and fertility, to ensure not just the survival, but continuous in situ improvement of heterogenous materials over time, contributing to climate resilient agriculture and food/nutritional security (Kochupillai, 2019a).

(iii) Sub-optimal focus on needs of organic and sustainable farming: A focus on high yield-oriented uniform seed multiplication and sales also undermines efforts, including within the EU, of promoting organic and sustainable farming. Unlike conventional farming, natural and organic farming requires intricate knowledge of local biotic and abiotic conditions and local solutions tailor made for these conditions. Lack of this knowledge and associated local solutions reduces the effectiveness of heterogeneous seeds. Under current regulatory regimes, information collection and management systems that compile and disseminate relevant information from marginal environments to support organic agriculture with heterogeneous seeds, have failed to evolve.

(iv) Inadequate importance given to traditional ecological knowledge (TEK)-based farming systems and associated know-how: A growing body of research indicates that successful farming with heterogenous seeds is closely linked with optimal in-soil biodiversity (e.g. microbial diversity in the soil). TEK and farming systems that utilize this knowledge, contain the lost art of maintaining in soil biodiversity such that heterogenous seeds can perform well and even outperform uniform seeds (Kochupillai & Köninger, 2022). While the CBD aims to protect both in and on soil diversity, neither the CBD nor the seed treaty give adequate importance to TEK and associated farming systems. Neither do they recognize or grant economic incentives for farmers to maintain, use and share their know-how of such farming systems or best practices for the cultivation of heterogenous seeds in diverse marginal environments. The lack of recognition and incentives leads to further erosion of agrobiodiversity, and of diverse (sustainable) farming systems that support in situ conservation and improvement of agrobiodiversity.

(v) A culture of IPR infringement: Focus on sales of certified uniform seeds has led to farmer-abandonment of traditional farming systems and local heterogeneous seeds. It has also (inevitably) led to farmer-dependence on uniform, certified seeds. This dependence, perhaps ironically, leads to infringement (by farmers) of plant breeders' rights and patents under the slogan of "seed sovereignty". Countries like India legalize seed saving (including of proprietary seeds) by farmers, leading to the adoption, by seed companies, of business models focused on the creation and sale of F1 hybrids that do not reproduce true to type (Kochupillai, 2016). Seed corporations also increasingly adopt Genetic Use Restriction Technologies (GURTs) for the production of seeds that do not produce any (viable) seeds.

(vi) Overemphasised top-down education: Current agricultural extension services almost exclusively focus on top-down education of farmers in conventional agriculture. More recently, these educational efforts have included the use of blockchain and AI solutions. While these technologies can be used to collect and monetize a wide range of data, currently, data collected are used primarily to inform farmers when to apply chemical inputs and/or when to irrigate. Apart from a few (small) AI companies offering "cultural" solutions (see for example, Plantix), the vast majority of the "big data" collected, analysed, monetized and used for agriculture is controlled by a very few institutions and corporations.

To remedy the above inequities, it is necessary to diversify directions of knowledge and value flows – not just top-down, but also bottom-up, i.e., from farmers engaged in agriculture with agrobiodiversity to other farmers and to the formal sector (e.g. research centres). A bottom-up flow of knowledge would facilitate a more equitable and sustainable exchange of data. In such a scenario, farmers could (i) contribute their valuable knowledge of sustainable agricultural practices and in situ agrobiodiversity conservation to one another and to the research community; (ii) actively engage as innovators in the seed innovation ecosystem instead of being uninvolved consumers of agricultural innovations devised by top-down producers that prioritise repeated consumption and up-selling.

A bottom-up flow requires, first, systems incentivising an increase in supply and demand for agrobiodiversity, e.g. through concrete monetary benefits. Second, such systems should also be capable of overcoming hurdles that long disuse (or misuse e.g. biopiracy) of agrobiodiversity have created. Third, the created system needs to be based on a comprehensive understanding of the components of agrobiodiversity that can be monetized. Monetizable components of agrobiodiversity, and hurdles preventing the emergence of a marketplace therefor, are described in the following sub-sections. Finally, the created system should permit collective governance by all stakeholders or by representatives of diverse stakeholders, all of who can participate actively, and be incentivized to participate, in the equitable system.

2.2 Hurdles Preventing the Emergence of a Marketplace for Agrobiodiversity

The following hurdles lie in the way of creating a robust international marketplace for agrobiodiversity:

1. Sources lost or unknown: There is a concrete problem in identifying existing (remaining) *in situ* sources of agrobiodiversity. Indeed, interactions with academic researchers reveal that when these researchers go back to regions from where acquisitions of local agrobiodiversity were previously made, they find that farmers have abandoned this heterogenous agrobiodiversity and replaced it with "improved" homogenous materials.[7]

---

[7]E.g. interaction with Dr. Charlotte Allender, Warwick School of Life Sciences, during the Conference: "Plant Variety Protection Debates: Connecting Law, Science and Social Science" Warwick University, 14 June 2018. See https://www.floraip.com/2018/07/10/plant-variety-protection-debates-connecting-law-science-and-social-science/ (last accessed 01 April 2022).

2. Concentrated agro-seed industry: Four companies currently control 56% of the proprietary seed supply and 70% of the global seed market (Howard, 2009). These companies have amassed a huge proprietary reserve of germplasm and do not consider it necessary to deal with complex regulatory hurdles to acquire materials cultivated and improved *in situ* by the informal sector.

3. Lack of incentives: Significant collections of seed agrobiodiversity are in *ex situ* seed banks and are considered to be in the public domain, also giving no (monetary) incentive to the farmer-providers of the genetic resources to continue cultivating and improving these *in situ*. Provisions of the seed treaty that do not mandate benefit sharing unless these materials are used for the creation of proprietary varieties, also leave farmers with no incentives to share their materials. While pockets of efforts are seen in government initiatives to recognize plant genome saviour communities (see India's Plant Variety Protection and Farmers' Rights Act, 2001), the system leaves several farmers and farmers' communities who have not been recognized (so far), feeling disillusioned. Moreover, the one-time recognition does not provide farmers with the opportunity to share their innovations in return for continuing economic benefits or to further develop means of benefiting from their innovations. Sub-optimal incentives for *in situ* agrobiodiversity conservation/improvement also prevent the sharing of associated materials and know-how.

4. Lack of incentives (continues): Acquisitions of 'samples' of agrobiodiversity (soil microbial diversity and crop diversity), are regularly made by research centres and private entities from farmers or other informal sources. These acquisitions are often made without giving any meaningful monetary or other benefit/ recognition to communities that have provided the material and associated information. When made through the 'back-door' to avoid existing regulatory hurdles within national biodiversity protection laws, such acquisitions are labelled as biopiracy, and provided they are caught/identified, are punishable offences. Biopiracy, whether or not it is caught and punished, leads to loss of trust amongst stakeholders, further reducing farmer-incentives to share their agrobiodiversity and associated know-how.

5. Lack of knowledge for reconversion: Adoption of chemical intensive farming encouraged by most of the agro-seed industry, while enhancing yields, has also led to unaccountable loss of in-soil (microbial) diversity. Heterogeneous seeds are not designed to perform in soils treated with chemical inputs that display reduced soil microbial diversity. Regions where farmers adopt conventional agriculture, may, over time, need to abandon local/heterogeneous seeds as chemical fertiliser residues have been known to interfere with the performance of local heterogeneous seeds. The concrete problem here is lack of knowledge on how to convert back to traditional/organic/sustainable agriculture while maintaining income/ profit levels. Perhaps ironically, Traditional Ecological Knowledge (TEK) based farming systems such as Natural Farming in India, contain know-how for rapid reconversion of conventional soils and farmers to agrobiodiversity preserving systems. Unfortunately, such systems are not so far recognized or known in

Europe and several of the necessary farm-made formulations are likely to be outlawed by existing (top-down) agricultural regulations.

6. Loss of diversity in farming systems: Recent scientific research reveals that farming methods that promote seed/soil health and diversity also enhance crop yields (Zhou et al., 2017). Yet, regulatory regimes have, until recently, not only out-lawed the sale of non-uniform (local, heterogeneous) seeds and materials for agriculture, but have also failed to support the parallel evolution of diverse, including traditional farming systems (e.g. "Natural Farming" and other systems of Agroecology). The lack of diversity in education, and lost knowledge of traditional, sustainable farming systems and practices, results in mono-cultures and uni-directional scientific progress. As discussed above, it also excludes farmers who are actual and potential innovators, from participating in the seed innovation ecosystem, and equitably benefiting from such participation.

7. Lack of trust and of systems supporting traceability of materials to the source: Recent EU legislation (EU 848/2018) attempts to reverse the above trends by facilitating easy marketing and registration of "heterogeneous materials", a term that is broad enough to include indigenous or locally adapted, non-uniform, farmers seeds. Yet, more fundamental issues remain unaddressed, particularly issues linked to the lack of trust in sharing agrobio resources (fear of biopiracy), and the absence of means that facilitate their traceability to source. These issues, again, disincentivize wider dissemination (sharing, selling) of agrobiodiversity and agricultural best practices associated with their cultivation and improvement *in situ*. This leaves farmer-innovators who generate and improve agrobiodiversity economically weak and socially marginalized.

To overcome these hurdles and take concrete steps for the creation of an equitable marketplace for agrobiodiversity, it is necessary to first understand the monetizable components of agrobiodiversity. This has been done in the following sub-section.

2.3 Identifying Monetizable Components of Agrobiodiversity

Plant genetic materials co-evolve with their surrounding microorganisms, forming the holobiont. These assemblages between plants and soil microbes are critical for plant health, e.g. by helping suppress diseases. In sustainable farming systems, the soil surrounding the plant root, called the rhizosphere, is particularly rich in beneficial microbiological activity. The more diverse the rhizosphere, the better the symbiotic exchange between plants and microorganisms, such as nutrient exchange (Van Der Heijden et al., 2016), resulting in higher nutrient content in the plant, vegetable or crop (Sangabriel-Conde et al., 2014). The beneficial microbial population in the soil, called the soil microbiome, is also influenced by the plant genotype.

While heterogeneous plant materials are more dependent on microbial synergies for nutrient access, improved varieties are more dependent on external inputs (Cobb et al., 2016). According to an expert in soil microbiology, "there is currently great interest in developing sustainable farming systems in which the plant microbiome is utilised to support plant nutrition and health, replacing the use of fertilisers and pesticides. This may include manipulating the microbiome through crop genotype or by using microbial inoculants with specific functional traits. Wild genotypes and cultivars adapted to low input systems are vital tools for the development of these resources".[8]

Therefore, it is not just the commercial potential of heterogeneous seeds and PGRs that is growing but also the commercial potential of soil microbial communities associated with specific crops and crop species that can be used to create microbial rather than chemical fertilisers (Velmourougane et al., 2017). It also appears likely that soil microbial populations that get optimized with traditional and sustainable farming systems may be unique to each location and to each crop. Soil systems that are linked to traditional farming systems are, therefore, like potential goldmines.

Accordingly, the following components of agrobiodiversity are monetizable (at least potentially):

(i) Heterogeneous seeds, including those with any locally or traditionally known qualities and characteristics (e.g. disease resistance or drought/flood resistance) of the seed;
(ii) Produce/food resulting from cultivation of heterogenous seed, including knowledge of any locally or traditionally known qualities and characteristics (e.g. nutrient content, medicinal properties) of the food/produce;
(iii) Beneficial soil microbial populations and diversity therein corresponding with specific seeds/plants, soil types and farming methods;
(iv) Farmers' know-how, which has evolved from local Traditional Ecological Knowledge (TEK)-based farming systems, on best practices for soil management, on-farm (in situ) seed storage and seed selection/conservation/ improvement.

Yet, under current legal and regulatory regimes, none of these categories are directly monetizable by farmer-custodians of agrobiodiversity.

Current and ongoing research suggests that the revival and wider application of traditional agricultural systems is necessary to ensure food security in the face of climate change (Nyong et al., 2007), and especially so in marginal conditions facing unique biotic and abiotic stresses (Roberts & Mattoo, 2018). There is, therefore, an urgent need to establish systems that incentivize:

(i) the use of agrobiodiversity in agriculture;
(ii) the identification of the value inherent in such agrobiodiversity, including local, heterogeneous seeds, associated farming systems and soil microbial diversity;
(iii) the transparent and traceable transfers/sharing of this agrobiodiversity (and associated data and knowledge) to diverse end-users for diverse end purposes

---

[8]Email correspondence from Prof. Gary Bending, soil microbiologist, University of Warwick, dated 14 July 2019.

(e.g. to other farmers for successful migration to organic/sustainable agriculture, to researchers for further R&D, to consumers for consumption etc.), and
(iv) fair, equitable and assured transfer of monetary benefits to farmers and rural
communities that share local agrobiodiversity and associated data and know-
how.

In the following section, we discuss whether and how blockchain and AI-based
technologies can help in reviving and (re)establishing such systems.

3 Diversifying Directions of Knowledge, Data and Value Flows: Can Blockchain and AI-Based Solutions Help?

The General Data Protection Regulation (GDPR), as well as EU's general policy on
(big) data, demand fair data collection and processing, for justified, limited purpose,
without sacrificing privacy. The regulations also aim to facilitate data trading. The
rights guaranteed to those who contribute data or permit its collection and use
include the right to revoke permission and/or grant permission at will in return for
monetary compensation (data as currency) (Duch-Brown et al., 2017). There is also
an EU wide effort to create a market for non-personal data. Several technological
solutions can support the creation and growth of a European (or global) Data
Economy, including data emerging from farmers' fields, managed and transferred
for the benefit of the grassroots (small and subsistence farmers) and for the benefit of
the environment (agrobiodiversity protection). To collect, manage and monetize at
least the four major monetizable components of agrobiodiversity (Sect. 2.3), the
hurdles identified in Sect. 2.2 need to be overcome.

3.1 Blockchain/DLT for In Situ Innovations with Agrobiodiversity: Creating Incentives

Blockchain technology, or the more generic distributed ledger technologies (DLTs),
permit secure data collection, arrangement and storage, as also the transfer of data in
an immutable or change sensitive manner (Drescher, 2017). DLTs can, therefore,
help create a decentralized system that is much more trustworthy than a 'centralized'
system managed by a third-party intermediary. Additionally, "smart contracts" can
be appended to blockchain/DLT solutions to automatically trigger a series of digital
occurrences as soon as a pre-determined set of conditions is fulfilled.[9]

Blockchain/DLTs also supports anonymization of users, ensuring that Personal
Identifiable Information (PII) of farmers and other stakeholders can be secured. At

---

[9]For a simple explanation of blockchain, see Kochupillai 2020a and Drescher, 2017.

the same time, the system can also be designed to facilitate limited or conditional disclosure of identities of contributing parties, either on their choice or, in case of need (e.g. for purposes of legal enforcement, facilitation of payments/encashment and/or correction of technical glitches).

The code underlying the system can help reduce transaction costs, delays, and a host of other problems, including problems that may be a result of corruption or breakdown of one of the nodes/computers in the system.[10] These features of blockchain/DLT technology[11] can help overcome issues of lack of trust and trace-ability in transfer (or sharing) of agrobiodiversity and its monetizable components. However, to provide stronger economic incentives for in situ conservation and innovation with agrobiodiversity and associated research, it is also necessary to acknowledge and identify the 'value' of agrobiodiversity (material and knowledge), and adopt systems that help identify (or 'mine') and monetize this 'value'.

Blockchain has been called the internet of 'value' (Tapscott & Tapscott, 2017).[12] The width and ambiguity of the term 'value' itself is responsible for the diversity of use cases and business models that blockchain (potentially) facilitates and promotes. At a very fundamental level, the 'value' of anything is very subjective: anything has 'value' because people believe it to have value. As the number of people who believe or place subjective value on anything increases, the apparent objective value of the thing also increases (Tapscott & Tapscott, 2017; Swan, 2015). For example, when people demand more bitcoins, this demand adds (monetary) value to bitcoins. When people demand more agrobiodiversity or produce derived from heterogenous seeds, its value also increases.

Currencies, including digital currencies, are essentially a medium of storing and trading 'value" (Yermack, 2015; Allee, 2008). Digital currencies currently exist in the form of digital payment systems (such as credit cards), in the form of digital reward points (such as airline mileage points, grocery and other marketplace pur-chase points or customer loyalty points), and in the form of cryptocurrencies. Blockchains or DLTs, together with cryptocurrencies or other point/reward systems linked to them, can help store and transfer the "value" associated with any underly-ing data, know-how or material being transferred or traded via a DLT backbone. In fact, blockchain/DLTs can also be designed as an incentive system, and preliminary research suggests that blockchain based systems may be well suited to incentivize and promote in situ conservation, research and innovation with agrobiodiversity (Kochupillai et al., 2021). This can be done, for example, by building smart contracts that work with the DLT system to automatically transfer cryptocurrencies, points or

---

[10] See also Harrison & Caccamo in this volume: "The use of distributed ledger technology (DLT) in certification could lead to less cumbersome processes and could in fact increase adoption of certification systems."

[11] For a simple explanation of all major features of blockchain technology, see KOCHUPILLAI, M. 04/2020. Blockchain for Equitable and Sustainable Agriculture. *European Seed*.

[12] For a more detailed understanding of the meaning „value" and how it is „mined" in the blockchain sense, see KOCHUPILLAI, M. 2019. 'Mining' the 'Value' of 'Work': Can Blockchain Incentivize Agrobiodiversity Conservation & Improvement? *Preprint*.

rewards to specific "nodes" or participants in the system, when specific criteria are met (without the need for an intermediary).

For example, in order to trigger a transfer of reward points to:

(i) a farmer, the criteria to be fulfilled by the farmer can include the contribution or transfer (into/via the blockchain based system) of a pre-defined "packet" of know-how, traditional knowledge or materials. Farmers can also add to or build on the knowledge contributed by other farmers on the system by adding insights relevant to their own unique agro-climatic, biotic and abiotic farming conditions.

(ii) a researcher, the criteria to be fulfilled by the researcher can include the contribution or transfer (into/via the blockchain based system) of research results testing and verifying (or rejecting) the validity or usability of the know-how or materials contributed by farmers from specific agro-climatic contexts. Researchers can also win reward points by contributing test results indicating or verifying specific unique features of seeds (e.g. resistance to specific biotic or abiotic stresses) or specific unique nutritional, medicinal or other properties of the produce resulting from indigenous/heterogenous seeds contributed by farmers to the system. Such research based verifications of properties of agrobiodiversity (seeds or produce) can add additional objective "value" to agrobiodiversity, creating a more robust marketplace therefor.

From the perspective of governance, especially governance systems that are inclusive, it is noteworthy that smart contracts can be designed based on conditions decided or dictated by farming communities participating/contributing to the system. Indeed, smart contracts designed based on needs/demands of local (farming) communities are more likely to facilitate participation in the DLT/blockchain based system.

Further, in previous works, we have suggested that a smart contract facilitated automated payment system (e.g., a one-time payment or automatic payment of royalty to originators of seed innovations, namely, farmers) be supported by a point-based reward system (instead of a cryptocurrency) (Kochupillai et al., 2019). This suggestion is important, not least because of the volatility of cryptocurrencies, high energy-consumption and costs associated with confirming/checking transfers of cryptocurrency, and the diversity of national laws that currently support or outlaw cryptocurrencies. Accordingly, it is worth investigating whether blockchain/DLT facilitated mechanisms to incentivize research and *in situ* innovation with agrobiodiversity can be linked with a simple (non-cryptographic and low energy consuming) automated point granting systems similar to existing systems that reward "carbon points". Such systems can then support point trading similar to "carbon trading" or "emission trading."[13]

---

[13] Trading of BioPoints was also announced by ComBank to boost biodiversity. See Palmer, D. 2019. CommBank Develops Blockchain Market to Boost Biodiversity. Available from: https://www.coindesk.com/commbank-develops-blockchain-marketplace-to-boost-sustainable-development-biodiversity

Collected points can also be exchanged for real cash (fiat currencies) from one or more of several possible sources, such as:

(i) established funds like the 'Gene Fund' or the 'Biodiversity Fund' under National laws (see, for example, the Indian Biodiversity Act, 2002)
(ii) from a fund maintained through the collection of a possible 'biodiversity tax' from sellers of uniform/homogenous seeds and chemical inputs that contribute to the depletion of agrobiodiversity; or
(iii) from exchanging points for money from industries that would want to acquire biodiversity points to avoid paying a possible 'biodiversity tax'.

In relation to (iii) above, industries that do not support the *in situ* conservation and improvement of agrobiodiversity could be required to pay a biodiversity tax unless they can show legitimate acquisition of "biodiversity points". Such points can be acquired, for example, by supporting institutions and farmers engaged in research and *in situ* conservation and innovation with agrobiodiversity.

Adopting the above envisaged system could help create incentives for downstream users and researchers to use rather than avoid the blockchain/DLT facilitated traceability and incentive system in two ways:

(i) by ensuring that points collected can be exchanged for cash from one (or more) of the above-suggested sources, and
(ii) by permitting the specific contribution of each farmer/farmer community and research institutions to be immutably recorded and known to the rest of the world.

Therefore, if any research and innovation with agrobiodiversity is done 'outside' the system (e.g. through illegally acquired PGRs), neither will the farmer contributor of the agrobiodiversity (or associated know-how) get royalties for his/her contribution, nor will the downstream researchers (whether these be other farmers or scientists) get point-based rewards for (research-based or new product development based) value addition or verification.

3.2   Incentivizing 'Work' on and with Agrobiodiversity and Associated Know-How

To illustrate how a blockchain/DLT-based incentive system might look like, we take the example of 'Sona Moti' – an ancient indigenous wheat seed recently (re)-discovered and named in India by the Art of Living Foundation (an international NGO) (Kochupillai, 2020a, b; Kopytko, 2019). Sona Moti is an ancient Emmer wheat variety that was re-discovered in rural Punjab. After its rediscovery, Sona Moti was found to have a particularly high folic acid content.[14] Its survival in cultivation depends on demand, and price can fluctuate significantly based on supply.

---

[14] For a more detailed account of the story of Sona Moti, see https://idip.leeds.ac.uk/category/ farmers-stories/

Further, Sona Moti farmers claim that its unique characteristics (taste, texture, nutritional content, quality of yield etc.) are associated with the method of farming they adopt for its cultivation. Specifically, these farmers use "Natural Farming", a compilation of farming practices sourced from Indian TEK systems. Natural Farming, as popularly practised in India, combines organic farming practices (no mineral fertilisers, pesticides, etc.) with the treatment and preparation of several formulations that recycle nutrients from farm waste and animal (particularly cow) manure to create organic fertilisers (the effect of which is similar to that of plant biostimulants), natural pest repellents, and seed germination enhancers.

Let us imagine a farming community cultivating Sona Moti, supported by an NGO, wants to sell Sona Moti via a blockchain based platform. We can expect the following to be the most relevant stakeholder categories in the system:

(i) **the public and private sector research community** that may wish to buy Sona Moti via the blockchain platform and can be given incentives to add or 'mine' value from it either through research on its characteristics or through research aimed at creating improved varieties through breeding activities,

(ii) **end consumers**, who may wish to buy Sona Moti for its specific nutritive value, i.e. for consumption;

(iii) **other farmers and seed multipliers** who may wish to buy Sona Moti as seed for their own cultivation and sale purposes. Such farmers may also be given incentives to contribute additional location specific know-how on best practices for the cultivation of Sona Moti;

(iv) **governmental bodies such as seed certifiers, organic certification agencies, the Biodiversity (Protection) Authority** etc., could be given incentives to test and certify the seeds;

(v) **non-governmental organizations** could be given incentives to help identify regions where farmers cultivate with heterogenous Sona Moti seeds, help farmers with data entry and help avoid or minimize the problem of fake entries;

(vi) **corporations**, who may wish to buy Sona Moti can either be given incentives to "mine" value through downstream research, or simply to support the packaging, labelling, (organic) certification and sales of seeds and produce.

The benefits blockchain could bring to such an environment include the following:

(a) Price discrimination based on stakeholder-categories and envisaged end uses: For example, farmers as well as seed multipliers, based on their landholding size, can be asked to pay lower prices or be permitted to buy their first Sona Moti seed samples free of cost. Corporations and research centres, can be required to pay a high(er) fee for each packet of information/know-how or seeds. Further, farmers can also receive payments under the following categories:

   (i) bulk payments from research purpose acquisitions,
   (ii) royalties every time multiplied seeds of Sona Moti or of improved varieties thereof are sold by downstream farmers, seed multipliers, or innovators, and
   (iii) per bag sale value from end consumers.

(b) Enhancing transparency and traceability to avoid biopiracy: Blockchain's capability to immutably record transaction history can be used together with regulatory frameworks such as "Know Your Customer" frameworks (Michael et al., 2018) (that are already mandated in countries like the US) to ensure that every category of user can be accurately identified and every sale of Sona Moti (as seed or grain) is recorded. This would help ensure that illegitimate/illegal transactions and use of agrobiodiversity or PGRs are minimized.

(c) Incentivizing honesty and legitimate use of the blockchain system: As discussed above, to prevent the appearance of parallel black markets, all stakeholders can be incentivized to use rather than avoid the blockchain based system. This can be done by granting automatic payments/rewards with the help of smart contracts to relevant stakeholders, whenever they have legitimately sourced the know-how or materials from the blockchain and added "value" to it – e.g. by confirming or testing the farmers' contributions through research, or by packaging or certifying farmer seeds/produce, etc. Those who acquire the materials or know-how off chain, would naturally be unable to access or claim such reward points or payments.

(d) Incentivizing continuous value addition: Given the nature of heterogenous seeds, their features are likely to vary based on the location of cultivation. This would permit each farmer-buyer of seed to further 'mine' 'value' by recording additional knowledge on cultivation best practices, local features etc. on the blockchain/DLT backbone, based on *in situ* cultivation and innovation in their own local areas. Over time, the knowledge base would be rich enough to permit farmers from various regions to successfully cultivate Sona Moti by accessing the rich knowledge base from the blockchain based system, and (optionally) contributing more localized know-how back into the system.

(e) Incentivizing and funding further R&D through ICOs: Based on the envisaged expansion of demand for Sona Moti seeds, the general public may be interested in buying and trading Sona Moti coins linked to the blockchain system, via initial coin offerings (ICOs). In addition to the point-based reward system, such ICOs can also be used to support further R&D on and with agrobiodiverstiy, thereby creating incentives for the *in situ* conservation and improvement of local heterogenous varieties. Such systems can, in this way, also support the creation of marketplaces for agrobiodiversity by enhancing demand for it among researchers and end consumers, and creating a transparent and trustworthy means of ensuring supply.

3.3   AI and ML-Based Searching of Data Managed and Governed Under a Blockchain/DLT Based System

Artificial Intelligence (AI) technologies, especially machine learning systems can utilize the data collected and managed by a blockchain/DLT based system to design apps that bring unique and custom-made information and solutions to farmers and researchers alike. For example, imagine a farmer is engaged in the cultivation of Sona Moti. The farmer shares information about the unique properties of the seed and resulting grain, namely, high folic acid content and low glycaemic index via a blockchain app. Scientists and plant breeders in search of similar properties can search and access this information using an AI based search app built on top of the blockchain system. Researchers can also test the applicability (or even the veracity) of the information shared by farmers in diverse farming conditions and share back their findings through the AI app.

Thus, AI apps, in combination with blockchain/DLT based systems, can not only help solve problems of trust and traceability but also support equitable data collection, meaningful querying of the collected data, and transparent downstream data usage, making the data monetizable and its usage controllable by those who contribute it. As AI and DLT based solutions can facilitate the collection of and access to disciplinary or geographic area-specific information from and by farmers, researchers and other stakeholders, governments across the globe, including in the EU and India, are looking at such solutions for sustainable agriculture.

4 Implementing the AI/Blockchain Solution: Legal and Ethical Considerations

To accomplish the goals of diversifying directions of knowledge and value flows and overcoming existing regulatory hurdles and inequities with the help of Blockchain/ DLT and AI solutions, it is necessary to revisit and partially amend existing regulatory schemes.

4.1 Blockchain and AI for Agrobiodiversity: Necessary Regulatory Amendments

(i) Disengage Benefit Sharing from Downstream IPR Protection

In the context of the Seed Treaty (and the Indian PPV & FR Act, 2001), it is necessary to re-think the current legal provisions that mandate benefit sharing only if the downstream research with PGRs is protected by IPRs (Patnaik et al., 2018). This current limitation may result from the (outdated) scientific understanding that the Mendelian "genes for traits" (and associated management of biotic and abiotic stresses) approach is the only way to accomplish food security (Radick, 2016; Kochupillai & Köninger, 2022). Today, the understanding has evolved. Heterogeneous seeds are recognised as crucial not only for food and nutritional security, but also for sustainable agriculture in the face of rapid climate change (as also noted in the preamble of new EU organic regulations) (Ficiciyan et al., 2018; Martínez-Nieto et al., 2020).

Further, farmer-custodians of agrobiodiversity are not just technology-takers but indispensable partners for the long-term continuation of formal innovations by the public and private sector seed industry and the organic fertiliser and pesticide industry. Ethics, equity, economics as well as common sense, therefore, dictate that farmer-contributors of agrobiodiversity get royalties in addition to significant initial (bulk) payments for sharing their agrobiodiversity and know-how. By incentivizing *in situ* agrobiodiversity conservation and improvement through long-term benefit sharing with farmers (such that both on-soil (crop/seed) and in-soil diversity is protected and enhanced), the research community as well as the private sector (breeders and corporations), would also sustainably benefit in the long run.[15]

Accordingly, it is necessary to re-think current laws limiting benefit sharing for PGR access only to cases where the downstream varieties are protected by IPRs or are utilized in a hybridization program.

(ii) Benefit Sharing for Access to Soil Microbial Diversity from TEK-Based Farming Systems:

For long-term food and nutritional security, incentives and monetary benefits must be secured not only for farmers cultivating locally relevant heterogeneous seeds but must also accrue to:

(a) farmers/communities who generate and share knowledge and information about *how best* to cultivate heterogenous seeds in specific local conditions to get the best results (vis-a-vis yields, nutritional quality, unique taste, aroma or medicinal properties), and
(b) those who generate and share knowledge/information about how to optimize beneficial microbial populations within specific soil types and in the context of specific crops.

Blockchain/DLTs, (together with AI applications) can facilitate secure and 'controllable' data sharing by farmer-generators of such know-how and data, while ensuring fair, inclusive and equitable economic benefits for those sharing the same.

Therefore, farming communities engaged with traditional farming practices that enhance seed and soil biodiversity may benefit if 'digital sequence information' associated with biodiversity is brought within the scope of the Nagoya Protocol (Kupferschmidt, 2018).[16]

Coupling such a move with the parallel adoption of concrete means (such as DLT/Blockchain-based solutions) that support the legitimate and traceable transfer of digital information linked to seed and soil biodiversity, will prevent any envisaged

---

[15] See also, Kochupillai (2016) 9–13.

[16] For the integration of digital sequence information into the ITPGRFA of the seed treaty, Manzella et al. suggested a solution in this volume using DOIs (digital object identifiers) to track plant genetic resources. They emphasize the interoperability among data systems in order to implement future legal solutions for benefit-sharing.

slow-down of globally beneficial research (Ibid.) while disincentivizing illegal/inequitable transfers of data/information.

(iii) Permitting farmer level (collective) branding of heterogeneous (local) seeds

Unlike several European countries, countries like India never banned the sales of farmers' heterogeneous seeds and materials. Under its Seed's Bill (PRS India, 2004) that has been pending since 2004, India plans to establish systems that can facilitate the emergence of regional, national as well as international markets for heterogeneous seeds. For example, the mandatory seed certification requirement can help farmers and farmer groups get their varieties quality tested and help them get a brand/denomination for their locally unique seeds (such as Sona Moti). However, this mandatory seed registration requirement has been opposed by farmers and farmer groups because it can create a heavy bureaucratic and financial burden on them. Accordingly, the pending Seeds Bill, based on the recommendations of the last Standing Committee Report (Parliamentary Research Service India, 2004), while seeking to make varietal registration mandatory, still excludes farmers' varieties from mandatory registration and certification (Pal et al., 2007).

However, the Bill bans farmers from selling branded seeds and farmer-to-farmer seed sales and exchanges can only take place in brown bags devoid of brands or other means of recognizing their source (Murdoch et al., 2000; Moschini et al., 2008). This mandate counters the ideal of traceability and prevents the emergence of profitable markets for heterogeneous seeds. It will also place small and subsistence farmers who wish to sell their seeds in the seed/agrobiodiversity trade market, at a disadvantage.

The Indian Seeds Bill, 2004 also "requires every person in the value chain to keep track of the preceding person, so that a faulty lot can be withdrawn." (Parliamentary Research Service India, 2004) To accomplish this goal while eliminating any chance of corruption or human error, and to increase accountability, DLT/Blockchain technologies are not only useful, but may be necessary to ensure meaningful and accurate traceability. DLT/Blockchain technologies may also help tackle the problem of affordability and feasibility of registration while still giving farmers and farmers' association the right to (collectively) brand and sell their seeds if they so desire. Such a system should first be tested at a small scale and then slowly expanded if pilot projects are found to be successful.

Aside from a market for 'uniform', non-variable varieties, it is necessary to permit, in parallel, 'True Labels' that declare the fact of heterogeneity and variability, together with the specific benefits and characteristics the cultivation of such seeds brings to farmers and biodiversity. Supported by digital traceability, distributed certification systems, smart-contract based automated payments, and biodiversity token/point awards, the parallel emergence of a market for agrobiodiversity can be facilitated. Such a marketplace would not only service farms engaged in organic or traditional agriculture, but would also bring both environmental and economic benefits for small and subsistence farmers, while facilitating equitable research and innovation with heterogenous seeds.

Here again, it is noteworthy that DLT/blockchain based systems support decentralized governance models (Zwitter & Hazenberg, 2020) – i.e., all stakeholders or stakeholder representatives can and must contribute to the creation of a governance framework for the system to be successful. This is a key reason why Blockchain systems are expected to facilitate democratization and re-distribution of power structures.

(iv) Re-thinking 'uniformity' and 'genetic purity' requirements in existing regulations

Existing regulations that mandate specific standards of efficacy (such as genetic purity) need to be revised. Genetic purity and uniformity are no longer considered valuable in all circumstances, and especially not in marginal environments. For example, in India, to the extent that the Seeds Bill mandates "genetic and physical purity" of seeds and specific 'limits of variability', it is worth looking into emerging scientific evidence that recommends using genetically diverse seeds (rather than uniform varieties) for sustainable agriculture (Gruber, 2017; Thrupp, 2000; Esquinas-Alcázar, 2005; Jacobsen et al., 2013).

This is also relevant in the context of the new EU organic regulations.[17] The EU Regulation clarifies that 'heterogeneous materials', unlike current proprietary seeds, need not be uniform or stable. Further, the EU's Farm to Fork Strategy aims "to facilitate the registration of seed varieties, including for organic farming, and to ensure easier market access for traditional and locally-adapted varieties." (European Commission, 2020, p. 8)

(v) Adopting regulations for sharing digital sequence information

Both the seed treaty and the CBD have failed to adequately trace exchanged PGRs (Martins et al., 2020; Kamau et al., 2015) and to integrate rules for sharing digital sequence information (Tsioumani, 2019; Prathapan et al., 2018). The creation of robust, fair and transparent digital marketplaces for agrobiodiversity and associated know-how can help overcome current regulatory loopholes. However, such a system can operate more effectively if equitable regulations for sharing digital sequence information (DSI) are adopted by concerned international and national regulatory authorities.

Regulations governing the sharing or transfer of DSI need, particularly, to be aimed at bringing benefits to farmers whose materials and knowhow is responsible for the creation and maintenance of underlying PGRs from where the DSI is mined. Absent such regulations, stakeholders in research centres and corporations can continue to avoid benefit sharing, thereby further disincentivizing in situ agrobiodiversity conservation and improvement.

---

[17] Regulation (EU) 2018/848 of the European Parliament and of the Council of 30 May 2018 on organic production and labelling of organic products and repealing Council Regulation (EC) No 834/2007.

(vi) The inadequacy of schemes that "subsidise" organic farming and conservation of biodiversity.

In the EU, the Common Agricultural Policy (CAP) allows countries to subsidise practices that enhance agricultural diversity (in situ and ex situ) which are then voluntary for farmers to implement (European Commission, 2019a, p. 24).

Although the implementing regulation has been in place since 2014 and 14 Member States have implemented subsidies (European Commission, Article 28, pillar 2, M 10.2), these voluntary mechanisms have failed to bring about notable increase in *in situ* agrobiodiversity conservation. Indeed, the legislation does not envisage mechanisms to document and share relevant hands-on knowledge with farmers wishing to migrate to such practices. European research has, in recent times, aimed to finance means to enhance engagement with agrobiodiversity. For example, EU's Horizon 2020 financed a two-million EUR project, "FarmersPride" (2018–2020), to establish a durable structure for in situ conservation of PGRs (European Commission, 2019b). Yet, to the knowledge of the authors, no existing system establishes a meaningful and trustworthy track and trace system and a concrete equitable incentive system.

4.2 Blockchain and AI for Agrobiodiversity: Flagging Ethical Concerns

A point-based rewards system coupled with blockchain/DLT's ability to capture, store and monetize various categories of 'value' can be a major boon for agrobiodiversity research by both the formal and informal sectors. It can incentivize unbiased, comprehensive research on all aspects of agrobiodiversity use and cultivation, including on human and animal health and environment. Despite numerous potential benefits of DLT/blockchain backed solutions, particularly its capacity to help 'mine' the 'value' of agrobiodiversity and any 'work' linked to it, one cannot overlook the fact that the subjectivity of 'value' can be an asset as well a liability (Palminteri et al., 2017). For example, DLT/Blockchain technology cannot, on its own, prevent the problem of 'garbage in, garbage out'. To minimize any misuse of the technology, in addition to the involvement of trusted third parties, the identification of ethical issues linked to blockchain and AI frameworks becomes crucial.

Some of the major ethical issues that can arise in the context of any blockchain and AI facilitated solution aiming to incentivize research and in situ innovations with agrobiodiversity include:

(i) Fairness, bias and inclusion. To avoid unintended biases and exclusions, checks and balances need to be built into AI and blockchain applications aiming at equitably promoting research and *in situ* innovation with agrobiodiversity by all stakeholders. Further, empirical research is needed to identify what is considered 'fair and inclusive' by contributors (farming communities), vis-à-vis the use of agrobiodiversity by downstream players and what constitutes fair remuneration/royalty for accessing the same. Laws do not regulate the sharing of agricultural data. However, various codes of conduct on agricultural data sharing by contractual agreement (e.g. by the EU) aim to raise awareness about the importance of transparency and rights linked to data by providing guidelines concerning privacy, the security of data and benefits for the data owner.[18] Guidelines for AI/blockchain based applications may be included in such codes of conduct.

(ii) Trust and privacy. Trust enhanced by DLT/Blockchain can enhance the number of legitimate transfers of agrobiodiversity and its monetizable components. However, the degree of trust depends on the design of the blockchain governance model. Adequate social science research engaging all stakeholders must precede the adoption of any governance model. Further, as mentioned previously (above), the governance models of DLT/blockchain systems need to be inclusive – taking into account views and demands of local (farming) communities participating in the system (e.g. when designing smart contracts) and also of all stakeholders (or stakeholder representatives) when designing the overall system's governance framework.

Building trust by inclusive governance frameworks, and ensuring privacy and security, are major concerns, especially at the start of any initiative that is likely to disrupt established systems. This is especially true in countries of the Global South where law enforcement may be problematic. There might be violent retaliation by powerful intermediaries. Skilful deployment of this platform will therefore be necessary, taking local governments as well as law enforcement agencies into confidence and keeping the identity of those contributing to the system confidential as long as necessary.

(iii) Transparency and traceability. While blockchain/DLT solutions enhance transparency and traceability to source vis-à-vis digital data, they are not the best suited to permit traceability of physical goods (such as seeds and soil samples) that are likely to get transformed soon after transfer (Xu et al., 2019; Perboli et al., 2018; Agrawal et al., 2018; Imeri & Khadraoui, 2018). For meaningful traceability of physical material transfers aimed at creating downstream products or information (e.g. materials used to generate DSI), additional technologies such as biomarkers, DNA barcodes etc. will need to be used alongside the blockchain backbone. Assignment of DOIs (Digital Object Identifiers) to such materials as recommended by Manzella et al. in this volume can also support traceability.[19] In fact, solutions such as DOI for agrobiodiversity are also more likely to be used by stakeholders, when

---

[18] See Zampati in this volume pointing out that these codes often neglect the rights of smallholder farmers. The international alliance Global Open Data for Agriculture and Nutrition (GODAN) aims to overcome those shortcomings protecting equally the rights of data contributors.

[19] Manzella et al. propose DOIs as an identifier technology to establish linkages between information, also because they are well-known and well established in research. Collisions that may be caused by locally assigned identifiers could be overcome by a Global Information System (GLIS),

implemented in combination with a DLT/Blockchain-based incentive system. Regulatory and multi-disciplinary issues linked to such technological combinations will, however, need to be investigated.

(iv) Governance, regulation and sustainability. The development and implementation of an AI and blockchain based marketplace for agrobiodiversity will require active interaction with existing governance structures and regulations. This is necessary, inter alia, to ensure a sustainable and seamless transition that maintains and secures meaningful and continuing interaction between human and autonomous actors. Further, some types of blockchain (notably, those that utilize 'proof-of-work' algorithms) are not sustainable due to their use of large quantities of energy.

However, in the context of this article, it is necessary to not reduce blockchain technology to bitcoins or other cryptocurrencies, the verification of transfers of which is what involves the most amount of "computing power", and therefore, of high energy consumption. Indeed, even in the context of cryptocurrencies and their transfer, recent advances in blockchain technology, which utilize consensus mechanisms based on proof-of-stake rather than proof-of-work, consume much less energy (Gallersdörfer et al., 2020; Kang et al., 2018). Algorand, for example, is a blockchain that utilizes proof of stake (Platt et al., 2021). Also, the bloxberg blockchain designed by the Max Planck Institute reduces energy consumption by replacing proof of work by proof of authority (https://bloxberg.org/). Further, as discussed above, the blockchain/ DLT based system envisaged by us is not (necessarily) reliant on cryptocurrencies for rewards or monetary incentives. Instead, we envisage a system that awards points similar to carbon points which are awarded when specific actions are performed. These carbon points need not be made transferable (although, a market for trade of such points can be created on a blockchain backbone), but may be made exchangeable for fiat currencies, thereby avoiding high energy consumption.

(v) Consumer Protection. The volatility of cryptocurrencies has moved several governments to consider banning them and limiting the operation of public permissionless blockchains, inter alia, to protect consumer/investor interests. It is necessary that a broader conceptual understanding of 'mining' the 'value' of 'work' in a blockchain sense, be considered when designing the system envisaged herein.[20] This would permit a shift in perspective, going beyond crypto-tokens to digitized systems that grant points for 'work' (e.g. research, sharing of new data and know-how) that incentivizes *in situ* conservation of and innovation with agrobiodiversity, and the creation of a robust and equitable digital marketplace for it.

---

which assigns DOIs to plant genetic resources to increase the security and accuracy of the exchange of information.

[20] For a more detailed explanation of the this conceptual understanding of "mining" the "value" of "work", see Kochupillai, M. 2019. 'Mining' the 'Value' of 'Work': Can Blockchain Incentivize Agrobiodiversity Conservation & Improvement? *Preprint*.

(vi) Equitable Participation by Stakeholders and Regulators. The benefit of any distributed DLT or Blockchain technology increases with the number of users and contributors; the greater the number of those who are engaged in contributing or testing seeds, soils, cultivation methods etc. on a blockchain, the higher the chances that any user of the system will be able to get an accurate view of the quality of the products and know-how being offered via the blockchain facilitated marketplace. To ensure that the system is not overtaken by vested interests, a large number of users (farmers, researchers, end consumers, government bodies etc.) must be a part of the blockchain network. It would also be necessary to determine which government agencies, NGOs and private players would need to act like check-posts in the system. Further research is also necessary to identify the most appropriate blockchain architecture (public permissioned, public permissionless or other architecture) and governance model for enhancing trust and securing privacy in the short and long term.

(vii) Legal liability. With the emergence of "code" based governance, it is also necessary to see how issues of liability would be reconciled. While blockchain technology can support private ordering and self-governance, in fields as sensitive and important as agriculture, blockchain codes must not be privately ordered.[21] However, semi-private ordering of codes, after consulting farmers, NGOs, scientists and government agencies, may be the best way forward. This can entail the creation of ethical codes via multi-disciplinary research engaging all stakeholders in consultations, or self-regulation by farmers supported by broad legislative guidelines and regulatory check-posts (e.g. mandatory government body nodes in any blockchain architecture created for promoting research and in situ innovation with agrobiodiversity or for the sale/purchase of agrobiodiversity).

(viii) Cultural Diversity. Finally, in a diverse world, fair and inclusive DLT/blockchain governance models must consider cultural diversity, equity, and practical usability for the benefit of farmers, researchers and the environment.

5 Conclusion

There is little doubt that all existing legal rules and regulatory frameworks operating in the sphere of agriculture are established with the best of intentions, also because these laws were likely passed based on the then prevailing or dominant scientific understanding (Louwaars, 2002).

---

[21] See also Harrison & Caccamo who advise the need of public funds to establish economically viable data standards.

However, with the scientific community in a state of flux about what kind of farming system is truly sustainable (from an economic, socio-cultural, environmental and a continuing innovation perspective), laws and policies governing agriculture and associated seed related innovations must also be revisited. In particular, legal regimes, policies and scientific research must not be skewed in favour of one type of farming system over and above other (re)emerging systems that protect, improve and conserve agrobiodiversity *in situ*. Emerging directions of research and policy (e.g. those linked to agroecology, ecosystem services and biodiversity in circular or sustainable farming systems) are particularly relevant; they can help identify or confirm the effectiveness and utility of traditional ecological knowledge-based low-cost and low-tech farming inputs and approaches, incentivize sustainable seed innovations, and support the emergence of an equitable and trustworthy marketplace for agrobiodiversity.

Existing regulations primarily envisage a top-down transfer of knowledge and materials, i.e. from corporations and/or research institutions down to farmers. TEK and agrobiodiversity-based farming systems, however, require local know-how and local seed and soil microbial diversity to enhance overall quality and diversity of crop/produce. Blockchain/DLT and AI-based systems provide an opportunity for bottom-up transfer of knowledge and materials, thereby supporting the diversification of directions of knowledge and value flows, reviving a plurality of knowledge systems and overcoming epistemic injustice.[22] By facilitating access to diverse sources of knowledge and materials, we can also help expand the market for know-how and materials (soil and seed diversity) emerging from these sources, enhancing (small) farmer incomes and protecting agrobiodiversity.

Efforts are underway to revive interest in farming using heterogeneous materials through various legislations and subsidization schemes. These efforts (Winter, 2010; Ewens, 1999) have met with little or sub-optimal success, perhaps due to the previously described lack of incentives and lack of means of transparently and equitably sharing farmer know-how. The European Commission under the Green Deal and its Farm2Fork Strategy is also looking to promote research on and with agrobiodiversity conserved *in situ* (Westengen et al., 2018), as also means of directly marketing produce and products derived from their cultivation to end consumers. (European Commission, 2017). A Blockchain/DLT based solution, as envisaged herein, can support the accomplishment of these goals.

This article attempts a conceptual discussion on how blockchain/DLT and AI-based solutions can help diversify directions of knowledge and value flows by incentivizing the capturing, storing, enhancement and (optional) trading of value in agrobiodiversity. In the agri-food sector, such solutions can also help diversify diets and revive (almost) lost local traditions and cultures linked to food. Most importantly, perhaps, these systems can also help bring back the pride associated with farming as a profession. They can help (small) farmers reclaim their position as innovators that are engaged in improving heterogeneous seeds and soils through

---

[22] See also Louafi et al. in this volume.

their keen observation and, in return for small but significant monetary payments, providing their insights to other farmers and researchers worldwide. At the same time, it is necessary that the research and rollout of such technologies be preceded by thorough multi-disciplinary, multi-stakeholder and multi-cultural research, guiding legal and policymakers globally. Additional empirical research will also be necessary to delve into appropriate governance models for the envisaged blockchain/DLT backed system.

References

Agrawal, T. K., Sharma, A., & Kumar, V. (2018). *Blockchain-based secured traceability system for textile and clothing supply chain.* Springer.

Allee, V. (2008). Value network analysis and value conversion of tangible and intangible assets. *Journal of Intellectual Capital, 9,* 5–24.

Borowiak, C. (2004). Farmers' rights: Intellectual property regimes and the struggle over seeds. *Politics & Society, 32,* 511–543.

CBD. (1992). *Convention on biological diversity.*

Cobb, A. B., Wilson, G. W., Goad, C. L., Bean, S. R., Kaufman, R. C., Herald, T. J., & Wilson, J. D. (2016). The role of arbuscular mycorrhizal fungi in grain production and nutrition of sorghum genotypes: Enhancing sustainability through plant-microbial partnership. *Agriculture, Ecosystems & Environment, 233,* 432–440.

Drescher, D. (2017). *Blockchain basics: A non-technical introduction in 25 steps* (1st ed.). Apress.

Duch-Brown, N., Martens, B., & Mueller-Langer, F. (2017). *The economics of ownership, access and trade in digital data.*

Dulloo, M. E., Hunter, D., & Borelli, T. (2010). Ex situ and in situ conservation of agricultural biodiversity: Major advances and research needs. *Notulae Botanicae Horti Agrobotanici Cluj-Napoca, 38,* 123–135.

Esquinas-Alcázar, J. (2005). Protecting crop genetic diversity for food security: Political, ethical and technical challenges. *Nature Reviews Genetics, 6,* 946.

European Commission. (2017). Networking, partnerships and tools to enhance in situ conservation of European plant genetic resource*s* [Online]. *European Commission.* Available: https://cordis.europa.eu/project/rcn/215955/factsheet/en

European Commission. (2019a). *Evaluation of the impact of the CAP on habitats, landscapes, biodiversity.* https://ec.europa.eu/info/sites/default/files/food-farming-fisheries/key_policies/documents/ext-eval-biodiversity-final-report_2020_en.pdf

European Commission. (2019b). *AGRIresearch factsheet. Genetic resources and breeding.* https://ec.europa.eu/info/sites/default/files/food-farming-fisheries/farming/documents/factsheet-agri-genetic-resources_en.pdf

European Commission. (2020). *Farm to fork strategy – Green deal.* https://ec.europa.eu/info/sites/default/files/communication-annex-farm-fork-green-deal_en.pdf

European Commission, Article 28, pillar 2 of the CAP, M 10.2 Genetic resources in agriculture: "*Support for conservation and sustainable use and development of genetic resources in agriculture and forestry*". https://enrd.ec.europa.eu/sites/default/files/rdp_analysis_m10-2.pdf

Ewens, L. E. (1999). Seed wars: Biotechnology, intellectual property, and the quest for high yield seeds. *Boston College International and Comparative Law Review, 23,* 285.

FAO. (1999). *What is happening to agrobiodiversity?*

FAO. (2011). *Plant genetic resources and food security Christine Frison* (F. L., & Esquinas-Alcázar, J. T., Eds.).

Ficiciyan, A., Loos, J., Sievers-Glotzbach, S., & Tscharntke, T. (2018). More than yield: Ecosystem services of traditional versus modern crop varieties revisited. *Sustainability, 10*, 2834.

Fraleigh, B., & Davidson, C. G. (2003). Overview of the international treaty on plant genetic resources for food and agriculture with emphasis on its significance for horticultural crops. In P. L. Forsline, C. Fideghelli, K. Richards, A. Meerow, H. Knupffer, J. Niens, A. Stoner, E. Thorn, A. F. C. Tombolato, & D. Williams (Eds.), *Plant genetic resources: The fabric of horticulture's future*.

Fusi, F., Welch, E. W., & Siciliano, M. (2019). Barriers and facilitators of access to biological material for international research: The role of institutions and networks. *Science and Public Policy, 46*, 275–289.

Gallersdörfer, U., Klaaßen, L., and Stoll, C. (2020). Energy consumption of cryptocurrencies beyond bitcoin. *Joule, 4*(9), 1843–1846.

Gruber, K. (2017). Agrobiodiversity: The living library. *Nature, 544*, S8–S8.

Halewood, M., Chiurugwi, T., Sackville Hamilton, R., Kurtz, B., Marden, E., Welch, E., Michiels, F., Mozafari, J., Sabran, M., & Patron, N. (2018). Plant genetic resources for food and agriculture: Opportunities and challenges emerging from the science and information technology revolution. *New Phytologist, 217*, 1407–1419.

Henry, C., & Stiglitz, J. E. (2010). Intellectual property, dissemination of innovation and sustainable development. *Global Policy, 1*, 237–251.

Howard, P. (2009). Visualizing consolidation in the global seed industry: 1996–2008. *Sustainability, 1*, 1266–1287.

Imeri, A., & Khadraoui, D. (2018). The security and traceability of shared information in the process of transportation of dangerous goods. In *9th IFIP international conference on new technologies, mobility and security (NTMS), 2018* (pp. 1–5). IEEE.

Jacobsen, S.-E., Sørensen, M., Pedersen, S. M., & Weiner, J. (2013). Feeding the world: Genetically modified crops versus agricultural biodiversity. *Agronomy for Sustainable Development, 33*, 651–662.

Kamau, E. C., Winter, G., & Stoll, P.-T. (2015). *Research and development on genetic resources: Public domain approaches in implementing the nagoya protocol*. Routledge.

Kang, J., Xiong, Z., Niyato, D., Wang, P., Ye, D., & Kim, D. I. (2018). Incentivizing consensus propagation in proof-of-stake based Consortium Blockchain Networks. *IEEE Wireless Communications Letters*, 1–1. https://doi.org/10.1109/lwc.2018.2864758

Kochupillai, M. (2016). *Promoting sustainable innovations in plant varieties*. Springer.

Kochupillai, M. (2019a). *'Mining' the 'Value' of 'Work': Can Blockchain incentivize agrobiodiversity conservation & improvement?* Preprint.

Kochupillai, M. (2019b). *Is UPOV 1991 a good fit for developing countries?* Innovation Society and Intellectual Property.

Kochupillai, M. (2020a). *Blockchain for equitable and sustainable agriculture*. European Seed.

Kochupillai, M. (2020b). *Blockchain for biodiversity: The benefits for the environment and for farmers*. Available from: https://european-seed.com/2020/05/blockchain-for-biodiversity-the-benefits-for-the-environment-and-for-farmers/

Kochupillai, M., & Köninger, J. (2022). Cast into the stones of international law: A critique of the UPOV standards and the underlying welfare and scientific assumptions they globalize. In A. Metzger & H. G. Ruse-Khan (Eds.), *Intellectual property ordering beyond borders*. Forthcoming with Cambridge University Press.

Kochupillai, M., Radick, G., Rao, P., Kopytko, N., Köninger, J., & Matthiessen, J. (2019). *Incentivizing & promoting sustainable seed innovations in india: A three-pronged approach*. Position Paper Submitted to the Government of India. https://doi.org/10.13140/RG.2.2.12670.74568/1

Kochupillai, M., Gallersdörfer, U., Köninger, J., & Beck, R. (2021). Incentivizing research & innovation with agrobiodiversity conserved in situ: Possibilities and limitations of a blockchain-based solution. *Journal of Cleaner Production*, 127155.

Kopytko, N. (2019). *The sustainable seed innovations project: The story of 'Sona Moti'*. Available from: https://spicyip.com/2019/07/the-sustainable-seed-innovations-project-the-story-of-sona-moti.html

Kupferschmidt, K. (2018). Biologists raise alarm over changes to biopiracy rules. *Science.* 61(6397): 14.

Louwaars, N. P. (2002). Seed policy, legislation and law: Widening a narrow focus. *Journal of New Seeds, 4*, 1–14.

Martínez-Nieto, M. I., Estrelles, E., Prieto-Mossi, J., Roselló, J., & Soriano, P. (2020). Resilience capacity assessment of the traditional Lima Bean (Phaseolus lunatus L.) landraces facing climate change. *Agronomy, 10*, 758.

Martins, J., Cruz, D., & Vasconcelos, V. (2020). The Nagoya Protocol and its implications on the EU Atlantic Area countries. *Journal of Marine Science and Engineering, 8*, 92–92.

Mcguire, S., & Sperling, L. (2016). Seed systems smallholder farmers use. *Food Security, 8*, 179–195.

Mekonnen, D., & Spielman, D. (2018). Changing patterns in the international movement of crop genetic material: An analysis of global policy drivers and potential consequences. In *30th international conference of agricultural economists*. International Food Policy Research Institute (IFPRI).

Michael, J., Cohn, A., & Butcher, J. R. (2018). Blockchain technology. *The Journal, 1*, 7.

Moschini, G., Menapace, L., & Pick, D. (2008). Geographical indications and the competitive provision of quality in agricultural markets. *American Journal of Agricultural Economics, 90*, 794–812.

Murdoch, J., Marsden, T., & Banks, J. (2000). Quality, nature, and embeddedness: Some theoretical considerations in the context of the food sector. *Economic Geography, 76*, 107–125.

Nyong, A., Adesina, F., & Elasha, B. O. (2007). The value of indigenous knowledge in climate change mitigation and adaptation strategies in the African Sahel. *Mitigation and Adaptation Strategies for global Change, 12*, 787–797.

Pal, S., Tripp, R., & Louwaars, N. P. (2007). Intellectual property rights in plant breeding and biotechnology: Assessing impact on the Indian seed industry. *Economic and Political Weekly*, 231–240.

Palmer, D. (2019). *CommBank develops blockchain market to boost biodiversity*. Available from: https://www.coindesk.com/commbank-develops-blockchain-marketplace-to-boost-sustainable-development-biodiversity

Palminteri, S., Lefebvre, G., Kilford, E. J., & Blakemore, S.-J. (2017). Confirmation bias in human reinforcement learning: Evidence from counterfactual feedback processing. *PLoS Computational Biology, 13*, e1005684.

Parliamentary Research Service India. (2004). *The Seeds Bill, 2004*. Available: https://www.prsindia.org/uploads/media/1167468389/legis1167477737_legislative_brief_seeds_bill.pdf

Patnaik, A., Jongerden, J., & Ruivenkamp, G. (2018). Rights or ability: Access to plant genetic resources in India. *The Journal of World Intellectual Property, 21*, 157–175.

Perboli, G., Musso, S., & Rosano, M. (2018). Blockchain in logistics and supply chain: A lean approach for designing real-world use cases. *IEEE Access, 6*, 62018–62028.

Platt, M., Sedlmeir, J., Platt, D., Tasca, P., Xu, J., Vadgama, N., & Ibañez, J. I. (2021). Energy footprint of blockchain consensus mechanisms beyond proof-of-work. *arXiv preprint. arXiv*, 2109.03667.

Prathapan, K. D., Pethiyagoda, R., Bawa, K. S., Raven, P. H., & Rajan, P. D. (2018). When the cure kills – CBD limits biodiversity research. *Science, 360*, 1405–1406.

PRS India. (2004) *The Seeds Bill* [Online]. PRS India. Available: https://www.prsindia.org/billtrack/the-seeds-bill-2004-104

Radick, G. (2016). Teach students the biology of their time. *Nature News, 533*, 293.

Roberts, D., & Mattoo, A. (2018). Sustainable agriculture – Enhancing environmental benefits, food nutritional quality and building crop resilience to abiotic and biotic stresses. *Agriculture, 8*, 8.

Saatkamp, A., Cochrane, A., Commander, L., Guja, L. K., Jimenez-Alfaro, B., Larson, J., Nicotra, A., Poschlod, P., Silveira, F. A., & Cross, A. T. (2019). A research agenda for seed-trait functional ecology. *New Phytologist, 221*, 1764–1775.

Sangabriel-Conde, W., Negrete-Yankelevich, S., Maldonado-Mendoza, I. E., & Trejo-Aguilar, D. (2014). Native maize landraces from Los Tuxtlas, Mexico show varying mycorrhizal dependency for P uptake. *Biology and Fertility of Soils, 50*, 405–414.

Swan, M. (2015). *Blockchain: Blueprint for a new economy*. O'Reilly Media.

Swanson, T., & Goeschl, T. (2000). Property rights issues involving plant genetic resources: Implications of ownership for economic efficiency. *Ecological Economics, 32*, 75–92.

Tapscott, D., & Tapscott, A. (2017). How blockchain will change organizations. *MIT Sloan Management Review, 58*, 10.

Thrupp, L. A. (2000). Linking agricultural biodiversity and food security: The valuable role of agrobiodiversity for sustainable agriculture. *International Affairs, 76*, 265–281.

Tsioumani, E. (2019). ITPGRFA GB-8: A missed opportunity for multilateralism. *Environmental Policy and Law, 49*, 320–323.

Van De Wouw, M., Kik, C., Van Hintum, T., Van Treuren, R., & Visser, B. (2010). Genetic erosion in crops: Concept, research results and challenges. *Plant Genetic Resources, 8*, 1–15.

Van Der Heijden, M. G., De Bruin, S., Luckerhoff, L., Van Logtestijn, R. S., & Schlaeppi, K. (2016). A widespread plant-fungal-bacterial symbiosis promotes plant biodiversity, plant nutrition and seedling recruitment. *The ISME Journal, 10*, 389.

Velmourougane, K., Prasanna, R., & Saxena, A. K. (2017). Agriculturally important microbial biofilms: Present status and future prospects. *Journal of Basic Microbiology, 57*, 548–573.

Westengen, O. T., Skarbø, K., Mulesa, T. H., & Berg, T. (2018). Access to genes: Linkages between genebanks and farmers' seed systems. *Food Security, 10*, 9–25.

Winter, L. (2010). Cultivating farmers' rights: Reconciling food security, indigenous agriculture, and TRIPS. *Vanderbilt Journal of Transnational Law, 43*, 223.

Xu, X., Lu, Q., Liu, Y., Zhu, L., Yao, H., & Vasilakos, A. V. (2019). Designing blockchain-based applications a case study for imported product traceability. *Future Generation Computer Systems, 92*, 399–406.

Yermack, D. (2015). Is Bitcoin a real currency? An economic appraisal. In *Handbook of digital currency*. Elsevier.

Zhou, X., Liu, J., & Wu, F. (2017). Soil microbial communities in cucumber monoculture and rotation systems and their feedback effects on cucumber seedling growth. *Plant and Soil, 415*, 507–520.

Zwitter, A., & Hazenberg, J. (2020). Decentralized network governance: Blockchain technology and the future of regulation. *Frontiers in Blockchain, 3*, 12.

Part IV
Challenges from/for Communities: Data Linkage Across the Food System

Preface

The fourth and final part of the book focuses on questions of participation in data linkage effort. The significance and meaning ascribed to plant data typically changes depending on who handles them and for which purposes. Recognition and debate around different perceptions and uses of data is therefore crucial to responsible data practices, which play an essential role in connecting stakeholders and facilitating communication across communities. Setting up data linkage systems is not just a technical challenge, but rather a platform to imagine and enact models of transdisciplinary collaborations, which facilitate dialogue among plant researchers, breeders, data scientists, data curators, farmers as well as consumers of crops and other stakeholders in seed and food systems. The contributors to this part propose various such models, including strategies to increase farmer engagement in efforts to link and re-use crop data and improve benefit-sharing mechanisms, and 'communities of practice' for sustained cooperation among stakeholders to improve the scientific and social impact of existing data tools. The chapters highlight the immense opportunities and advantages offered by focusing on heterogeneous communities of practice as well as the financial, administrative, ethical and conceptual obstacles to such engaged, multi-lateral approaches.

Ethical and Legal Considerations in Smart Farming: A Farmer's Perspective

Foteini Zampati

Abstract Smart farming contributes to exponential income growth, enhanced decision making, better services and products, as well as productivity and profitability. Nowadays, numerous agricultural technology providers are entering the market, focusing on aggregating farmers' data. But many farmers, especially smallholders, do not benefit from the sharing and exchange of this data, which leaves them feeling disempowered. Until today, ethical considerations were often side-lined because gathering more data was seen as necessary, and concerns about how data might be abused or misused were only subsequently considered. However, with the increase of big data in smart farming, it is more essential than ever to focus on the ethical aspects of data governance (access, control, consent) and practices. Therefore, these ethical questions will provide valuable insights into how data is being collected and used, for what purposes, how to bridge the digital divide, and how to create transparency and build trust between stakeholders. This chapter will focus on farmers' perspectives and how they could actively participate in a more equitable data sharing and exchange in the agri-food value chain by contributing to the design of a fairer data governance framework. The adoption of agricultural codes of conduct is the example that will be explored.

1 Introduction: The Challenges Posed by Digital Technologies for Farmers

It is evident that the adoption of digital technologies in agriculture has marked the start of a major transformation: Better services and products, innovations, enhanced decision making and increased profitability and productivity (Zampati, 2019). But do smallholder farmers really benefit equally, or even at all, from the benefits of data sharing? Moreover, do all stakeholders in the agricultural sector have the same

F. Zampati (✉)
Chapman Freeborn Airmarketing GmbH, Frankfurt, Germany
e-mail: foteini.zampati@chapmanfreeborn.aero

H. F. Williamson, S. Leonelli (eds.), *Towards Responsible Plant Data Linkage: Data Challenges for Agricultural Research and Development*,
https://doi.org/10.1007/978-3-031-13276-6_13

257

access and control to these insights? What concerns do farmers have on such issues as data ownership, access and control, security and privacy?

Lack of transparency around the above-mentioned issues and whether farm data should be considered 'personal' or not, are some of the data challenges faced by all agricultural stakeholders, particularly farmers. Moreover, data transactions are currently governed by contracts and licensing agreements, in which the terms and agreements are complex. This leaves smallholder farmers with very little negotiating power and it is obvious that a lack of trust dominates these relationships (Wiseman et al., 2019a, b).

The lack of awareness about these rights or the use of data (mostly for farmers) has contributed a lot to an unfair distribution of wealth in the agricultural sector. This perception of inequitable distribution of advantages and disadvantages in the world of production, collection, distribution and use of data is something quite common to agriculture. Global power imbalances have been identified relative to the limited access of some farmers to digital technologies and or to the data they generate (Kshetri, 2014; Rodriguez et al., 2017; De Beer, 2016; Maru et al., 2018; Ferris & Rahman, 2016). This is the so-called digital divide between the developed and the developing countries, which is caused by a lack of means to buy the technologies required for digital farming (Kshetri, 2014: 2; Maru et al., 2018; Ferris & Rahman, 2016) and a lack of scientific data skills among farmers (Ferris & Rahman, 2016: 6). Another reason for this unfair distribution of power is the fact that only large farms are able to pay for the costs of accessing the information based on data, while this is expensive for small-scale farms in developing countries (Ferris & Rahman, 2016: 8; Chaves Posada, 2014), and that recommendations made on the basis of data are not always well suited to the needs of small farms (Rodriguez et al., 2017; Kamilaris et al., 2017; Maru et al., 2018; Ferris & Rahman, 2016). Furthermore, the power imbalance between data contributors and data aggregators is evidenced by the inability of farmers to negotiate the standard terms of the large agri-businesses' data licences that govern the agricultural technology (Carbonell, 2016; Jakku et al., 2018). These ethical concerns have a definite impact in society. As Van de Burg et al. state, "in discussions about if and with whom data ought to be shared, or the desirability of different power (re)distributions, a lot is pre-supposed about the desirability of different impacts of smart farming on society" (Van der Burg et al., 2019).

Overall, farmers are perceived to have little control on data flows and use, information and knowledge. On the one hand their data sharing is hindered by unclear data governance and risks of data misuse, and on the other hand they face the challenge of gaining access to necessary data provided by others. Both types of challenge are linked to inequitable data flows.

In this chapter, we will explore first the challenges that farmers, and specifically smallholder farmers, face in the adoption of digital technologies. What specifically are the reasons that farmers are not fully or at all included in the discussions, design and governance of digital agriculture, even though the benefits are well known? In Sect. 2, we will share some insights from the work of the Global Open Data for Agriculture and Nutrition (GODAN) Initiative to enable farmers to harness the power of data driven agriculture by defining responsibilities among the various stakeholders and by balancing the obvious benefits of data sharing with legitimate concerns in relation to privacy, security, community rights and commercial interests, mostly from the farmers' perspective. In Sect. 3 we will focus on the ethical aspects of data governance (access, control, consent) and practices.

Systems of governance that could support a fairer, equal distribution of benefits such as codes of conduct and the development of the GODAN/CTA/GFAR toolkit on agricultural codes of conduct are presented in Sects. 4 and 5, respectively. Section 6 provides some conclusions.

2 The GODAN Approach

The Global Open Data for Agriculture and Nutrition (GODAN) initiative[1] is an international alliance with voluntary membership that aims to promote the global availability of open data in agriculture in order to stimulate innovation and increase productivity in this important sector. Today, its global network includes over 1100 organizations (spanning governments, international organizations, the private sector and academia) from 118 countries across the world.

More specifically GODAN supports global efforts to make agricultural and nutritionally relevant data available, accessible, and usable for unrestricted use worldwide. The initiative focuses on building high-level policy, and public and private promotion of open data. Its focus is to increase awareness of ongoing activities, innovations and good practices. Another focus is to guide and assist both private and public sector bodies on open data and open access policy, by promoting capacity development and diversity among open data users for more effective accessibility, use, engagement and understanding of open data. In order to do so, GODAN works closely to define actors' responsibilities to respect the rights of all those affected by the release and use of open data, by balancing the obvious benefits of open data with legitimate concerns in relation to privacy, security, community rights and commercial interests.

Through its enormous network of different international stakeholders, GODAN has enhanced the dialogue and actions between key organisations in order to achieve a common understanding and a consensus around the ethical and legal issues concerning agricultural data, data ownership, data rights, privacy, responsibilities and ethics. More specifically, in 2018 GODAN, the Global Forum on Agricultural Research (GFAR), the Technical Centre for Agricultural and Rural Cooperation ACP-EU(CTA) and the Küratorium für Technik und Bauwesen in der Landwirtschaft (KTBL) worked together on the ethical, policy and legal aspects of open data affecting smallholder farmers by engaging with various stakeholders (from governments, the private sector and academia). This collective action included

---

[1] https://www.godan.info/

organising workshops and webinars, participation in major conferences, supporting and advising on the development and implementation of data policies on a national and international level, and work on capacity building for farmers and farmers' organisations. Overall, the focus of this work was in relation to farmers' concerns about data sharing and how to promote food security and sustainability through the opening of access to farm data, primarily in developing countries.

This chapter reports on some of the insights acquired by GODAN in the course of these activities and engagements, with an emphasis on practical concerns and questions of interest both to farmers' rights and to ways forward for researchers aiming to re-use data acquired from farmers.

3 Relevant Ethical Questions

Until today, ethical considerations were often side-lined because gathering more data was seen as necessary, and concerns about how data might be abused or misused were only subsequently considered. However, with the increase of big data in smart farming, it is more essential than ever to focus on the ethical aspects of data governance (access, control, consent) and practices. Big data can contribute to improved profitability and productivity in the agriculture sector. However, there are some challenges to be considered such as access, distribution of benefits, equity, inclusion, data ownership and data rights as well as data governance. It is essential to address questions such as, what opportunities do the digital technologies provide? Do all the actors in the value chain receive the same information and insights? How is trust established between big agribusinesses and farmers? It is well known that whoever owns data may control data insights. By addressing the correct ethical questions and by engaging all stakeholders in the agricultural sector (specifically farmers) in an open dialogue, this will provide valuable insights into how data is being collected and used and for what purposes, how to bridge the digital divide, and how to create transparency in order to build trust between stakeholders.

Since the world of agriculture is quite diverse, it consists of different types of agricultural methods and farming realities. In order to maximize their potential, it is important that digital solutions are designed with a view to the farming communities' needs.[2] This is especially true in African countries with very low literacy levels and limited knowledge of digital technologies, yet where the highest untapped agricultural potential remains. Smallholder farmers are not harnessing the power of data and must overcome challenges and risks to ensure that investments benefit them. In this case, there are two main challenges that need to overcome: first, to gain access to

---

[2] See also Kochupillai and Köninger (this volume): "Artificial Intelligence (AI) technologies, especially machine learning systems can utilize the data collected and managed by a blockchain/DLT based system to design apps that bring unique and custom-made information and solutions to farmers."

relevant data and services provided by others and, second, to make sure that any data they share does not actually weaken their positions.

Data asymmetries and imbalances as well as monopolies are quite present/ dominant in the agricultural sector. These data asymmetries arise when smallholder farmers with rather limited resources reveal their most personal farm data in order to gain access to benefits of technology, while those who can transform the collected data into useful information reveal little to nothing about the back-end processes or how or where the information will be kept or used. Therefore, there is a need to address the question of the balance between the cost of introducing the technology versus the expected benefits for the farmers (Kritikos, 2017). Farmers need to feel and be engaged in the decision process of how collectors will use their data. They also need assurances of their privacy and control; they seek transparency and trust in their interactions with providers; and they would like to receive the benefits of their data and to have access to relevant data.

Because of the above-mentioned situation, important questions and issues have arisen:

- Who owns data?
- Who is entitled to the value of the data?
- How will that data be used or potentially shared?
- What about data protection? What do we mean by farmers' rights to data?
- What is the state of recognition of these rights at the national and international level?
- What is the role of the General Data Protection Regulation (GDPR) in the agricultural sector?
- How should these rights be implemented in local and international laws, guidelines and policies and how can they be protected?
- What should be done to include farmers in the mechanisms of data collection, evaluation, transmission and use?

These issues of course aren't new to people in the agricultural sector. But there is a big need today to address them right and quickly to ensure farmers' rights. It is already well known that ownership as a legal concept is rather complex, and farming data is not traditionally recognised as a type of property that is subject to ownership. The currently available ownership-like rights of data are limited to intellectual property rights (copyright, patents, database rights, trade secrets, plant breeders' rights and trade secrets). However, none of these provide adequate protection of data ownership. In particular, copyright is one way in which data can be owned but data is not always or even normally copyright protected by default. Facts – for example, statistics, formulas, geo-information and news – are not copyrightable (De Beer, 2016).

The European Parliament's 1996 Database Directive establishes sui generis, i.e. unique, rights in databases that fall short of the standard of an intellectual creation required by copyright law (De Beer, 2016). Database creators have the right to prevent extraction and/or reuse of the whole or of a substantial part of the contents of a database. To gain this protection, the database creator must establish that there has been a substantial investment in the obtaining, verification, or presentation of the contents. The term of protection is 15 years, but it is renewable whenever the database holder makes any substantial change to the contents of the database (Wiseman et al. 2019a, b). Patents and plant breeders' rights do not protect data directly, but can nonetheless limit the ability to use data related to innovations in agriculture (De Beer, 2016).

It is quite difficult to define data ownership in farm data. For example, even where data is protected by copyright law the ownership of copyright can be varied by contract (Wiseman et al. 2019a, b). Therefore, it is better to ask the question who has control and access to farm data, rather than asking who owns data. In many cases, farmers own data generated from their farms but they have little control over who or how their data is going to be used.

4 Practical Solutions: Codes of Conduct

While laws and regulations that govern personal data (such as The European General Data Protection Regulation or GDPR[3]) are becoming increasingly common, there is a lack of legislation covering the collection, sharing and use of data in agriculture (Zampati, 2019).

It is worth mentioning though the recent EU Regulation on the free flow of non-personal data, and its relevance for digital agriculture. Just one year after launching the General Data Protection Regulation, the EU launched a new regulation about the control of non-personal data, which defines data on precision farming as non-personal data. This highlights the need for more analysis to achieve a clearer distinction on personal versus non-personal farm data, which would help alleviate privacy concerns going forward. Appropriately, the new regulation emphasizes the importance of self-regulation within the data economy: It encourages the development of industry-specific codes of conduct, allowing for transparent, structured and seamless sharing of data between service providers.

Consequently, to steer a new paradigm of agricultural data governance, there is a need to develop transparent data sharing codes of conduct, and self-regulation that responds to the situation and needs of communities and balances the distribution of benefits between actors in the agricultural value chain. Therefore, codes of conduct have started to emerge to fill the legislative void and to set common standards for data sharing contracts: codes provide principles that the signatories/subscribers/ members agree to apply in their contracts. Farm data is an example of such sensitive data flows. Farm data flows go from the farm to many other actors (extension agents/ advisory service providers/agri-tech companies, farmers' associations, financial service providers, government, etc.) and then – aggregated and combined and in the form of services – back to the farm. Such flows potentially open up data that

---

[3] https://eugdpr.org/

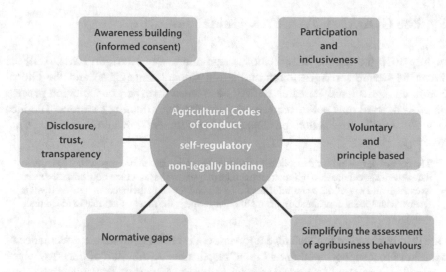

Fig. 1 Key characteristics of agricultural codes of conduct (Zampati, 2019)

should only be shared with specific actors under specific conditions, or should be anonymised in order not to harm the farmer's interests and privacy. This is especially true in the case of smallholder farmers whose farm data often coincides with household and personal data and who are in the weakest position to negotiate their data rights (Wiseman et al. 2019a, b).

Currently, there are five main agriculture data codes: the US American Farm Bureau Federations' Privacy and Security Principles for Farm Data,[4] the New Zealand Farm Data Code,[5] the EU Code of Conduct on Agricultural Data Sharing by Contractual Agreement,[6] the French *Charte sur l'utilisation des données agricoles* (French Charter on the use of agricultural data) and very recently in 2020 the Australian Farm Data Code. The existing codes of conduct cover central issues such as terminology, data ownership, data rights (including right to access, data portability, and the right to erasure/right to be forgotten), privacy issues, security, consent, disclosure and transparency. In general, these codes of conduct attempt to harness the benefits of ag-data while protecting producers' privacy and security. Even though they are not legally binding (they are a form of self-regulation that relies on the goodwill and social responsibility of industry and agribusinesses), these codes help build awareness around data use and sharing and the importance of transparency in agricultural data flows, they change the way agribusinesses view data, and they make data producers – primarily farmers – more aware of their rights (Wiseman et al. 2019a, b; Sanderson et al., 2018) (Fig. 1).

---

[4] https://www.agdatatransparent.com/principles

[5] http://www.farmdatacode.org.nz/

[6] https://cema-agri.org/images/publications/brochures/EU_Code_of_conduct_on_agricultural_data_sharing_by_contractual_agreement_2020_ENGLISH.pdf

5   The GODAN/CTA/CFAR Guidelines

In July 2018 the Global Open Data for Agriculture and Nutrition (GODAN), the Technical Centre for Agricultural and Rural Cooperation (CTA) and the Global Forum on Agricultural Research (GFAR) convened an expert consultation process **on ethical, legal and policy aspects of data sharing affecting farmers**. The idea was for a collective action on *Empowering Farmers through Equitable Data Sharing*.

> The core of our vision for the collective action is that farmers can be empowered to harness data driven agriculture through **inclusive data ecosystems that nurture equitable sharing, exchange and use of data** and information by all and for all participants in agri-food value chains, with special consideration of smallholder farmers, the most vulnerable to inequitable data flows.[7]

One key requirement for such an empowerment is enabling equitable governance of data flows that support a fairer and more responsible distribution of benefits, where transactions are based on mutual interest and trust. As part of this collective action GODAN/CTA/GFAR decided to focus on developing better data management practices through the adoption and implementation of agricultural data codes of conduct, voluntary guidelines and principles. The suggestion of the development of codes of conduct, voluntary guidelines and principles specifically comes at a time where there is no clear legal framework for farm data sharing, as mentioned above.

GODAN, CTA and GFAR launched in May 2020 an online tool on agricultural codes of conduct. This tool was created by initially reviewing existing codes of conduct, voluntary guidelines and principles relevant for farm data sharing. It was drafted as part of the consultative process taking place in the GODAN/CTA Sub-Group on Data Codes of Conduct. The aim and purpose of this tool is to provide a guide to best data management practice to farmers and agri-businesses and associations who collect, manage and share their data. It has an additional practical purpose: to provide the conceptual basis for general scalable guidelines for everyone dealing with the production, ownership, sharing and use of data in agriculture. An interesting point from GODAN and its partners' farmer-oriented perspective is that, as the review showed, the existing farm data codes do not have farmers or farmers organisations as their primary target audience – not to mention smallholder farmers – but rather agribusinesses and ag tech companies that work with farmers and use their data. Codes of conduct are an instrument for these companies to ensure data sharing by gaining the trust of farmers through transparent documentation of good practices. So, while being prepared by bodies that represent also farmers and indirectly raising farmers awareness of their data rights, they are not written primarily for farmers and so far surely not for smallholder farmers (Wiseman et al. 2019a, b). Therefore, with these general scalable guidelines, this tool provides guidance for associations of smallholder farmers in developing countries on how to use/adjust/negotiate/set up a

---

[7] https://www.gfar.net/documents/vision-and-strategic-plan-collective-action-empowering-farmers-through-equitable-data

farmer-centred farm data sharing code. A key point of these guidelines is the essential role of trusted organisations like farmers' cooperatives in interpreting/ contributing to/negotiating the code for their farmers (Wiseman et al. 2019a, b; Sanderson et al., 2018).

These Guidelines help to produce a guidance list to consider when sharing or collecting agricultural data with partners. They aim to:

- Raise awareness around the collection, use and sharing of farm data.
- Improve transparency, clarity and honesty in the way farm data is collected, used and shared.
- Encourage the fair and equitable collection, use and sharing of farm data in a way that benefits farmers.
- Build trust and confidence in the way farm data is collected, used and shared so that, where appropriate, farm data can be utilised in ways that bring benefits to agriculture.
- Allow flexible implementation, so that providers can establish appropriate practices around farm data collection, use and sharing.

The tool features 17 clauses from which the users should be able to select a clause relevant to their situation and proceed to a checkout where the selected clauses can be used as a document. The clauses are as follows:

1. **Definitions**: A list of definitions that are relevant to the agricultural sector (e.g. exactly what types of data are going to be collected, what is considered personal data, agricultural data, individual farm data, raw data, aggregated data, data originator, data provider, etc.).
2. **Ability to control and access**: Farmers, in particular whoever has produced/ collected data on their farming operations, cither by technical means or manually, or who has commissioned data providers for this purpose, have a leading role in controlling the access to and use of data from their business and the right to benefit from sharing the data with any partner that wishes to use their data. Providers should preserve the ability of the farmer to determine who can access and use individual farm data. However, it would be good for the farmer to agree upon data use and sharing with the other stakeholders who share an economic interest, such as the tenant, landowner, cooperative, owner of the precision agriculture system hardware, and/or an Ag Tech Provider (ATP), etc.
3. **Consent for collection, access, control**: Collection, access and use of farm data should be granted only with the affirmative and explicit consent of the farmer. Via a contractual arrangement the collection, access, storage and use of agricultural data can occur only with the explicit informed permission of the data originator. Consent must be freely given, specific, informed and unambiguous. In order to obtain freely given consent, it must be given on a voluntary basis. The element "free" implies a real choice by the farmer. Any element of inappropriate pressure or influence which could affect the outcome of that choice renders the consent invalid. For consent to be informed and specific, the farmer must at least be notified about the provider's identity, what kind of data will be

processed, how it will be used, to whom it will be disclosed and the purpose of the processing. The farmer must also be informed about his or her right to withdraw consent anytime.

4. **Purpose limitation**: Providers must only collect, use and share farm data for the purposes that have been made clear to the farmer. No reuse of data is allowed for different purposes to those that had been originally agreed.
5. **Notice**: Data originators (farmers) must be notified that their data is being collected and about how and to whom the farm data will be disclosed.
6. **Transparency and consistency**: Data originators (farmers) should be notified about what types of farm data is being collected, as well the purposes for which agribusinesses and ATPs collect, use and share data in a more transparent way (e.g. algorithms). In addition, information should be provided about how farmers can contact e.g. the ATPs with any inquiries or complaints, and also become aware of the third parties to whom their data is disclosed and any risks that may affect farmers who share data with the providers.

   All agribusinesses' and Ag Tech Providers' policies, principles and practices should be transparent and consistent with the terms and conditions in the legal contacts. No contract change can be effective without the other party's agreement.
7. **Rights of the data originator**: Within the context of the agreement and retention policy, the data originator (farmer) should be able to have the following rights:

   - Right to portability: Data providers should be responsible for making individual farm data easily available to farmers.
   - Farmers should be able to retrieve their individual farm data in both processed (cleaned) and unprocessed form for storage or use in other systems, with the exception of data that has been anonymised or aggregated and is no longer specifically identifiable.
   - Right to remove, destroy, erase (the right to be forgotten) or return data to the data originator.

8. **Right to benefit**: providers should recognise the originator's right to benefit or be compensated for the use of data they originated.
9. **Disclosure, use and sale limitation**: An agribusiness or an Ag Tech Provider will not sell and/or disclose individual farm data to a third party without first securing a legally binding commitment to be bound by the same terms and conditions as the ATP has with the farmer. Farmers must be notified if such a sale is going to take place and have the option to opt out or have their data removed prior to that sale. An Ag Tech Provider will not share or disclose original farm data with a third party in any manner that is inconsistent with the contract with the farmer. If the agreement with the third party is not the same as the agreement with the ATP, farmers must be presented with the third party's terms for agreement or rejection.
10. **Data retention and availability**: Each agribusiness or Ag Tech Provider should provide for the removal, secure destruction and return of individual farm data

from the farmer's account upon the request of the farmer or after a pre-agreed period of time. The ATP should include a requirement that farmers have access to the data that an ATP holds during that data retention period.

11. **Contract termination**: Farmers must be given the possibility to opt out of the contract and terminate the collection and usage of their data provided that it's stated in the contract and the data originator is informed about the consequences. Procedures for termination of services should be clearly defined in the contract.

12. **Unlawful or anti-competitive activities**: Data should not be used for unlawful or anti-competitive activities, such as the use of farm data by agribusinesses or Ag Tech Providers to speculate in commodity markets (e.g. price discrimination).

13. **Data protection safeguards**: The contract should mention responsibilities and measures for farmers' privacy, security and confidentiality that data users/providers should take. Farm data should be protected with security safeguards against risks such as loss or unauthorised access, destruction, use, modification or disclosure. Notification policies and measures in the event of a breach should be established.

14. **Liability and Protection of IP rights**: Terms of liability should be defined. The contract should also acknowledge the rights of all parties to protect sensitive information via restrictions on further use or processing. Protection of sensitive data such as personal/financial data, confidential information, trade secrets or intellectual property rights against tampering should be ensured.

15. **Simple and Understandable Contracts**: Providers should be responsible for making a clear contract that is easily understandable to farmers. Contracts for ag data should use simple and plain language. In addition, contracts will clearly specify: (1) important terms and definitions, (2) the purpose of collecting, sharing, and processing data, (3) rights and obligations of parties related to data, (4) information related to storage and use of ag data, (5) verification mechanisms for the data originator, and (6) transparent mechanisms for adding new uses.

16. **Certification Schemes**: Data certification schemes develop transparency and trust around data uses. Codes or accreditation requirements can be monitored through the establishment of an independent Supervisory Authority to evaluate whether contracts comply with these principles. Compliance with the codes of conduct should be rewarded. All stakeholders who respect these principles should submit their contracts and policies for evaluation by an audit team of an independent accredited organization. Upon evaluation a certificate of compliance will be issued.

17. **Compliance with the National and International Laws**. All stakeholders who work and develop Codes of Conduct shall comply with local and international laws.[8]

---

[8]The content of the clauses is available at: https://www.godan.info/codes

These principles and guidelines intend to set common standards for data sharing contracts by touching on significant topics such as definitions, access and control, consent, data rights, and certification schemes. For example, clause 2 refers to access and control. It is actually stated that data collected from farmers should remain farmers property. This clause aims to ensure that only the data for which farmers have given permission is used and shared and that the farmer continues to own all data created by his or her operations. In clause 3, collection, access and use of farm data should be permitted only with farmers' consent. Farmers should be granted appropriate and easy access to their own data unless the aggregated data is not linked to farmer ownership. There is a need to ensure that farmers get a return from sharing their data and that they are informed in a clear and unambiguous way when their data is being collected, used and shared. In this way, making farmers owners of their data and providing opportunities to control the flow of their data to various stakeholders should help build trust with farmers for exchanging data and harvesting the benefits of big data. In addition, when third parties are involved in data collection on farming the third party should have an agreement with the farmer to ensure farmers data availability, access and control of his data. As such, the farmer would be more involved in the discussions and would be able to better control who gets the data produced by his or her technology devices or machines and what exactly can be done with it. More importantly, it should also be recognized that farmers have the right to benefit from the use of data produced on the farm during farming operations where different stakeholders are involved, and the benefits of data sharing should be returned to the farmers (Kritikos, 2017). Furthermore, another topic that is dealt with is the need for simple and plain language within the contracts: "*All contracts should use simple and understandable language and clearly define the purposes for which the data can be used, ensuring that any transfer or change to the data is traceable.*"

In clause 16 another important aspect that is promoted with the codes of conduct is the enhancement of the development of certification schemes. Certifications allow farmers to identify technology providers whose data management practices adhere to certain criteria set out by a standard setting or accreditation body. These standards are geared towards ensuring open and transparent data practices, particularly around data collection, processing and sharing, and data storage and security (Jouanjean et al., 2020).

An example is the Ag-Data Transparency Evaluator. A process was developed in 2016 to certify the Ag Tech Providers whose contracts complied with the 13 principles of the American Farm Bureau Federation's Privacy and Security Principles ("Principles for Farm Data", 2014). This tool, in which ATPs voluntarily submit their data contracts to a ten-question evaluation, was created by the American Farm Bureau Federation and is backed by a consortium of farm industry groups, commodity organisations and technology providers. The Evaluator allows ATPs to assess themselves against the Principles for Farm Data in regard to compliance.

Answers to these questions, plus the ATP's contracts and policies, are submitted to, and reviewed by, an independent third-party administrator (the law firm Janzen Agricultural Law LLC). Once reviewed, the results are posted on a website for farmers and other agricultural stakeholders to consult and review. If ATPs receive approval from Janzen Agricultural Law LLC, they can use the "Ag-Data Transparent" seal. The use of the seal informs farmers that the ATP's approach to data management is in line with the Principles for Farm Data.

From January 2020 the "Ag Data Transparent" accreditation and evaluation process has been updated to reflect the growing awareness of the need for data rights that protect the individual.[9] The scope of Ag Data Transparency evaluation was expanded to include the farm financial sector. It was decided that extending the Seal to the farm financial sector provided some additional protection of farmers' privacy (Jouanjean et al., 2020). Tech providers seeking accreditation now need to answer eleven new and updated questions about how they collect, use, share, and safeguard farmers' data. The updates contain precisions such as the types of data, or the nature of the data user. The issue of user data ownership and consent, including whether a user can opt out, is also explored and the practice of companies selling data to third parties is also taken into consideration.

These clauses developed with the GODAN/CTA/GFAR tool are not intended to be exhaustive and are no substitute for a robust institutional framework to guide and operationalize decision making concerning privacy, ethics, and so on. Overall, these guidelines for farm data sharing provide a voluntary framework. They are designed for use and consultation within national legislation. This online tool aims to describe the shared responsibility of many sectors, addresses the need for a cooperative effort, recognizes the need for capacity-strengthening for its implementation and describes the standards of conduct for fairer and more responsible data management, complementing the existing legally binding instruments.

It is also an evolving tool, with recommendations for a general, scalable and further customisable code of conduct template that best addresses farmers' needs around fair and responsible data sharing. GODAN in collaboration with Youths in Technology and Development Uganda (YITEDEV) is going to work together on the toolkit on codes of conduct in order to empower smallholder farmers and specifically youth and women, and to raise awareness about their rights when negotiating with various stakeholders. This project follows the proposal during the Expert Consultation on ethical, legal and policy aspects of data sharing affecting farmers in Bonn in 2018 to take Uganda as a pilot case. A virtual workshop will be organised to introduce to farmers the concepts of open data, privacy and data rights and to increase understanding. A second workshop will then be organised focused on the codes of conduct toolkit and to get feedback from the farmers.

---

[9] https://www.aglaw.us/janzenaglaw/2021/1/26/ag-data-transparent-updates-for-2021

6 Conclusion

No one can really doubt the power and potential of modern technologies in agriculture. Digital developments such as AI, internet of things (IoT), blockchain and autonomous systems are contributing to efficiency, productivity and profitability. The rate of technological advance is accelerating and doesn't look set to slow down. Yet data issues are altogether more complex in agriculture as there are many and varied actors throughout the agri-food data value chain, with their own specific needs. Many farmers, especially smallholders, do not personally benefit from the sharing and exchange of data that epitomizes the digital age, leaving them feeling disempowered. Monopolies, data asymmetries, discrimination and the lack of transparency and trust as well as lack of legislation and regulation on data ownership, data rights and privacy issues are some of the basic challenges that farmers face in relation to digital agriculture. Therefore in many cases they are often reluctant to share their data because they either feel it might be unsafe, or are unaware of its value. The fact that they don't get any benefit from it only serves to increase their reluctance.

The lack of inclusion of farmers in the design and governance of digital agriculture processes contributes to widening the gap between bigger and smaller stakeholders, and in reality smallholder farmers don't fully embrace the benefits of digital agriculture. As stated in this volume by Devare et al., "an effective data governance strategy must recognize that governance is primarily about people and not directly about tools or technologies. These latter are important but are far from the sole determinants of process and organizational efforts".

In this chapter we addressed farmers' concerns in relation to the adoption of digital technologies in agriculture. We looked at systems of governance that could support a fairer, equal distribution of benefits, where transactions could be based on mutual interest and trust. One such system is the development of agricultural codes of conduct. These principles and guidelines are a means of improving transparency and fairness in agricultural data contracts, and, as such, they can be a viable option to support farmers in their relationship with technology providers and foster trust around digital technologies (Wiseman et al. 2019a, b). It is worthwhile mentioning that despite agricultural codes of conduct being voluntary and not legally binding, they can nevertheless contribute to major cultural shifts, as they provide a solid framework for best practice in data management through the engagement of stakeholders at every level (including and especially farmers) in open dialogue to find solutions that address their differing needs and concerns. This approach can also serve to strengthen trust throughout the data value chain.

References

Carbonell, I. (2016). The ethics of big data in agriculture. *Internet Policy Review, 5*. https://doi.org/10.14763/2016.1.405

Chaves Posada, J. (2014). *Rights of farmers for data, information and knowledge*. Global Forum on Agricultural Research. http://www.gfar.net/sites/default/files/rights_of_farmers_for_data_infor mation_and_knowledge.pdf Accessed 28 Sept 2021

De Beer, J. (2016). *Ownership of open data: Governance options for agriculture and nutrition*. Global Open Data for Agriculture and Nutrition.

Ferris, L., & Rahman, Z. (2016). *Responsible data in agriculture*. GODAN. Available via F100 Research. https://f1000research.com/documents/6-1306. Accessed 28 Sept 2021

Jakku, E., Taylor, B., Fleming, A., Mason, C., Fielke, S., Sounnes, S., & Thorburn, P. (2018). "If they don't tell us what they do with it, why would we trust them?" Trust, transparency and benefit-sharing in smart farming. *NJAS – Wageningen Journal of Life Sciences*. https://doi.org/10.1016/j.njas.2018.11.002

Janzen, T. (2021). Ag data transparent: Updates 2021. *Janzen Ag Tech Blog*. https://www.aglaw.us/janzenaglaw/2021/1/26/ag-data-transparent-updates-for-2021. Accessed 28 Sept 2021.

Jouanjean, M., Casalini, F., Wiseman, L., & Gray, E. (2020). *Issues around data governance in the digital transformation of agriculture: The farmers' perspective* (OECD Food, Agriculture and Fisheries Papers, No. 146). OECD Publishing. https://doi.org/10.1787/53ecf2ab-en

Kamilaris, A., Kartakoullis, A., & Prenafeta-Boldú, F. X. (2017). A review on the practice of big data analysis in agriculture. *Computers and Electronics in Agriculture, 143*, 23–37. https://doi.org/10.1016/j.compag.2017.09.037

Kritikos, M. (2017). *Precision agriculture in Europe: Legal, social and ethical considerations*. Scientific Foresight Unit (STOA).

Kshetri, N. (2014). The emerging role of big data in key development issues: Opportunities, challenges, and concerns. *Big Data & Society, 1*(2), 1–20. https://doi.org/10.1177/2053951714564227

Maru, A., Berne, D., De Beer, J., Ballantyne, P., Pesce, V., Kalyesubula, S., Fourie, N., Addison, C., Collett, A., & Chaves, P. (2018). *Digital and data-driven agriculture: Harnessing the power of data for smallholders*. GFAR, GODAN, CTA. Available via F100 Research. https://f1000research.com/documents/7-525. Assessed 28 Sept 2021

Rodriguez, D., de Voil, P., Rufino, M. C., Odendo, M., & van Wijk, M. T. (2017). To mulch or to munch? Big modelling of big data. *Agricultural Systems, 153*, 32–42. https://doi.org/10.1016/j.agsy.2017.01.010

Sanderson, J., Wiseman, L., & Poncini, S. (2018). What's behind the ag-data logo? An examination of voluntary agricultural data codes of practice. *Rural Law and Policy*. https://doi.org/10.5130/ijrlp.1.2018.6043

Van der Burg, S., Bogaardt, M., & Wolfert, S. (2019). Ethics of smart farming: Current questions and directions for responsible innovation towards the future. *NJAS – Wageningen Journal of Life Sciences*. https://doi.org/10.1016/j.njas.2019.01.001

Wiseman L, Pesce V, Zampati F, Sullivan S, Addison C, Drolet J. (2019a) *Review of codes of conduct, voluntary guidelines and principles relevant for farm data sharing*. CTA Working Paper 19/01. https://cgspace.cgiar.org/handle/10568/106587. Accessed 28 Sept 2021.

Wiseman, L., Sanderson, J., Zhang, A., & Jakku, E. (2019b). Farmers and their data: An examination of farmers' reluctance to share their data through the lens of the laws impacting smart farming. *NJAS – Wageningen Journal of Life Sciences*. https://doi.org/10.1016/j.njas.2019.04.007

Zampati, F. (2019). Providing an ethical approach to big data in agriculture. *CTA Spore Magazine*. https://spore.cta.int/en/opinions/article/codes-of-conduct-providing-an-ethical-approach-to-big-data-in-agriculture-sid0700a1b64-6f93-4c70-aa77-186c152b2b1d. Accessed 28 Sept 2021.

Communities of Practice in Crop Diversity Management: From Data to Collaborative Governance

Selim Louafi, Mathieu Thomas, Frédérique Jankowski, Christian Leclerc, Adeline Barnaud, Servane Baufumé, Alexandre Guichardaz, Hélène Joly, Vanesse Labeyrie, Morgane Leclercq, Alihou Ndiaye, Jean-Louis Pham, Christine Raimond, Alexandrine Rey, Abdoul-Aziz Saïdou, and Ludovic Temple

Abstract Establishing linkage among data of diverse domains (e.g. biological, environmental, socio-economical, and geographical) is critical to address complex multidimensional issues such as food security or sustainable agriculture. The complexity of this challenge increases with the level of heterogeneity of the data but also with the social context of production of datasets, a dimension usually less considered. Building on the experience of a transdisciplinary project on the diversity of crop diversity management systems in West Africa (CoEx), this chapter reflects on the importance to better account for agency for more meaningful, responsible and

S. Louafi (✉) · M. Thomas · C. Leclerc · H. Joly · A.-A. Saïdou
Agricultural Research Centre for International Development (CIRAD), UMR Genetic Improvement and Adaptation of Plants Institute (AGAP), Montpellier, France
e-mail: selim.louafi@cirad.fr; mathieu.thomas@cirad.fr; christian.leclerc@cirad.fr; helene.joly@cirad.fr

F. Jankowski · V. Labeyrie
Agricultural Research Centre for International Development (CIRAD), UMR Savoirs, Environnement et Sociétés (SENS), Montpellier, France
e-mail: frederique.jankowski@cirad.fr; vanesse.labeyrie@cirad.fr

A. Barnaud · J.-L. Pham
Diversity – Adaptation – Plant Development (DIADE), University of Montpellier, French National Research Institute for Sustainable Development (IRD), Montpellier, France
e-mail: adeline.barnaud@ird.fr; jean-louis.pham@ird.fr

S. Baufumé · A. Rey · L. Temple
Agricultural Research Centre for International Development (CIRAD), Montpellier, France
e-mail: servane.baufume@cirad.fr; Alexandrine.Rey@cirad.fr; ludovic.temple@cirad.fr

A. Guichardaz
Independent Consultant, Montpellier, France

M. Leclercq
Faculté de Droit, Université Laval, Québec, Canada
e-mail: morgane.leclercq.1@ulaval.ca

© The Author(s) 2023
H. F. Williamson, S. Leonelli (eds.), *Towards Responsible Plant Data Linkage: Data Challenges for Agricultural Research and Development*,
https://doi.org/10.1007/978-3-031-13276-6_14

efficient plant data linkage. The chapter addresses sequentially the cognitive and political challenges related to data work and the way they could be addressed simultaneously within the same social unit. To do this, we rely on the concept of community of practice (CoP) which gained enormous popularity in relation to data and knowledge management. More than simply a social mechanism for community knowledge management, we show in this contribution that CoP needs to be approached as a social experiment and a terrain of collective situated learning in order to address each challenge and their linkages with respect to data work.

1 Introduction

Establishing linkage among data from diverse domains (e.g. biological, environmental, socio-economical, and geographical) is critical to address complex multidimensional issues such as food security or sustainable agriculture. As illustrated by Rawlings and Davey (among others) in this volume, many technical solutions exist to link heterogeneous datasets. However, the complexity of these technical solutions increases with the level of heterogeneity of the data. It also increases with the social context of production of datasets, a dimension less frequently considered.

Dataset production may be carried out by scientists only. In this case, the difficulty for establishing data linkages would increase with the level of heterogeneity of disciplines and conceptual frameworks involved. The problem becomes even more acute in transdisciplinary contexts in which dataset production is carried out not only by scientists, but also by farmers themselves, or by other stakeholders. In such a complex social context of production of datasets, people may not necessarily share the same background and cognitive references, and they may not necessarily follow the same normative orientations about the way to produce, access, exchange and use plant data.

The diversity of people involved in dataset production implies a diversity of practices. The way people interact and value the knowledge they produce, as well as the rules they adopt about how this knowledge can be used by other people, directly relate to the ethical and political dimensions of data work. Data, datasets and databases do not exist only as a material, numerical or technical product, but also as a result of social processes at work before, during, and after the data production. In addition, different levels of responsibility are engaged before, during, and after the data production: why collect new data? To what extent are the diversity of actors and

A. Ndiaye
Senegalese Association of Peasant Seed Producers (ASPSP) and of the West-African Committee of Peasant Seeds (COASP), Thiès, Senegal

C. Raimond
PRODIG, Pôle de Recherche pour L'Organisation et La Diffusion de L'Information Géographique, Université Paris 1 Panthéon-Sorbonne, Paris, France
e-mail: christine.raimond@univ-paris1.fr

practices considered in data work? Are these actors able to interpret the data they contribute to producing or collecting? Are data production processes and modalities for using and exchanging the knowledge produced collectively defined?

Unfortunately, the technical aspect of data production often tends to quickly take precedence over these social, political and ethical dimensions (Boeckhout et al., 2018). Diversity of practice in data work, and different levels of responsibility, are too often overlooked in discussions about plant data linkage. This chapter asks, to what extent does enhanced understanding and recognition of the practice of data work and the people involved help to make plant data linkage more meaningful, responsible and efficient?

Building on the experience of a transdisciplinary project on the diversity of crop diversity management systems in West Africa, this chapter reflects on the importance of better accounting for agency to achieve more meaningful, responsible and efficient plant data linkage. We argue that this cannot be done in isolation from the technical challenges or – even worse – only once the technical challenges are solved, as their resolution has a direct impact on data quality (metadata) and their actual and legitimate linkages.

This chapter distinguishes between the cognitive and political challenges related to data work. The cognitive challenge refers to how data is produced and interpreted. From a technical point of view, this challenge consists in developing standards for metadata (data about data) and data annotation that are meaningful and computer-readable (Arnaud et al., 2020). Such efforts aim to enhance epistemic accuracy through the production and linkage of multidisciplinary data, which spans genetics, environment, agroecology, biology, and socioeconomics (Arnaud et al., 2020). If such an approach certainly enriches plant knowledge representation, it leaves aside the difficult issue of collective data-making in multi-stakeholder contexts characterized by a great heterogeneity of actors with diverse backgrounds and cognitive references, beyond academic disciplines. Responding to this socio-cognitive challenge obliges us to move away from a vision of the epistemic activity of data production as a passive contemplation of the 'world out there' (Popa et al., 2015) in which each discipline brings additional descriptors to enrich what is implicitly defined as the same entity. We show in this chapter that the concept of community of practice (CoP), by contrast, can help by considering the creation of meaning as a collective production process, negotiated through participation and social interactions. Attention in this paper is paid to the way objects (here seeds) get their meaning and reality in the course of practical activity that involves the relations amongst humans and between humans and non-human entities. In this context, ontologies are no longer about modes of knowing pre-existing entities, but the way objects are enacted in practice (Woolgar & Lezaun, 2013).

The political challenge classically refers to both normative and procedural issues. By normative issues, we understand the underlying conflicting logics, values and assumptions that arise among heterogeneous actors with regard to data content. The procedural issue refers to the various ways in which power and participation are constructed and enacted in data practices (Couldry & Powell, 2014). Plant science crystallizes a significant number of issues related to divergent visions about marketing, quality and certification of seed, intellectual property and access and benefit sharing legislations, and risk management, among others. These legislations impact many actors (farmers and their organizations, NGOs, breeders in the public or private sectors, researchers, genebank managers, policy makers, etc.). The complex landscape of rights and responsibilities and associated institutional frameworks generates tensions among these stakeholder groups, which in turn affect plant data exchange and use practices. Hence, addressing both normative and procedural political challenges requires a critical stance towards the understandings, values and assumptions of the various stakeholders as well as towards the institutional and power structures that shape the current organization of data work.

This chapter addresses sequentially these two challenges related to responsibility and agency in plant data linkage, but is also interested by the way they could be addressed simultaneously within the same social unit. To do this, we rely on the concept of community of practice. CoP has gained enormous popularity in relation to data and knowledge management as a way to cultivate expertise and foster learning. However, more than simply a tool and social mechanism for community knowledge management, we show in this contribution that CoP need to be approached as a social experiment and a terrain of collective situated learning in order to address each challenge and their linkages with respect to data work. We argue that responsible data production and linkages require not only consideration of the diversity of knowledge systems and practices (socio-cognitive challenge) but also the need to enhance the ability of various stakeholders to contribute to meaning production in a context of strong heterogeneity among actors (political/normative challenge). More importantly, we argue that responsible data production and linkages cannot be fully achieved if the political/procedural challenge is not addressed simultaneously with the two others to translate the recognition of the socio-cognitive and political/normative challenges into concrete changes in everyday practices and organization of data work. Addressing this last challenge, which is too often overlooked or reduced to its managerial dimension, requires enhancing our understanding of the way the community of practice acts on itself to manage collective data work.

The paper is broken down into four sections. In the first section, we present the case study of the CoEx project and the way data work has been organized and conducted within this project. We then present the versatile concept of CoP and describe how it could apply to the collaborative context of the CoEx project. The following sections describe how CoEx has addressed respectively the three challenges. We conclude by discussing the relationship between these three challenges and the extent to which they offer a way to combine in a fruitful way both the managerial and situated learning dimensions of CoP.

2   The CoEx Project

CoEx is a 4-year collaborative (2016–2020) project funded by Agropolis Foundation and constructed as a collective and multi-actor inquiry on crop diversity management systems in West Africa. This collective gathered researchers from various disciplines as well as farmers' organizations and NGOs in Burkina Faso, Canada, France, Mali, Niger and Senegal with the overall objective of providing a more accurate picture of actual practices surrounding seed acquisition, uses and exchange, beyond the usual "formal" and "informal" binary division that still predominates international and national legal and policy frameworks.

The so-called formal system is based on breeding programs and is organized around the release of genetically uniform certified seeds in a market in which farmers are end users. The so-called informal system covers genetically heterogeneous seeds selected, produced and distributed by farmers and their organizations, generally in a subsistence economy (Almekinders et al., 1994).

The reality of farming practices regarding seed management is not so clear-cut, with interactions and a continuum between these two systems (Louwaars, 2007). Faced with the diversity of farmers' and consumers' demands, the diversity of production contexts and the diversity of crop types, the diversification of seed supply sources is an essential strategy for food security and sustainable agriculture. Indeed, by promoting the diversity of plants and cultivated varieties, the diversification of supply sources also favors the resilience of agricultural systems through stabilization of yields over time, nutritional improvement or adaptation to climate change (Labeyrie et al., 2021). Hence, reconciling the legal and policy frameworks surrounding crop diversity management systems with the diversity of actors, rules, standards and practices is a key challenge for sustainable agriculture.

However, abandoning the binary vision in order to characterize the plurality of crop diversity management practices with a more refined approach represents a definite methodological challenge: How to characterise the most diverse situations throughout the world while being as faithful as possible to what farmers are experiencing, without falling into analyses that are too context-specific?

To this end, a conceptual framework general enough to accommodate a diversity of knowledge systems and specific enough to provide relevant applicable knowledge for various stakeholders (Popa et al., 2015) was established. This framework was based on the concept of social-ecological systems (SES) that accounts for the intertwined social and biological dimensions of resource management (Berkes & Folke, 1998; Anderies et al., 2004; Ostrom, 2009; Young et al., 2006; Folke et al., 2005). The social-ecological seed system focuses on the relationships and interplay between resources (seed, crops), actors (farmers, sellers, community seed bank managers, researchers, breeders, etc.) and institutions (understood as the rules, accepted norms, standard procedures according to which individuals and organizations think and act) within a specific environment defined as a socio-ecological context (Labeyrie et al., 2021). All kinds of relationships between resource systems, resource units, users and governance systems, acting at multiple levels within the same system, were considered without any established hierarchy. For example, in a classical unidirectional human/ecological interaction model, communities establish rules to (sustainably) manage resources but this framework also allows accounting for the other way around, i.e. the way resources (in this case, seeds) 'create' a community with specific types of attachments and relationships.

Three major farmer-crop relationships were considered within CoEx: seed access, crop choice and use of the harvest. Data work to characterise these relationships involved collecting quantitative and qualitative information and covered different disciplines and protocols, some of them co-constructed with farmers' organizations. Moreover, specific attention was paid within CoEx to agreeing on data sharing and management practices as part of a collaborative governance approach of the project and in which any output was considered as a commons.

3   Community of Practice in the Context of CoEx

Communities of practice have gained enormous popularity in relation to knowledge management. They have emerged as a powerful governance mechanism to address complex problems by fostering spaces for collaboration and opportunities across wide areas of expertise, geographies and actors. Building on information and communication tools and coupled with the movement of big data, they are today presented as a way to facilitate connectivity and leverage maximum impact from data by accelerating linkages (see https://bigdata.cgiar.org/communities-of-practice/ and Bertin et al., this volume). Tracing the genealogy and providing a critical review of the notion of CoP are out of the scope of this paper. We rather build upon existing reviews of this concept (see in particular Bolisani & Scarso, 2014; Cox, 2005; Gherardi, 2009; Handley et al., 2006; Li et al., 2009) to point to issues of relevance to the CoEx project.

Community of practice has become an "umbrella term" since its inception by Lave and Wenger (1991). This notion was initially coined to reflect on the collective and socially situated dimension of learning in opposition to the dominant cognitive and individual approach of learning. The active, experimental and collective character of knowledge building departs from a conception of learning as a matter of individual construction and acquisition. As Stahl noted (2003: 523, quoted in Allert, 2004: 6), "*meaning making is not understood as a psychological process which takes place in individuals' minds but as an 'essentially social activity that is conducted jointly – collaboratively – by a community, rather than by individuals who happen to be co-located*". The focus shifts from outcomes and products such as 'knowledge' or 'data' to activities such as "knowing" as a process of participation in shared learning activities and social processes of knowledge construction (Allert, 2004).

Over time, the CoP concept has been taken up in management literature and the focus shifted rapidly from CoP as a terrain of social learning to CoP as an organisational tool to manage knowledge teams in a more effective way (Li et al., 2009). In this stream of literature, CoPs are approached as a mechanism through which knowledge is held, transferred and created (Gherardi, 2009). Expertise is seen as the most crucial resource in CoPs and skills and knowledge interdependencies need to be effectively managed through technology-mediated tools, standards and protocols (Gherardi, 2009). The prevailing notion among knowledge management scholars is the one that considers a CoP as an entity in itself that requires managerial efforts to initiate, develop or cultivate. Wenger himself departed from his initial view and further reinforced this 'managerial turn' in his book published with colleagues in 2001 (Wenger et al., 2001), which formalised the three main elements of community of practices (Bolisani & Scarso, 2014): the domain (i.e., the area of knowledge that brings the community together); the community (i.e., the group of people for whom the domain is relevant, the relationships among them and the boundaries); and the practice (i.e., the body of knowledge, methods, tools, stories, documents which members share and develop together).

The original meaning, in which CoP is a way to consider learning and knowing as situated in social practices, is much more relevant to describe how concretely knowledge building and data work has been conducted within CoEx. In effect, in our willingness to producing new data and knowledge about seed systems that move away from pre-conceived categories and truly reflect what actors on the ground were experiencing, 'practice' rather than 'community' is what mattered most. As previously described, the SES framework focuses on the relationship between human and non-human entities rather than on representation of fixed entities. This allowed us to describe how entities such as seeds are enacted in practice, i.e. get their particular reality in a specific context. Such an approach gives more importance to the diversity of practices producing knowledge about seeds and recognizes the dynamic relationship between what can be deduced from the object itself on one hand and collectively constructed in social situations to give its meaning to the knowledge and data produced on the other hand.

4   Socio-Cognitive Challenges in Crop Diversity Management Characterisation

Two classifications coexist within the scientific biological community interested in crop diversity. The Linnaean (botanical) system of classification creates fixed categories based on agreed (though arbitrary) criteria based on morphological traits and biological characteristics to which all scientists subscribe: Class, Order, Family, Genus, Species, Sub-Species. Genetic information is increasingly used to classify or to update previous classification.

In addition to this botanical classification, researchers working on crop diversity use different categories to describe species and sub-species: wild relatives of crop plants; local varieties and primitive cultivars; obsolete ancient cultivars; advanced breeding lines, mutations and other products of plant breeding programs; and high-yield elite modern cultivars (Wilkes, 1988; National Research Council, 1993). This whole set of categories constitutes what is called genetic resources (or germplasm) in a generic way, which implicitly refers to the breeding (use) value. In effect, it describes a spectrum that holds a vision of genetic progress in which modern cultivars constitute the end goal and become a variety that is meant to be certified and sold in the market for production purposes (Bonneuil, 2019; Fenzi & Bonneuil, 2016). Unlike the botanical Linnaean system, this categorisation is not based on agreed criteria and thus, this information could be reported in various way by researchers.

In parallel to this biological classification systems, ethnobotanists and anthropologists have described a diversity of autochthonous classification systems used by farmers around the world to describe crop diversity, using vernacular name based on morphological, origin or symbolic characteristics.

CoEx discussed and investigated to what extent these existing categories and descriptors were able to accurately reflect existing seed acquisition, use and exchange practices with the overall objective to move away from the linear vision of genetic progress underpinning the binary formal/informal vision that predominates the characterisation of crop diversity management systems.

To this end, specific efforts were made to focus on the way seeds were being enacted in practice. Within the CoEx's community of practice, our object of inquiry was the coming into existence of entities (here seed "varieties") in the course of a practical activity rather than the modes of knowing of given entities. For example, during the field research carried out in one village of the Thiès Province in Senegal, we were able to observe, as notably and already described in other contexts (Leclerc et al., 2014; Labeyrie et al., 2019), a lack of consensus among farmers from the same village regarding the history and names of certain seed varieties, a fact presumably linked to their different life trajectories. This diversity of classification systems within the same village is made even more complex if we introduce a temporal perspective. In effect, our observations show that the categories used to designate the status of seed varieties used by farmers rapidly evolve over time. This semantic shift in our Lissar case study was particularly noticeable when seed lots move from the farmers' individual stores to the community 'seed bank' (a collective crop storage and conservation place). Through this movement of seeds that took place, we observed a change of the status of a given variety, from improved variety to collective variety. This suggests an ontological reconfiguration in the perception of farmers towards their plant material during these physical flows of seed lots in the village. Besides, lots designated as "farmers' varieties" or "local varieties" covered many varied origins and genetic compositions. This dynamic character, associated with the unstable character of variety qualifiers, underlined not only the diversity and combinatorial nature of variety perception and seed management by farmers, but also challenged the unquestioned use of these denominations for data collection and research purposes (see Fullilove, this volume).

This also brought into question the very notion of variety from a practice-oriented view. In our attempt to characterize seed acquisition strategies, we were interested in documenting whether privileged forms of acquisition were associated with varietal status, something that required aggregating data collected by seed lots at the scale of varietal status. However, such a task proved to be difficult since these categories, as noted above, were dynamic and not fixed over time and among farmers. Repositioning the analysis at the level of the physical entity managed by the farmer, i.e. the seed lot he/she sows or harvests in his/her field, was perceived as a promising way to avoid this caveat. This was especially the case when the description of seed lots mobilized several different descriptors known and used by farmers including seed variety name, varietal status and also other morphological or agronomic characteristics. Using an unweighted multi-dimensional description of the seed lot offered the possibility of diluting the impact of the heterogeneity of perceptions on all the descriptors when clustering methods are applied to aggregate the different seed lots.

In order to circumvent the difficulty of producing data without referring to pre-existing categories and better reflecting those categories experienced by the different actors involved, some CoEx members have also explored the specific attachments of farmers to their seeds (Lewicka, 2011). This was done through mixed quantitative and qualitative methods based on farmers' surveys to clarify the dimensions involved in the characterisation of seeds from the farmers' point of view. Such a relational approach to seed characterisation offers a way to grasp what really matters for farmers besides the instrumental values and the plurality of status and values associated with seeds within each community. This includes in particular emotional dimensions (such as pride, hope, collective, emancipation) and moral values (such as faithfulness, loyalty, reliability, solidarity), two aspects seldom explored despite their critical importance to the perception of fairness and equity in the management of crop diversity (Jankowski et al., 2020). Such an approach has underlined the way in which any characterisation is embedded and defined through the social relations that engage the farmers to their seeds.

By focusing on how entities such as seeds get their particular reality in a specific context and in the course of a relationship between human and non-human entities, rather than on representation of fixed entities, the CoEx's CoP addressed the socio-cognitive challenge of producing data on objects/dimensions that better reflect the diversity of actual practices in the field and are usually left out in existing knowledge systems and database about seed and seed systems. However, producing equitable and responsible data not only means taking into account the plurality of ways of knowing about seed but also tackling the political challenges related to the background values and assumptions guiding research, and to the socio-institutional structures supporting particular norms and practices.

5 Political Challenges in Crop Diversity Management Characterisation

In the context of CoEx, learning did not only occur in regard to the ontological and semantic status of a particular entity (seed). It also took place by encouraging various processes of critical assessment and social learning in regard to the different values and assumptions as well as the institutional and power structures that shape the current organization of research. Such learning is deemed essential to overcome the fact that knowledge and data production and sharing take place within a political context of strong inequalities among actors in their ability to contribute to meaning production (Bezuidenhout, 2020; Godrie et al., 2020; Fricker, 2007).

Knowledge actors rarely share complementary or compatible motivations and objectives. Various institutional logics usually coexist within a CoP without necessarily being made explicit or transparent to others in the course of knowledge production. In the specific context of plant science and breeding, strong divergence of views and power differentials among actors requires paying particular attention to collective meaning production.

The situated learning approach of CoP, as approached within CoEx, helps address the political challenge in both normative and procedural ways.

5.1   The Political Challenges from a Normative Point of View

As previously noted, seeds, genetic resources and associated knowledge crystallize a significant number of issues related to the divergent views about their legal status (see Manzella et al., this volume). Seeds are critical to the conventional paradigm of industrialization of agriculture, while at the same time they are at the core of the food sovereignty movement as part of farmers' autonomy and diversification and adaptation strategies in a rapidly changing environment. Recognizing and accounting for the diversity of normative orientations of the different stakeholders within CoEx involved paying careful attention to strengthening the capacity of partners to engage in the research design and to contribute to the meaning of the data collectively produced.

In this regard, CoEx established a research process that tried to ensure that all partners could benefit from the data produced in the course of the research. This was achieved by promoting the appropriation of scientific methods by and sharing results with farmers' organizations' members and rural communities. Besides simply sharing intermediary findings, such a process also permitted the project to integrate multiple legitimate perspectives into the scientific analysis and to ensure better linkages between scientific and societal problems.

One CoEx task was particularly amenable to this approach. Cognizant about the 'local trap' that accompanies collecting comprehensive data in one specific location, to provide as accurate a picture as possible about the so-called informal seed systems, CoEx proposed to characterize the diversity of seed systems at a larger scale. The project did this by enabling the collection of data over a large geographical area. In order to obtain an overview of the diversity of seed systems, a spatial uniform distribution was used as sampling strategy to describe – without any geographical a priori – the diversity of crops (species and varieties via their morphological traits), the variety of their uses, and the different modalities of seed acquisition. Surveys were carried out in 144 villages, spread over four countries in the 1.5 million square kilometres of the Sahelian strip, from Senegal to Niger, passing through Mali and Burkina Faso moving from West to East.

The surveys documented the presence or absence of 32 crop species, with a particular focus on pearl millet, cowpea, sorghum, gumbo, groundnut and maize to further characterize varieties, seed acquisition patterns and uses at multiple spatial – farm, local, regional, national, and multi-country – and temporal scales. These crops are the staple food of the population, i.e. reaching about 330 million people.

In order to reconcile the scientific robustness and objectivity of the surveys on the one hand, and the social relevance of the knowledge produced on the other hand, an approach covering all stages of the collaborative research process was developed, ranging from the co-construction of research questions and data collection protocols to the joint analysis of results.

A first workshop entitled 'common research protocol' took place in Ouagadougou (Burkina-Faso) for 1 week, bringing together partners from five universities and five farmers' organizations from Niger, Mali, Burkina Faso, Senegal and France. The survey protocol and questions were discussed in order to agree on what data and information will specifically be collected. The participants also considered ways to address cultural difference between languages and countries, and the associated diversity of crop management practices. All participants collectively ran a testing phase of the survey in the field to ensure a higher level of common understanding despite these specificities.

The survey was implemented simultaneously in the four countries with a total of eight field survey teams. Each team used a touch screen tablet with Kobo Tools software. Touch screen tablets allowed the teams to conduct data entry directly in-field. Data was uploaded each day in a server, which was common to all partners. This offered the possibility for each survey team to see the progress of other teams in real-time, on a map with dots representing villages that were already surveyed in different countries. Each team had access to the results obtained by the others. This form of data management and access has made "shared data" a common and central value among the partners.

A second workshop entitled 'collective results interpretation' took place in Montpellier (France) for 1 week. Questions to be analysed first were formulated before the meeting by each partner, according to their priority. During the meeting, results were analysed first separately by country, and then transversally across country. The interpretation of the results was based on the local knowledge and practices of each partner. Thus, a given observation was considered of broader significance when different (and independently formulated) interpretations from different teams converged. This form of collective interpretation contributed to enhanced appropriation of the results by the partners.

Moreover, the formation along the way of a CoP built as a group of partners sharing a common research frame of reference made it possible to forge a common understanding about the value of the knowledge being produced. The whole challenge of the approach consisted in ensuring that the facts observed by researchers in the different countries during the surveys were collective facts, not only shared in materiality (via the computer tools used) but also in meaning (via co-constructed protocols and collective analysis). By simultaneously asking the questions "what do we do" and "how do we do it" and by conceiving the production of knowledge as process of co-participation and social interactions, the CoEx's CoP made it possible to produce collective meaning from the huge amount of data collectively collected.

This allowed overcoming the opposition often presented as irreconcilable between the criteria of scientific robustness on the one hand and those of social relevance and legitimacy on the other hand (see Leonelli and Williamson, this volume). Not only did the mobilization of farmers' organizations make it possible to carry out a large number of surveys, but they also provided valuable elements of understanding of the context. This allowed an informed interpretation of the results complementary to those of the researchers (particularly in statistics) that made it possible to compare the different situations and give them a general scope.

With this knowledge, farmers, researchers and public decision-makers have valuable information to better valorise the solutions experimented with locally to adjust to their environment. The partnering farmers' organizations are now in a position to build a discourse based on scientific evidence they contributed to generate, and even to participate in gaining better recognition of farmers' right to cultivate their seeds and, more generally, of their role in the management of agrobiodiversity. This could also be made possible by the opening of new public space for enhanced interactions between the various stakeholders and policy makers (Nlend Nkott & Temple, 2021).

5.2   The Political Challenges from a Procedural Point of View

Our approach to participation within CoEx was not only limited to the data production and interpretation process. It also included participation in decision-making processes about the use of this (collectively produced) data. This decision-making process has not only been approached from a managerial perspective, but rather as a knowledge area in itself that became part of the collective inquiry process. CoEx has indeed been conceived as a collective experiment that has taken up the question of the conditions of its own collaboration and defined its own modus operandi, objectives and means according to the specific problems to be solved in the course of the project.

To do this, and in line with the situated learning approach, CoEx established the two following activities: a collective analysis of past research collaborations experienced by members; and a reflexive process in regard to the CoEx conditions of collaboration as the research continues to develop.

The first activity consisted of collecting agreements made in past multi-stakeholder research collaboration and surveying farmers' organization members and researchers about their perceptions and perspectives about what worked or not in these agreements. Two workshops were organized in 2017 and 2018 to collectively organize this work and analyse the findings. A third workshop in 2019 lead to the drafting of a manual on multi-stakeholder research collaboration that listed different points to be considered and monitored within such complex collaborative contexts and options to address them. This learning process enhanced the collaborative capacity of the participants in multi-stakeholder contexts.

The second activity consisted in establishing a project governance structure that reflected the multi-stakeholder nature of the collaboration process and examined specific collaborative issues as they occurred in the context of the project implementation. Rather than relying on existing resources and administrative routines to the research organization in charge of the coordination of the project (the French CIRAD, in this case), this reflexive process increased the involvement of the community in its own collective functioning. In relation to data work, three topics were specifically covered: one related to the ethical aspects of collecting survey data from farming communities; one related to data management and the status of data and results within collaborative projects; one on the type and status of knowledge made available by project participants, in particular concerning the practices and know-how disclosed by producer members of farmers' organisations participating in the project. To address this issue, CoEx recognized that all partners have equal rights over the data, dataset and data analysis jointly produced, considering these outputs as a commons. This was materialized by the recognition of the right to participate in any decision regarding publication, utilization for various purposes, transfer to third party, or application of any intellectual property right. One concrete experience during the project that offered an opportunity to test these ideas related to the willingness of one researcher to engage in a new collaboration with a US university that would use data collected in farmers' surveys on seed acquisition practices and sources,. A 'Data Provision agreement' was prepared and negotiated between CIRAD and the US university. This involved many back and forth exchanges between CIRAD and the CoEx members on the one hand, and CIRAD and the US university on the other hand, as many clauses proposed to protect the integrity of the (social) context in which such data had been collected, as well as the collective validation process of data use in publications through the CoEX multi-stakeholder steering committee, were perceived at odds with established practices in inter-academic collaborative practices. The partners also undertook to treat with the utmost vigilance, and in accordance with the various national legislations, any local or traditional knowledge associated with seed and genetic resources that may be transmitted to them by farmers in the course of the surveys, in order to prevent any kind of misappropriation.

6   Conclusion

In considering the notion of community of practice in its initial conceptualisation, which involves blurring the distinction between knowledge and practice and between production and use of knowledge, this paper revisited both socio-cognitive and political challenges related to data production and responsible data linkages. The concrete combination within the same project of these three dimensions offers a perspective on CoPs quite different from the managerial angle through which most of the discussions on data work have apprehend them. This in turn has consequences for the way to approach the agency issue in plant data production and linkages.

Regarding the socio-cognitive challenge, beyond accounting for the plurality of ontologies or knowledge systems (often reduced to a diversity of classificatory modes), we have shown that producing data that "make sense" to the different members requires also to get a better understanding of their respective epistemologies or ways of knowing. More specifically, we showed that the dynamic character of seed circulation and the social relations that engage farmers with their seeds ultimately impact their characterization.

The recognition of the various ways of knowing is also crucial to address political/normative challenges in interpreting the data and creating collective meaning and learning. The establishment of a group of partners sharing a common research frame of reference and the mobilization of the different set of available interpretative resources provided by the different actors made it possible to forge a common understanding of the value of the knowledge being produced. Through the collectively produced knowledge, the partnering farmers' organizations are now in a better position to build a discourse based on scientific evidence that they contributed to generating, and even to participate in gaining better recognition of their rights.

Beyond this collective learning process on the (technical) topic of research, the establishment of modalities for collective organization and decision-making itself was also part of the learning process. Very often reduced to its managerial aspect, this governance dimension is crucial to ensure full participation in addressing the socio-cognitive and political/normative challenges in the first place, and to manage as best as possible the differences in power between groups in the use of data and knowledge jointly produced.

Acknowledgments The CoEx project was publicly funded through ANR (the French National Research Agency) under the "Investissements d'avenir" programme with the reference ANR-10-LABX-001–01 Labex Agro and coordinated by Agropolis Foundation.

References

Allert, H. (2004). Coherent social systems for learning: An approach for contextualized and community-centred metadata. *Journal of Interactive Media in Education 1*(2004), 2004(2).

Almekinders, C. J., Louwaars, N. P., & De Bruijn, G. H. (1994). Local seed systems and their importance for an improved seed supply in developing countries. *Euphytica, 78*(3), 207–216.

Anderies, J. M., Janssen, M. A., & Ostrom, E. (2004). A framework to analyze the robustness of social-ecological systems from an institutional perspective. *Ecology and Society, 9*(1), 18.

Arnaud, E., Laporte, M. A., Kim, S., Aubert, C., Leonelli, S., Miro, B., et al. (2020). The ontologies community of practice: A CGIAR initiative for big data in agrifood systems. *Patterns, 1*(7), 100105.

Berkes, F., & Folke, C. (Eds.). (1998). *Linking social andecological systems: Management practices and social mechanisms for building resilience.* Cambridge University Press.

Bezuidenhout, L. (2020). Being fair about the design of FAIR data standards. *Digital Government: Research and Practice, 1*(3), 1–7.

Boeckhout, M., Zielhuis, G. A., & Bredenoord, A. L. (2018). The FAIR guiding principles for data stewardship: Fair enough? *European Journal of Human Genetics, 26*(7), 931–936.

Bolisani, E., & Scarso, E. (2014). The place of communities of practice in knowledge management studies: A critical review. *Journal of Knowledge Management, 18*(2), 366.

Bonneuil, C. (2019). Seeing nature as a 'universal store of genes': How biological diversity became 'genetic resources', 1890–1940. *Studies in History and Philosophy of Science Part C: Studies in History and Philosophy of Biological and Biomedical Sciences, 75*, 1–14.

Couldry, N., & Powell, A. (2014). Big data from the bottom up. *Big Data and Society, 1*(1), 1–5.

Cox, A. (2005). What are communities of practice? A comparative review of four seminal works. *Journal of Information Science, 31*(6), 527–540.

Fenzi, M., & Bonneuil, C. (2016). From "genetic resources" to "ecosystems services": A century of science and global policies for crop diversity conservation. *Culture, Agriculture, Food and Environment, 38*(2), 72–83.

Folke, C., Hahn, T., Olsson, P., & Norberg, J. (2005). Adaptive governance of social-ecological systems. *Annual Review of Environment and Resources, 30*, 441–473.

Fricker, M. (2007). *Epistemic injustice: Power and the ethics of knowing.* Oxford University Press.

Gherardi, S. (2009). Community of practice or practices of a community. In *The Sage handbook of management learning, education, and development* (pp. 514–530). SAGE.

Godrie, B., Boucher, M., Bissonnette, S., Chaput, P., Flores, J., Dupéré, S., et al. (2020). Epistemic injustices and participatory research: A research agenda at the crossroads of university and community. *Gateways: International Journal of Community Research and Engagement, 13*(1), 1–15.

Handley, K., Sturdy, A., Fincham, R., & Clark, T. (2006). Within and beyond communities of practice: Making sense of learning through participation, identity and practice. *Journal of Management Studies, 43*(3), 641–653.

Jankowski, F., Louafi, S., Kane, N. A., Diol, M., Camara, A. D., Pham, J. L., & Barnaud, A. (2020). From texts to enacting practices: Defining fair and equitable research principles for plant genetic resources in West Africa. *Agriculture and Human Values, 37*, 1083–1094.

Labeyrie, V., Kamau, J. I., Dubois, C., Perrier, X., & Leclerc, C. (2019). So close yet so different: Cultural differences among farmers in Central Kenya affect their knowledge of sorghum (Sorghum bicolor [L.] Moench) landrace identification. *Economic Botany, 73*(2), 265–280.

Labeyrie, V., Antona, M., Baudry, J., Bazile, D., Bodin, Ö., Caillon, S., Louafi, S., & Thomas, M. (2021). Networking agrobiodiversity management to foster biodiversity-based agriculture. A review. *Agronomy for Sustainable Development, 41*(1), 1–15.

Lave, J., & Wenger, E. (1991). *Situated learning: Legitimate peripheral participation.* Cambridge University Press.

Leclerc, C., Mwongera, C., Camberlin, P., & Moron, V. (2014). Cropping system dynamics, climate variability, and seed losses among East African smallholder farmers: A retrospective survey. *Weather, Climate, and Society, 6*(3), 354–370.

Lewicka, M. (2011). Place attachment: How far have we come in the last 40 years? *Journal of Environmental Psychology, 31*(3), 207–230.

Li, L. C., Grimshaw, J. M., Nielsen, C., Judd, M., Coyte, P. C., & Graham, I. D. (2009). Evolution of Wenger's concept of community of practice. *Implementation Science, 4*(1), 1–8.

Louwaars, N. P. (2007). *Seeds of confusion: The impact of policies on seed systems.* Wageningen University. ISBN 9789085047933 – 151.

National Research Council. (1993). *Managing global genetic resources, agricultural crop issues and policies.* National Academy Press.

Nkott, N., & Temple, L. (2021). Le secteur semencier céréalier au Burkina Faso: dépendance de sentier et trajectoires d'évolution depuis 1970. *Economie et institutions* (in press).

Ostrom, E. (2009). A general framework for analyzing sustainability of social-ecological systems. *Science, 325*(5939), 419–422.

Popa, F., Guillermin, M., & Dedeurwaerdere, T. (2015). A pragmatist approach to transdisciplinarity in sustainability research: From complex systems theory to reflexive science. *Futures, 65*, 45–56. S0016328714000391. https://doi.org/10.1016/j.futures.2014.02.002

Stahl, G. (2003). Meaning and interpretation in collaboration. In *Designing for change in networked learning environments* (pp. 523–532). Springer.

Wenger, E., McDermott, R., & Snyder, W. (2001). *Cultivating communities of practice: A guide to managing knowledge*. Harvard Business School Press.

Wilkes, H. G. (1988). Plant genetic resources over ten thousand years: From handful of seed to the crop-specific mega-gene banks. In J. R. Kloppenburg (Ed.), *Seeds and sovereignty: The use and control of plant genetic resources*. Duke University Press.

Woolgar, S., & Lezaun, J. (2013). The wrong bin bag: A turn to ontology in science and technology studies? *Social Studies of Science, 43*(3), 321–340.

Young, O. R., Berkhout, F., Gallopin, G. C., Janssen, M. A., Ostrom, E., & Van der Leeuw, S. (2006). The globalization of socio-ecological systems: An agenda for scientific research. *Global Environmental Change, 16*(3), 304–316.

The Research Data Alliance Interest Group on Agricultural Data: Supporting a Global Community of Practice

Patrícia Rocha Bello Bertin, Cynthia Parr, Debora Pignatari Drucker, and Imma Subirats

Abstract Efforts to address equity and inclusion in agricultural data infrastructures face numerous challenges. People and networks are widely distributed geographically. This means some solutions to data problems may arise regionally and independently, yet many people are not easily able to engage with their distant colleagues to learn about them or collaborate. In general, constraints on funding for such projects are often national rather than international, and travel funding is not equally distributed. Finally, the breadth of activity means interdisciplinary communication is important but difficult and hard to sustain. Addressing these challenges, the Research Data Alliance (RDA) has been a home for the Interest Group on Agricultural Data (IGAD) since 2013. In 2021, IGAD became the first example of a new type of RDA group – a Community of Practice. A future goal is to use this community of practice to put good regional or national work into practice via inclusive collaborations. This chapter reflects on the lessons learnt from the IGAD community of practice in its attempts to include new voices around the world.

P. R. B. Bertin (✉)
Empresa Brasileira de Pesquisa Agropecuária – Embrapa, Brasília, Brazil
e-mail: patricia.bertin@embrapa.br

C. Parr
U.S. Department of Agriculture – USDA, Agricultural Research Service, Washington, DC, USA
e-mail: cynthia.parr@usda.gov

D. P. Drucker
Embrapa Agricultura Digital, Campinas, Brazil
e-mail: debora.drucker@embrapa.br

I. Subirats
Food and Agriculture Organization of the United Nations – FAO, Rome, Italy
e-mail: imma.subirats@fao.org

© The Author(s) 2023
H. F. Williamson, S. Leonelli (eds.), *Towards Responsible Plant Data Linkage: Data Challenges for Agricultural Research and Development*,
https://doi.org/10.1007/978-3-031-13276-6_15

1 Introduction

Efforts to address equity and inclusion in agricultural data infrastructures face numerous challenges. People and networks are widely distributed geographically. This means some solutions to data problems may arise regionally and independently, yet many people are not easily able to engage with their distant colleagues to learn about them or collaborate. In general, constraints on funding for such projects are often national rather than international, and travel funding is not equally distributed. Finally, the breadth of activity means interdisciplinary communication is important but difficult and hard to sustain.

This chapter describes the ongoing transition of the Research Data Alliance (RDA) Interest Group on Agricultural Data (IGAD) into a Community of Practice. With practical examples, it explains how IGAD has helped identify and promote awareness of efforts around the world that may currently be restricted to one region but that have the potential to democratize participation in agricultural data management infrastructure initiatives and generally improve capacity for managing and leveraging agricultural data.

1.1 A Brief Introduction to the Research Data Alliance (RDA) and the Interest Group on Agricultural Data (IGAD)

The Research Data Alliance (RDA) is a community-driven initiative that was launched in 2013 by the European Commission, the United States Government's National Science Foundation and National Institute of Standards and Technology, and the Australian Government's Department of Innovation as a neutral space where its members could come together to develop and adopt infrastructure that promotes data-sharing and data-driven research (Berman & Crosas, 2020). As for today, the RDA has attracted over 12,000 members from 145 countries. The vision is: "researchers and innovators openly share and re-use data across technologies, disciplines, and countries to address the grand challenges of society" (Research Data Alliance, 2021).

The work of the RDA is conducted through self-organized Interest Groups (IGs) and Working Groups (WGs) that discuss solutions to real-world problems. Participation in one of the 97 existing groups is open to anyone who agrees to the RDA's principles – usually experts from academia, private sector and government, who are attracted to these groups as a means to identify and build the infrastructure that is needed to overcome their research data management challenges.

The Interest Group on Agricultural Data (IGAD)[1] was formed in 2013, as a forum for sharing experience and providing visibility to research and work with agricultural

---

[1] https://www.rd-alliance.org/groups/agriculture-data-interest-group-igad.html

data. Since then, it has grown in community strength to over 260 members, becoming one of the RDA's most prominent Thematic Groups, serving itself as a platform to the creation of specific Working Groups. In keeping with RDA's strategy, IGAD has supported the creation of five WGs: Wheat Data Interoperability, Rice Data Interoperability, Agrisemantics, On-Farm Data Sharing, and Capacity Development for Agricultural Data WGs.

2 Examples of Global Coordination in Previous IGAD Activities

The RDA holds a global plenary meeting every 6 months, in which the IGs and WGs participate to display and engage the wider community around their work, deliverables and outcomes. The IGAD and its associated WGs have played an active role at the RDA Plenaries, as a means to reach out and forge new alliances with other groups, as well as to create new offshoot groups aimed at specific challenges and solutions. During the plenary sessions, the IGAD has hosted a wide array of speakers and discussions, seeking to work alongside major international initiatives in agricultural research data management and interoperability from private and public organizations such as GODAN, CGIAR, FAO of the UN, INRAe, and Syngenta, among others. Prior to each of the RDA Plenaries, IGAD has also successfully organized pre-meetings to engage the agricultural data community in taking stock of existing issues and laying the groundwork for concrete future action.

To sustain engagement even through the Covid-19 pandemic, IGAD has conducted several webinars and virtual events. One of them focused on the theme 'IGAD/RDA: Sharing Experiences and Creating Digital Dialogues'. The week-long event (25–28 May 2020) brought together 350 IGAD members to discuss semantics, crop data interoperability and experiences and lessons learnt from Asia, Europe, Africa and Americas, producing many interesting results and interactions. In 2021, IGAD promoted 30 min 'Coffee Break' Webinars, a new kind of webinar series to support the exchange of experiences within the agricultural data community, which consisted of virtual 15-min presentations on topics of interest, followed by 15 min of discussion. With presentations coming from participants all over the world to share their experiences, the sessions were also recorded for those who could not attend live. Virtual meetings have the advantage of allowing anyone to participate from anywhere and helps inclusion as there are no travelling costs involved. In fact, the events attracted many hundreds of interested people that approached the IGAD community for the first time.

From all WGs that have been created under the IGAD umbrella, the Agrisemantics and the Wheat Data Interoperability (WDI) Working Groups were particularly successful, with consensus recommendations being approved for implementation (Caracciolo et al., 2020; Ycumo et al., 2016). The Agrisemantics Working Group produced a set of recommendations to facilitate the adoption of semantic technologies and methods for the purpose of data interoperability in the field of agriculture and nutrition. To achieve so, between 2016 and 2019 the group gathered researchers and practitioners to study all aspects in the life cycle of semantic resources: conceptualization, edition, sharing, standardization, services, alignment, long term support (Caracciolo et al., 2020). Beginning with a landscape study, a number of use cases for the exploitation of agricultural semantic resources were analyzed. The outputs of the WG were synthesized into 39 'hints' for users and developers of semantic resources, and providers of semantic resources' services. A wide range of applications of the recommendations of the Agrisemantics WG followed – AgroPortal, for example, represents the importance of domain-specific repositories and tools for mappings, and VocBench offers a web-based platform for the creation and maintenance of semantic resources according to best practices.

With regards to the WDI Working Group, by the time it was created, in 2014, the goal was to make the best use of existing genetic, genomic, and phenotypic data in fundamental and applied wheat science. Given the ever-growing data deluge coming from modern technologies such as DNA (Deoxyribonucleic acid) and RNA (Ribonucleic acid) sequencing, high throughput genotyping and phenotyping, high throughput imaging and satellite monitoring, data interoperability became a priority for the wheat research community (Yeumo et al., 2016).

The WDI WG was formed by data and information practitioners and scientists from different organizations and countries, with a clear standpoint, which was to avoid the creation of new standards, but to provide a common framework for describing and representing data with respect to existing open standards. In order to converge and agree on specific recommendations, the WDI WG began by surveying the practices of the wheat research community. The proposed guidelines were then endorsed by the RDA and early adopted by organizations such as the Australian Center for Plan Functional Genomics, the French Institute National de Recherche pour l'Agriculture, l'Alimentation et l'Environnement (INRAe), and the English Rothamsted Research. Recommendations are frequently revised to consider the evolving landscape of data practices and standards.

Replicating the methodology used in the WDI in the context of other crops was a challenge, though. It was noted that institutional support and the pre-existence of a well-structured and vivid community is an important prerequisite for the success of the WGs. The Rice Data Interoperability WG, for instance, had to be cancelled for not being able to sustain the effort needed to develop recommendations. The group is now in the process of being replaced by a more general Crop Data Interoperability WG. Soil experts are also committed to partnering with the IGAD.

3 Transitioning to a Global RDA Community of Practice

IGAD has helped create awareness about research data management within the food and agricultural community, linking with other communities to facilitate the adoption of RDA recommendations, inviting experts from different fields of expertise to join and enrich the dialogue and the sharing of knowledge, and encouraging researchers to share their experiences.

As a form of recognition of the IGAD's role to promote the RDA within the food and agricultural data community, the interest group was the first to become a Community of Practice (CoP) under the formal structure of RDA. Although the RDA is not particularly concerned with establishing a single unified concept for a 'community of practice', the notion clearly draws from the original work by Lave and Wenger (1991, p. 98), where a community of practice is "a set of relations among persons, activity, and world, over time and in relation with other tangential and overlapping communities of practice". The community of practice, according to the authors, would provide a proper social context for learning to take place.

In practice, an RDA CoP offers a discipline or domain the opportunity to create an open forum for the discussion, development and maintenance of specific and generic solutions to the data challenges faced by that community. By offering a forum to discuss data-related trends and challenges, CoP members will learn from one another experiences and collaborate on implementing solutions. It supports the RDA to attract new individual, organizational, and regional members, including researchers and stakeholders from low and medium-income countries, establishing connections with other international initiatives.

On a logistical level, one of IGAD's chief roles has been to serve as a platform that leads to the creation of domain-specific Working Groups. As a CoP, this role is strengthened, providing a neutral space for networking and blending ideas related to data management and interoperability. The IGAD CoP can use community building and capacity support as a means of ensuring working groups' success.

Recently approved by the RDA Technical Advisory Board, the CoP will maintain the IGAD acronym, which now stands for 'Improving Global Agricultural Data'. Each year, one specific objective or priority theme will be added as 'sub headers', for example: IGADs (Semantics), IGADm (Management), IGADw (Workforce), IGADs (Sovereignty), IGADc (Capacity Building), and IGADi (Infrastructure), and so forth. From a community perspective, agricultural data practitioners and the organizations they work for will benefit from participating in the IGAD CoP due to a better alignment with global practice, identifying opportunities to form partnerships on specific projects, better ability to impact stakeholders via improved data systems and practices, and mutual learning from exchanging experiences.

As to the operational mechanisms, the IGAD CoP will be coordinated by at least three professionals from the global agricultural data community, drawn from different geographic regions, whose role is to plan and operate by consensus. A communication plan will be developed to keep the community updated on the several engagement opportunities within the CoP, such as in-person or virtual RDA meetings, monthly webinars or longer events to happen at least annually.

The philosophical approach behind the IGAD CoP is to represent all geographic regions and increase the participation of the global south. Leadership and a process of chair rotation is expected to reflect this. Some of the challenges are related to inclusiveness. For instance, the times at which plenaries and meetings are often scheduled do not favor the engagement of participants from the global south.

Recording sessions and varying the times for the virtual encounters has proved to be a reliable method for wider engagement.

Members of the IGAD CoP include practitioners of agricultural data management in academia, government and industry, engaged in large part via the regional or disciplinary organizations that they have formed to support their efforts. They usually have skills in both their domains and in relevant aspects of data management, whether for research or for agricultural activities, but because the community expects to enhance skills and knowledge, no specific requirements are expected. Engaging key members of other relevant networks is expected to act as liaisons with their larger communities, in line with the participative approach of a community of practice.

3.1 Farmer Research Data Framework

The IGAD Community of Practice offers a valuable forum for sharing approaches to difficult issues such as how to protect data generated by farmers while ensuring that valuable research can be conducted to improve agricultural practices for both economic and natural resources stewardship. A recent workshop held in the United States, Big Data Promises and Obstacles: Agricultural Data Ownership and Privacy, was inspired by work in Europe on codes of conduct.

According to Zampati (2021), codes of conduct emerged to fill the legislative void and to set common standards for data sharing contracts. Farm data would be an example of sensitive data, which flows from the farm to many other actors (such as extensionists, agri-tech companies, farmers' associations, financial service providers, etc.) to be usually aggregated and combined in the form of services and sent back to the farm.

These topics have also been discussed at the RDA 11th Plenary. However, to become truly part of an actionable global framework accessible to everyone, the ideas will need to be brought again to the IGAD CoP. The participants from regional networks can discuss and consider how to reground and modify them to suit cultural and legal practices elsewhere.

3.2 CARE Indigenous Data Governance Principles

The FAIR (Findable Accessible Interoperable and Re-usable) data principles (Wilkinson et al., 2016) are becoming increasingly important in several disciplines, including within agricultural data. Devare et al. (this volume) advocate that FAIR agricultural data assets should be the norm rather than the exception, to foster a transition towards 'translational agriculture', a new agricultural system that would make use of powerful technologies to enable more effective data mining and use, making agrifood research and business more agile and responsive to user needs. The

FAIR principles were extensively discussed at the IGAD meeting prior to the RDA Plenary 11 and are now being put into practice by many IGAD participants.

A new, complementary set of data governance principles has recently emerged to balance indigenous rights and interests in data with the desire to honor the FAIR principles of supporting open, machine-readable data (Carroll et al., 2020). These CARE principles (Collective Benefit, Authority to Control, Responsibility, and Ethics) hold great promise for indigenous and local communities with agricultural data, ensuring that its re-use benefits those communities, but they have not yet been widely shared among IGAD participants. They take a very user and usage-centered approach, to complement the very data-focused FAIR principles. The new community of practice and its network of regional networks should provide a very effective means to increase awareness and discussion so that adoption of or refinement of these principles can happen more quickly.

3.3 Taxonomic Plant Data Linkage

IGAD would do well to engage a related community of practice, the Biodiversity Informatics Standards[2] community, previously known as the Taxonomic Databases Working Group (TDWG). While this group has already co-sponsored activity with the Research Data Alliance there is untapped opportunity to engage with the IGAD Cop. Biodiversity informaticists are acutely aware that linking plant data across datasets requires effective identification of the organism from which the data derives. A series of recent Biodiversity Informatics Standards symposia on agricultural biodiversity have made clear that standards must accommodate a wealth of valuable information about crop wild relatives and land races in agrobiodiverse regions. For example, in India, typical biodiversity data standards must be able to accommodate local names and smallholder cultivation practices in order to support analysis of crop phenotypes, genotypes, and their environmental influences and impacts beyond industrial western farming operations (Arnaud et al., 2016; Rajagopal et al., 2017).

Another relevant TDWG group, the Species Interaction Data Group[3] was established for developing a data standard to allow universal exchange of data and information that is relevant not only to biology but also to agriculture and ecosystems services such as pollination. Connecting both the IGAD and TDWG communities can increase awareness of the existence of such standards efforts, and the broad geographic representation in both communities can ensure that diverse use cases and cultural differences are accommodated in these standards.

---

[2] https://www.tdwg.org/

[3] https://github.com/tdwg/interaction

3.4 IGAD's Regional Outreach Efforts: The Brazilian Experience

It is noteworthy that the IGAD activities have contributed to the implementation of good data management policies and practices within agricultural research institutions all over the world. Very often, these actions are in support of openness and the adoption of standards to data repositories.

An example are the recent efforts by the Brazilian Agricultural Research Corporation (Embrapa) to incorporate the FAIR principles into its research data management processes and practices. Embrapa is a public agricultural research institution whose mission is to "provide research, development and innovation solutions for the sustainability of agriculture and for the benefit of Brazilian society" (Embrapa, 2021). Structured in 43 research centers geographically distributed throughout the country, the company generates a large volume of research data on the various strategic themes of agricultural research.

Aware of the volume, speed, variety and value of research data produced in the development of its activities, Embrapa has mobilized efforts to properly govern and manage these assets throughout their life cycle, in order to and to make them findable, accessible, interoperable and reusable. Among these efforts is the publication of the company's 'Data, Information and Knowledge Governance Policy', which establishes the principles, guidelines, attributions and responsibilities that will strengthen the mechanisms of generation, organization, treatment, access, preservation, recovery, disclosure, sharing and reuse of Embrapa's information assets.

The document is based on the premise that well-organized, documented, accessible and verified data are more easily shared and reusable, with several advantages to the organization. Knowledge exchange within IGAD informed the content of Embrapa's Data Governance policy, drawing upon other research institutes' experiences and guidelines, such as INRAe's Open Access and Open Data Policy (INRAe, 2016). Another important reference to Embrapa's policy is the FAIR principles (Wilkinson et al., 2016), a central pillar of the corporate 'Research Data Management Program'.

Adherence to the FAIR principles is crucial when data services are discussed, as interoperability plays a key role. Embrapa has implemented data services through APIs for different purposes that allow users from companies, startups, universities and students, among others, to solve real-world and real-time problems in agriculture. The AgroAPI Platform[4] offers Agritec API, for instance, which gathers useful information for crop production management. It includes data and models on: (i) ideal planting time for dozens of crops, based on agricultural zoning of climatic risk; (ii) ratio of the most suitable cultivars for 12 different crops (Rice, Cotton, Peanuts, Barley, Beans, Cowpeas, Sunflower, Castor, Maize, Soy, Sorghum and

---

[4] https://www.agroapi.cnptia.embrapa.br/

Wheat); (iii) indication of fertilization and soil correction as a result of previous soil analysis, productivity forecast and climatic conditions before and during the harvest for five crops (Rice, Beans, Maize, Soy and Wheat). These inform decision making on defining planting season with less risk of loss and fittest cultivars, productivity forecasts and water balance and climatic conditions before and during harvest. Another example is SATVeg API, which is derived from the Temporal Vegetation Analysis System (SATVeg), a web tool developed by Embrapa Agricultura Digital, aimed at generating and viewing temporal profiles of the NDVI and EVI vegetative indices for Brazil and all of South America, with the objective of supporting activities of territorial management and agricultural and environmental monitoring. Vegetative indices are generated from multispectral images provided by the MODIS sensor, on board NASA and Terra and Aqua satellites, covering data produced from 2000 until the last date then made available by its official repository, with a 16-day temporal resolution and spatial resolution of 250 m. SatVeg is being expanded to cover Sentinel products that will also be offered as a machine-to-machine data service through APIs.

The experience of Embrapa is serving as a basis for the construction of the GO FAIR Agro Implementation Network, benefiting the whole national agricultural Research, Development & Innovation system. The Brazilian regional GO-FAIR office is structured following the international GO-FAIR initiative[5] and currently embraces 7 thematic implementation networks. The regional office produced a letter of principles agreed by the participating organizations which exposes its functioning and rules of engagement.[6]

The agricultural data implementation network is coordinated by Embrapa and is supported by other relevant research institutions in the country. It is in the early stages of a bottom-up community effort and the experience of IGAD activities inspire its construction, considering the different approaches within agricultural data science, community facilitation tools, inclusivity regarding gender and minorities, and regional diversity in a continental country of great importance for food production. A manifesto was constructed by the agricultural data community in Brazil and was launched in November 2021, during the XIII Brazilian Conference on Agroinformatics. Its mission is to work in an articulated and collaborative way to encourage the sharing and reuse of data produced in the context of agricultural production systems and also those arising from research in agricultural sciences in Brazil, supported by the FAIR principles. It includes objectives related to agricultural data science, cultural change towards FAIR good practices, training activities, articulation and collaboration with the other GO FAIR Brazil National Thematic Implementation Networks and with the Food Systems International Implementation Network. The network was launched in April 2022 during a virtual event that brought together 130 professionals from the agricultural sciences, information science and information and technology domains, representing more than 40 public

---

[5] https://www.go-fair.org/

[6] https://www.go-fair.org/wp-content/uploads/2019/06/Declaration_GO-FAIR-Brazil_Jun2019.pdf

and private institutions such universities, research and development institutes, companies and startups.

Communities of practice in agriculture can encompass a multitude of subjects and one of them is related to preserving cultural and biodiversity heritages. Diverse agrifood products traditionally grown by local populations are also getting more attention worldwide and also in Brazil. Agrobiodiversity data standards are needed to properly represent and make sense of such data and that is being improved by collaborative work from several organizations. Collaboration is also the motivation behind the creation of a national GO-FAIR implementation network focused on agriculture in Brazil. All of this work will benefit if the IGAD CoP can include new voices from the field.

4 Concluding Remarks

Communities of practice in agriculture need to share information about regional developments in the use of data intensive activities such as Internet of Things embedded in agricultural machinery or irrigation devices, and the development of decision-making support tools that rely on climatic and remote sensing data sources. A community of practice can ensure that these developments are informed by local farmers' traditional knowledge and that they preserve and protect cultural and agrobiodiversity. The FAIR and CARE guiding principles help us to move forward towards linked data and bridging gaps that will allow many diverse communities to connect and share experiences for a more sustainable food production environment.

Addressing these challenges, the Research Data Alliance (RDA) has been a home for the Interest Group on Agricultural Data (IGAD) since 2013. This chapter reflected on the lessons learnt from the IGAD community of practice in its attempts to include new voices from around the world. As in Lave and Wenger (1991, p. 100), the focus of the community of practice is to provide the members with "access to a wide range of ongoing activity, old-timers, and other members of the community; and to information, resources, and opportunities for participation".

The convening power of the RDA provides many advantages, such as the ability to sustain multiple threads of interdisciplinary work, and worldwide networking. Several important working groups have been supported by IGAD such as an emerging crop data interoperability working group.

IGAD regularly convenes some meetings outside the RDA Plenaries to allow for participation from practitioners with fewer resources. FAIR data (Findable, Accessible, Interoperable, and Reusable) has been a frequent topic of discussion. In recent years, virtual sessions have expanded the conversations even more to enable global participation. For example, in the US, several workshops have addressed the need for progress on issues relating to farmer data ownership and privacy; these are informed by work happening in Europe, but ideas will need to be re-grounded and modified to cultural and legal practices elsewhere. For plant data in particular, ideas about land races and nomenclature from the Biodiversity Information Standards

(TDWG) could be combined with the work of the CGIAR institutes to provide more seamless access to indigenous knowledge.

In Brazil, several efforts to support data driven decision-making in the field could serve as models for other IGAD members. For instance, as we have discussed, the Brazilian Agricultural Research Corporation (Embrapa) has implemented data services through APIs that provide real-time data on climate, productivity and most favorable days for planting different crops. Diverse agrifood products traditionally grown by local populations are also getting more emphasis in Brazil and agrobiodiversity data standards are being improved by collaborative work from several organizations.

Collaboration is a keyword behind the creation of a Brazilian GO-FAIR Implementation Network focused on agriculture. Like the Brazilian example, geographic barriers should not prevent the global agricultural research data community from actively participating in the IGAD CoP.

References

Arnaud, E., Castañeda-Álvarez, N. P., Cossi, J. G., et al. (2016). *Final report of the task group on GBIF data fitness for use in agrobiodiversity*. Global Biodiversity Information Facility. http://www.gbif.org/resource/82283. Accessed 30 Sept 2021.

Berman, F., & Crosas, M. (2020). The Research Data Alliance: Benefits and challenges of building a community organization. *Harvard Data Science Review, 2*(1). https://doi.org/10.1162/99608f92.5e126552

Caracciolo, C., Aubin, S., Jonquet, C., et al. (2020). 39 hints to facilitate the use of semantics for data on agriculture and nutrition. *Data Science Journal, 19*(1), 47. https://doi.org/10.5334/dsj-2020-047

Carroll, S. R., Garba, I., Figueroa-Rodríguez, O. L., et al. (2020). The CARE principles for indigenous data governance. *Data Science Journal, 19*(1), 43. https://doi.org/10.5334/dsj-2020-043

Devare, M., Arnaud, E., Antezana, E., & King, B. (this volume). Governing agricultural data: Challenges and recommendations. In: S. Leonelli, & H. F. Williamson (Eds.), *Towards responsible plant data linkage: Data challenges for agricultural research and development*. Springer.

Empresa Brasileira de Pesquisa Agropecuária – Embrapa. (2021). *Mission, vision and values*. https://www.embrapa.br/en/missao-visao-e-valores. Accessed 30 Sept 2021.

INRAe. (2016). *Open access and open data policy*. https://ist.inrae.fr/wp-content/uploads/sites/21/2020/02/2016_Charte-libre-access_Inra-anglais.pdf. Accessed 1 Oct 2021.

Lave, J., & Wenger, E. (1991). *Situated learning: Legitimate peripheral participation*. Cambridge University Press.

Rajagopal, P., Vattakaven, T., & Dhandapani, B. (2017). Databasing crop plants from the People's Biodiversity Register of India. *Biodiversity Information Science and Standards, 1*, e19820. https://doi.org/10.3897/tdwgproceedings.1.19820

Research Data Alliance – RDA. (2021). *About RDA*. https://www.rd-alliance.org/about-rda. Accessed 30 Sept 2021.

Wilkinson, M., Dumontier, M., Aalbersberg, I., et al. (2016). The FAIR Guiding Principles for scientific data management and stewardship. *Scientific Data, 3*, 160018. https://doi.org/10.1038/sdata.2016.18

Yeumo, D., Fulss, R., Alaux, M., et al. (2016). *Wheat data interoperability guidelines, ontologies and user cases: Recommendations from the RDA Wheat Data Interoperability Working Group.* https://doi.org/10.15497/RDA00018

Zampati, F. (2021). Ethical and legal considerations in smart farming: A farmers' perspective. In H. Williamson & S. Leonelli (Eds.), *Towards responsible plant data linkage.* Springer.

Cultivating Responsible Plant Breeding Strategies: Conceptual and Normative Commitments in Data-Intensive Agriculture

Hugh F. Williamson and Sabina Leonelli

Abstract This chapter argues for the importance of considering conceptual and normative commitments when addressing questions of responsible practice in data-intensive agricultural research and development. We consider genetic gain-focused plant breeding strategies that envision a data-intensive mode of breeding in which genomic, environmental and socio-economic data are mobilised for rapid crop variety development. Focusing on socio-economic data linkage, we examine methods of product profiling and how they accommodate gendered dimensions of breeding in the field. Through a comparison with participatory breeding methods, we argue that the conceptual commitments underpinning current methods of integrating socioeconomic data into calculations of genetic gain can preclude the achievement of key social development goals, and that better engagement with participatory approaches can help address this problem. We conclude by identifying three key avenues towards a data-intensive approach to plant breeding that utilises the diverse sources of relevant evidence available, including socio-economic data, and maximises the chance of developing sustainable and responsible strategies and research practices in this domain: (1) reliable, long-term management of data infrastructures; (2) ongoing critical analysis of the conceptual foundations of specific strategies; and (3) regular transdisciplinary consultations including expertise in the social studies of agricultural science as well as participatory breeding techniques.

Keywords Plant breeding · Genetic gain · Responsible research practice · Participatory methodologies · Gender · Precision agriculture

H. F. Williamson (✉) · S. Leonelli
Exeter Centre for the Study of the Life Sciences (Egenis), University of Exeter, Exeter, UK
e-mail: h.williamson@exeter.ac.uk; s.leonelli@exeter.ac.uk

© The Author(s) 2023
H. F. Williamson, S. Leonelli (eds.), *Towards Responsible Plant Data Linkage: Data Challenges for Agricultural Research and Development*,
https://doi.org/10.1007/978-3-031-13276-6_16

1 Introduction: Data-Intensive Breeding for Accelerated Genetic Gain

As the previous chapters in this volume have made abundantly clear, the impact of data-intensive tools and methods on plant science and agriculture is extensive and motivated by a wide variety of goals ranging from reducing human labour to tracking dangerous pathogens, increasing yield, identifying agronomically promising plant varieties and understanding the impact of environmental and climatic changes on cultivation and food systems. Here we aim to reflect on some of the conceptual assumptions underpinning the implementation of data-intensive technologies and genetic insights in the agricultural domain. Specifically, we focus on one emerging trend for how plant breeding systems could be built to combine big data availability, including genomic, environmental and socio-economic data, with models for genomic prediction and specific selection methods. This trend is organised around the widespread adoption of genetic gain as a key indicator for evaluating and monitoring the outcomes of plant breeding, and for designing plant breeding strategies and seed system interventions for the future. In our view, this includes specific conceptual and normative commitments to a particular vision of agricultural development, which need to be explicitly drawn out to ensure that the strategies used to realise such a vision within specific situations are both scientifically reliable and socially responsible.

The rate of genetic gain is a statistical measure of the change in a population average for a given trait or set of traits that is due to selection, the use of which is increasingly being encouraged as a high-level performance indicator for plant breeding (Covarrubias-Pazaran, 2020). As a key indicator of biological (more specifically, quantitative genetic) change relative to selection practices, genetic gain bridges concerns over biological improvement of crops with concerns over the efficiency of breeding practice (thus reflecting a quest towards cost-efficiency comparable to that described by Curry, this volume, in relation to the rationalisation of genebanks). Previously to the introduction of genetic gain, breeding programmes have most commonly been evaluated by counting the number of varieties released, a measure which reflects neither the extent of trait improvement realised in new varieties nor their actual uptake among farmers. Genetic gain therefore provides an alternative and potentially more effective metric, increasingly used world-wide, to assess the success of breeding programmes and quantify the agronomic value of new varieties.[1]

Alongside its use as an evaluative measure, normative commitments to increase the rate of genetic gain in breeding programmes have been established as key policy goals for plant breeding in recent years, for example in the funder-led Crops to End Hunger strategy of the CGIAR. The adoption of increased (or "accelerated", as it is

---

[1] The case study presented here draws on a more detailed analysis of genetic gain as an indicator, how it is being implemented in international plant breeding networks, and some key implications (see Williamson & Leonelli, 2022).

often phrased) genetic gain as a policy objective has been linked to: improved cost efficiency; better improvement of complex, quantitative crop traits; and the adaptation of agriculture to climate change through faster development of new varieties targeted to rapidly changing environments (Atlin et al., 2017). This objective has come with new reporting requirements for breeders and managers, and has been incorporated into formal systems for evaluating breeding programs such as the Breeding Program Assessment Tool, developed at the University of Queensland, evaluation through which is now mandatory for any programs receiving funding from the Bill and Melinda Gates Foundation.

Reporting on genetic gain is encouraged not only through the retrospective calculation of rates on the basis of historical data or designated 'era' trials using stored germplasm, but also through the estimation of future genetic gains from a given breeding programme design. Such estimations have significant implications for how breeding programmes are designed, with choices about breeding source materials, selection methods and trial environments, among other factors, decided on the basis of their estimated contribution to genetic gain.

There are several means through which rates of genetic gain can be increased. These include reducing the length of time that breeding takes (i.e. fewer breeding generations), improving selection accuracy, and/or increasing selection intensity (Williamson & Leonelli, 2022). The methods available to achieve these goals frequently involve complex forms of data linkage that have emerged since the turn of the millennium. Indeed, new methods to collate and integrate disparate data sources have arguably driven the turn towards viewing rapid improvement of genetic gain as a feasible goal, progress against which can be precisely measured and quantified. The most prominent example of this is Genomic Selection, whereby molecular marker data taken from biological samples can be used to predict the performance of individual plants based on their genotype, using complex and highly tailored models, thus allowing selection decisions to be made well before the plants in a given generation reach maturity (Xu et al., 2019). Other methods include the use of environmental characterisation together with climatic data (including predicted data) and crop modelling to increase the accuracy of selection by better targeting evaluation and selection to the environmental conditions for which a new variety is being bred (Ramirez-Villegas et al., 2020; Chenu, 2015). Across these methods, there is a particular emphasis on those that increase the speed of breeding, which is often a particularly cost-effective way of increasing genetic gain (Cobb et al., 2019: 634).

Increasing genetic gain thus plays directly into one vision of data-intensive agriculture – what we might call precision breeding – in which maximal trait improvements can be realised in a rapid space of time through tightly integrated pipelines for data collection, integration and analysis, moving back and forth between the field, sequencing labs and computational facilities (cf. Cobb et al., 2019). Rather than discussing genomic or environmental data linkage, which have been discussed at length elsewhere in this volume, this chapter will primarily focus on the role and status of socio-economic data and knowledge in this data-intensive vision.[2] Understanding how such data is collected and used is critical to assessing the social dimensions of responsibility in plant and agricultural science. While privacy and data protection form one pillar of responsible practice in this domain, as discussed in the introduction to this volume and the chapter by Zampati, we are specifically thinking here about the possibilities that the integration of such data into plant breeding programs afford for achieving goals of socio-economic development and improved human wellbeing, for a diverse and inclusive constituency of actors. Our aim is to demonstrate how the conceptual and normative commitments that have accompanied the increased focus on maximising genetic gain, especially in the CGIAR, have significant implications for the kind of engagement with agricultural stakeholders that can be imagined and implemented. This in turn has implications for the kinds of social benefit that breeding programs can deliver.

The starting point for this discussion is a tension that has dogged appeals to accelerate genetic gain: Namely, that despite the range of potential benefits that have been attributed to this goal (such as improved yields or increased resilience of agricultural systems to climate change), it does not necessarily lead to greater adoption of new crop varieties by farmers (cf. Ceccarelli, 2015). Indeed, low rates of adoption of improved varieties among farmers in the Global South is a longstanding concern of breeders and managers working in international agricultural research (Atlin et al., 2017). In order to combat this problem, alongside calls to accelerate genetic gain there has been a recognition by breeders and funders that breeding needs to become more 'demand-led', responding more closely to the needs and desires of farmers and other actors in food systems, as part of a wider 'varietal replacement strategy' (Atlin et al., 2017; Cobb et al., 2019).[3]

Implementing demand-led breeding requires processes for accessing and utilising socioeconomic data that can inform breeding targets and selection decisions. The primary method being promoted for this task is product profiling. In the following section, we analyse what this involves and some of the conceptual implications and limitations that follow from it. We then discuss work that has been undertaken in recent years to overcome some of these limitations and ensure that product profiling is gender-responsive. Following this discussion of gender-responsive breeding, we compare data-intensive breeding methods based on product profiling to participatory methods. Using the example of the Mother and Baby Trial Design, we suggest that many of the principles and goals of gender-responsive breeding can be achieved more consistently and dynamically through the latter. Participatory methods have tended to be excluded from breeding programs focused on maximising genetic gain, however, in line with longstanding disputes about whether highly centralised

---

[2] On 'visions' or 'imaginaries' of socio-technical systems (including data systems) and their implications for science, society and the future, see Jasanoff and Kim (2015) and Leonelli (2021).

[3] Other solutions involve the use of policy levers to encourage or oblige seed companies and farmers to distribute and adopt new varieties, respectively (Spielman & Smale, 2017). These solutions have quite significant political, legal and economic implications regarding the control of seed systems, but we leave that topic aside for the purposes of this chapter (see Williamson & Leonelli, 2022).

breeding grounded in formal selection theory or decentralised, participatory breeding produce greater impact. We conclude the case study by looking at how new data infrastructures are being developed to facilitate dense data collection from participatory methods and their integration into breeding programs alongside other data-intensive methods such as Genomic Selection. We argue that these infrastructures point to alternative visions for breeding and agricultural development, but the prospects for wider adoption of such socially responsive, integrated programs will depend on the extent to which normatively entrenched goals such as accelerating genetic gain govern the distribution of resources and labour in international agricultural research.

2   Product Profiling and Gender-Responsive Breeding

Genetic gain is an indicator that can be assessed and realised for any given trait or index (set) of traits. While the selection of target traits for improvement has long been led by breeders, recently both public and private sectors have shifted to demand-led modes of breeding, where the choice of desirable traits is made through the collection and analysis of socio-economic data. A key method in that respect is product profiling (e.g. Persley & Anthony, 2017).[4] This method has been strongly promoted by proponents of genetic gain, alongside changes to biological breeding methods. This is not only because it facilitates a demand-led approach to breeding, but also because it allows a rapid, formalised delivery of socio-economic information that can be integrated into the tight timescales and optimised pipelines needed to increase genetic gain (e.g. Cobb et al., 2019; Atlin et al., 2017).

A product profile can broadly be defined as "a set of targeted attributes that a new plant variety is expected to meet in order to be successfully released onto a market segment" (Ragot et al., 2018, cited in Cobb et al., 2019: 628). In other words, a product profile describes a plant variety viewed as a desirable replacement for already established varieties within a particular market, thus establishing a key objective for breeders' work over the coming years. Indeed, product profiles are framed as a concise, formalised set of targets that can guide the design of a breeding programme and selection decisions throughout (cf. Ragot et al., 2018). They are assembled at the start of a breeding project by breeders in collaboration with market and socioeconomic researchers. Supporting the creation of product profiles are a set of techniques of market segmentation that allow the target constituency for a breeding programme to be identified and studied. These involve distinguishing distinct groups within a market, "segments" defined by "a relatively homogeneous demand for a commodity (here crop varieties or animal breeds)" (Gender &

---

[4]For a detailed case study of how product profiling methods are being incorporated into plant breeding (specifically cassava breeding) at one CGIAR centre, the International Institute of Tropical Agriculture (IITA), see Agbona et al. (this volume).

Breeding Initiative 2017, cited in Orr et al., 2018: 6). A target segment for breeding is then identified, based on the desired social and/or economic intervention of the breeding programme, and taking into account factors such as agroecosystems, demographics and technological skill. The target segment is the group (usually agricultural producers) who will adopt the resulting variety, although the actual beneficiaries of a breeding programme may be different, for example consumers of a food variety or other actors in the value chain. Product profiles can then be assembled by surveying the needs and desires of both the target segment and other stakeholders (see Orr et al., 2018; Ragot et al., 2018).

Product profiles are meant to facilitate demand-led breeding, where the demand primarily envisioned is market demand, focused on breeding crops that facilitate new commercial opportunities and advantages for farmers and other producers. This is a distinct and non-trivial commitment. While improving the economic position of farmers is a valuable goal that has ramifications for wellbeing and the fight against poverty, it is only one among several important objectives when considered from the wider perspective of social development, including climate action, responsible consumption and reduced inequalities (to pick the most relevant three objectives among the seventeen UN Sustainable Development Goals). The focus on market-led demand underpinning the construction of product profiles reflects a longstanding bias in development discourse and practice towards economic growth and commercialisation (cf. Escobar, 1995), a bias that was largely true of the Green Revolution (e.g. Harwood, 2020) and continues to be true of its legacy projects (e.g. Holt-Giménez, 2008). More ambitious sustainability-focused goals that don't necessarily contribute to market outcomes, such as supporting agroecological systems, tend to receive less support (cf. Rosset & Altieri, 2017).

This situation is problematic in several respects, and we shall here briefly discuss only one of them, concerning the intersection between product profiling, breeder communities and gender equality. It has been well documented that crop improvement focused on commercial value tends to favour men substantially more than women, especially in rural and underdeveloped agricultural settings (cf. Sachs, 2019). Gender differences provide an especially useful lens for thinking about social responsibility in relation to plant breeding, socioeconomic data and indicators such as genetic gain, so it is worth here turning to this topic in some detail.

In order to overcome some of the conventional biases in breeding towards forms of crop improvement that favour men, significant work has been undertaken to improve the gender-responsiveness of breeding in the CGIAR and related networks and institutions (for a history of this work, see Van der Burg, 2019, 2021). In recent years, this has included a significant push to design gender-responsive methods and principles for product profiling, organised through the CGIAR Gender and Breeding Initiative (e.g. Ashby et al., 2018; CGIAR Gender & Breeding Initiative, 2018). This work has been extensively documented by Ashby and Polar (2019), who also summarise some of the key differences in crop trait preferences and socio-economic position between men and women. Such differences are in practice highly variable, and there is no universal set of women's preferences as opposed to men's: In many cases, gendered preferences converge. Nevertheless, there are recurring themes that can be used to guide the design of gender-responsive agricultural research and plant breeding practice (see also Sachs, 2019). One such theme is the importance often placed on particular qualities of the crop rather than overall yield. Due to the distribution of labour in the household economy, women will frequently prefer qualities that reduce labour (such as cooking time, or ease of peeling roots and tubers), even at some cost to overall yield. Where men might primarily be concerned with the income that can be made from selling a harvest in larger commercial markets, women frequently have to consider trade-offs related to household work, the sustenance of their own community and the ultimate end use of a crop (such as household processing and consumption), whether by themselves or by other local women to whom they might sell in more informal markets. As Ashby and Polar observe, it is necessary to consider "the different ways in which resources, rights and responsibilities are shared among women and men engaged in small-farm production, processing and marketing" (2019: 28–9). This is especially so because increased commercialisation resulting from the introduction of new, "improved" varieties can in practice lead to a loss of control for women as cultivation of those more lucrative crops are taken over by men (2019: 23).

Participants in the Gender and Breeding Initiative have made major contributions towards developing methods for incorporating "gender screening" into product profiling in order to take account of gendered differences, such as specific weighting techniques and the differentiation between "niche" and "game-changing" traits. As Ashby and Polar's comment on the distribution of resources, rights and responsibilities indicates, understanding these matters requires in-depth socio-economic research on the relevant groups for whom breeding is targeted. This is where questions of data return to the fore. "There is a practical challenge, therefore", they note: "how to systematize relevant information about gender differences, especially men's and women's trait preferences, in a way that breeders can factor it into their trait prioritization and product profiles" (Ashby & Polar, 2019: 13). Unfortunately,

> much of the published information is inadequate for this task: it consists of a description of a trait preferred by women or ranked higher by women than by men, for example "earliness," without an explanation of the desired extent or level of the trait. This limits the usefulness of the information to breeders, who need to understand what producers consider the desired performance level of a trait. Trait preferences are also reported without analysis of the socioeconomic characteristics of respondents other than their gender and geographic location. Simple sex-disaggregation of preference data is not very useful for informing breeding objectives, because it is essential to understand what resource constraints or producers' objectives are associated with a given preference and whether there is an underlying gender inequality at work. In addition, data on gender differences in trait preference studies is too often reported without evidence that the respondents are representative of a clearly identified population of end users. This makes it difficult to draw general conclusions about the significance of a gender-differentiated trait preference at a scale that a breeding program can rely on, as predictive of widespread end-user acceptance. (2019: 29)

What this points to is the significant issues that remain around access to and integration of appropriately detailed socio-economic data and information. Indeed, as noted in the report on a CGIAR workshop on product profiling, "For some questions, good evidence may not exist, and until it can be obtained, best instincts and knowledge from the breeding team may need to be used as a starting point" (CGIAR Gender & Breeding Initiative, 2018: 18).

In part, this situation relates to difficulties surrounding data collection. Indeed, Almekinders et al. (2019) have argued that methods for researching farmer seed demand present an under-acknowledged bottleneck to attempts to redesign breeding pipelines and to increasing adoption of improved varieties (cf. McEwan et al., 2021).[5] Partly, however, it also relates to the structure of product profiling, which still places with breeders the responsibility for making critical decisions that have wide-ranging implications.

> In order to arrive at a final Product Profile, breeders evaluate, weight and prioritize the individual plant traits under consideration for inclusion in the product profile. Trait prioritization is highly selective, because the number of traits that can be included in any one profile is usually restricted to prevent the selection process from becoming unduly complex. The criteria breeders use for trait prioritization are often a mix of commercial, technical and business considerations, shaped by the goals of the breeding program. (Ashby & Polar, 2019: 13)

Socio-economic data and expertise, including gender data, are only incorporated at such key decision points, and often through very informal means. This is understandable where information is a limited resource. Stepping back, however, we might throw this situation into relief by comparison with some alternative modes of breeding available, specifically those that take a more systematic approach to the inclusion of socio-economic data and knowledge through participatory approaches.

3   Participatory Breeding for Dynamic Socio-Economic Data Flows

Participatory plant breeding methods, involving farmers directly in the selection process for new varieties, began to emerge in the late 1970s before taking root more substantially in the 1990s (Harwood, 2012: 142–3; Cleveland & Soleri, 2002; Westengen & Winge, 2020). Participatory methods provide a very different model of socio-economic responsiveness in comparison to conventional, centralised breeding.

Consider the Mother and Baby Trial Design method developed for participatory potato selection at the International Potato Center (CIP) in Peru (De Haan et al., 2019). This method utilises a combination of a centrally managed, experimental field trial (the 'mother') in which multiple varieties are grown and smaller trial plots (the 'babies') in farmers' fields that reflect the latter's own agronomic conditions. Participating farmers engage in evaluation of the different varieties at key stages,

---

[5] The history of social research in the CGIAR has long been marked by highly variable investment and integration with core plant breeding and research activities (Cernea & Kassam, 2006).

from flowering through to harvest, at both the managed and on-farm plots. These evaluations include standard yield assessment, but more importantly they include evaluation on the basis of selection criteria that are identified and ranked by farmers themselves at the time of evaluation. These criteria may include relatively conventional trait preferences such as resistances to blight, but also trait preferences that are more contextual and tangential to crop production, such as the adequacy of foliage for feeding livestock (2019: 26–27). A particularly important set of additional evaluations are those concerning the qualities of the crop, especially qualities relating to cuisine and organoleptic (i.e. sensory) traits such as appearance, taste and texture (2019: 45–6). Once participant farmers have chosen their preferred criteria, they rank plant varieties on that basis through simple voting methods involving placing seeds or other tokens in paper bags.

Gender-responsiveness is critical to the Mother and Baby Trial Design. This is achieved, first, through the focus on crop qualities, which as we saw is typically favored by women participants; second, by ensuring that women have the space to make their own contributions and decisions free from the influence of male farmers; and third, by designing participation such that the data collected from these trials can be disaggregated by gender. Ensuring space for women may require not only an equal balance of female and male participants, but also conducting discussions and voting with women separately from men (2019: 26). The ability to disaggregate data by gender can be achieved by providing men and women with different seeds/tokens for voting that can be counted separately (2019: 27–8).

Participatory breeding methods such as the Mother and Baby Trial Design have the advantage of providing socio-economic data collection and integration that is more *consistent* and more *dynamic*: Consistent, because they do not depend on highly variable and often heavily mediated flows of information; and dynamic, because the data collected from and opinions offered by farmers contribute directly to the shaping and reshaping of breeding and selection decisions throughout the whole process. As Almekinders et al. note, "The picture we create of the farmers' preferences is a snapshot taken from our perspective as researchers and devoid of trade-offs and considerations farmers have in a real-life situation" (2019: 17). This 'snapshot' quality is accentuated where socio-economic data is incorporated at a single decision point in the product profiling process.

On top of these advantages, and perhaps most critically, it has also been argued that participatory breeding leads directly to greater varietal adoption by farmers. Ceccarelli and Grando (2007) have observed that in conventional breeding "the entire process is supply-driven; as a consequence, in many developing countries many varieties are produced and released but only a small fraction of these are adopted. With [participatory plant breeding], decision[s] on which variety to release depend on initial adoption by farmers; the process is demand-driven" (2007: 356). This is quite a different model of demand-driven breeding to the idea of market demand discussed above, one in which demand is community-led and treated as demonstrable adoption by farmers rather than a 'snapshot' of preferences, thus building adoption itself into the breeding process. Moreover, Ceccarelli notes elsewhere that "in a conventional system, 5 to 6 [years] typically pass after official release before appreciable adoption commences [. . .], and during this time, farmers' priorities, agronomic conditions (e.g., availability of irrigation or fertilizer price), policy measures (e.g., introduction or removal of subsidies), and market demands may change, making the breeding objectives set at the beginning of the breeding program obsolete" (2015: 89). The dynamic engagement with farmer needs, priorities and growing conditions in participatory breeding directly responds to such issues, ensuring that varieties remain relevant to changing conditions.

Given these advantages, then, why are participatory breeding methods practically invisible in key discussions of genetic gain (e.g. Atlin et al., 2017; Cobb et al., 2019), despite the corresponding concern for varietal adoption? And why are questions of the social responsiveness of breeding limited to those information flows that can be condensed into a limited set of goals captured in a product profile? This situation is not new. As Harwood notes, participatory breeding has often been strongly resisted by breeders, with many considering it "an unnecessary alternative to conventional breeding (rather than an additional option)" (2012: 146). Indeed, one prominent proponent of accelerating genetic gain has asked of participatory plant breeding, "Why do we need it? We need it because we don't do good market research to really understand what farmers need, what millers need, what consumers prioritise" (Atlin, 2016). The attention to additional actors in food systems beyond farmers is important. But as we have indicated above, it is often market research flows that tend to be inconsistent by comparison to participatory methods. Moreover, it is debatable whether much of what is conventionally conducted under the rubric of market research addresses socio-economic concerns over gender relations and the distribution of resources, rights and responsibilities, which Ashby and Polar (2019) among many others have flagged as vital to addressing social and economic inequalities.

More broadly, we take this discussion of product profiling in relation to participatory breeding methods as exemplifying the critical role of the conceptual and normative dimensions of plant breeding for the design and implementation of data-intensive approaches. Specifically, our analysis highlights a tension between how data-intensive plant breeding is being imagined and the practical requirements of organising participatory breeding schemes. When implemented within breeding programs, the commitment to maximise genetic gain is typically accompanied by a commitment towards speed and efficiency in the collation of data and criteria underpinning the choice of product profiles (e.g. Cobb et al., 2019: 634; cf. Williamson & Leonelli, 2022): the CGIAR for instance is pushing for tightly integrated pipelines for data production, integration and analysis, such that selection decisions can be brought forward and the length of time from initiation of breeding to variety release reduced, potentially by up to 5 years depending on the crop species and methods used. Product profiling is attractive in relation to these commitments, because it provides a clear and limited set of target traits for improvement that breeders can use to make selection decisions under conditions of time pressure, in conjunction with molecular and evaluation data drawn from field trials. In comparison, participatory breeding programs fare much worse: they require significantly higher investments to set up, especially if large numbers of farmers and on-farm trials are involved; and collection and analysis of data from those on-farm trials and from participatory evaluation sessions takes considerable time, especially when compared to the possibilities of Genomic Selection to predict plant performance before it has even reached maturity. This can lead to drag on rates of genetic gain, by adding additional time and labour requirements into pipelines, making participatory breeding look unappealing despite the above-mentioned advantages in terms of supporting social equality and agrodiversity.

This in turn underscores a continuing tension between the commitment to accelerating genetic gain and the need to increase varietal adoption, goals which are practically and conceptually separated in current visions of data-intensive plant breeding. If the aim of public plant breeding is ultimately to deliver social as well as economic impact, then any accounting for the efficiency of breeding should factor in a combination of genetic gain, varietal adoption and agrodiversity assessment more broadly (cf. Ceccarelli, 2015). Focusing solely or even primarily on genetic gain and its delivery to farmers as key indicators of success for plant breeding risks perpetuating a situation of supply-driven breeding and market-led seed systems, where biotechnological improvement becomes a primary value and an end in itself, while the social impacts of breeding are shaped to accommodate this goal. When it comes to data-intensive breeding, it is not outlandish to suggest that responsible practice should invert this situation, with the social impact of breeding driving the choice and implementation of biotechnological improvement. We argue that this may require rethinking the maximisation of genetic gain as a *situated* rather than a universal objective: One that can be deployed in certain circumstances but should always take into account the potential conflicts this can produce with other commitments, rather than being imposed as a key objective across breeding programs at large and then onto seed systems, through a treadmill of variety release that is pushed onto farmers.[6]

4 Conclusion: Essential Components of Responsible Breeding Strategies

The eminent historian of agriculture James C. Scott has provided a provocative reading of efforts to improve agriculture through biotechnology, as follows: "if the logic of actual farming is one of an inventive, practiced response to a highly variable environment, the logic of scientific agriculture is, by contrast, one of adapting the environment as much as possible to its centralising and standardising formulas" (1998: 301). This controversial reading may be viewed as applying well to the current fixation on accelerating genetic gain, where the infrastructures and evaluative procedures supporting data-intensive breeding are constructed around highly centralised and standardised methods of product profiling, which do not admit –

---

[6] See footnote 2, and cf. Williamson & Leonelli, 2022 for a more detailed discussion of seed system issues.

through their commitment to speed and market-led understandings of varietal demand – of participatory approaches which may be slower and yet yield better outcomes in terms of social equality and support for agrodiversity. However, we do not think that it is necessary or even fully warranted to juxtapose conventional, data-intensive breeding focused on increasing genetic gain with participatory breeding methods, as if these two approaches were incompatible and intrinsically opposed to each other. What we have suggested is that there is a tension among some of the commitments explicitly or implicitly endorsed by these two approaches, which needs to be highlighted and critically discussed in order to successfully reconcile their respective advantages. In Scott's terms, there may be ways to reconcile the logic of actual farming with that of scientific agriculture, as long as a balance is sought between standardisation and speed on the one hand, and participation and inclusive data-intensive methods on the other.

This point has been made most thoroughly by Fadda et al. (2020), drawing on the example of the Bioversity International 'Seeds for Needs' project. What such projects indicate is not a necessary conflict between competing methods, "even though the two approaches are different from a conceptual and underlying philosophical point of views" (2020: 2), but the potential for an innovative and deeper *integration* of participatory methods with genomic and other data-intensive methods. Steps in this direction are also being taken by the cassava breeding programme at IITA, for instance through the use of the Tricot (triadic comparison of technologies) participatory methodology, which reflects similar goals to the Mother and Baby Trial Design method (see Agbona et al., this volume). In closing this chapter, we shall identify and discuss what we regard as three essential components to such an integrated approach to plant breeding.

The first component is **the development and reliable maintenance of digital infrastructures** that support the sourcing and integration of data from farmers and on-farm trials. This needs to include semantic standards that incorporate farmer and other local terminologies, such as the Crop Ontology (Arnaud et al., 2020; Leonelli, 2022). It also needs to include platforms for crowdsourcing participatory trial data directly from farmers, such as the ClimMob platform being developed to support the Tricot methodology (van Etten et al., 2020), which allow much greater scaling of participation, and thus greater efficiency and reliability of results (an aspect that has been the source of criticism by proponents of conventional breeding; e.g. Atlin et al., 2001). The appointment of 'quality champions' or similar designated experts to support the effective use of digital infrastructures, as has been undertaken for the BREEDBASE breeding data management system, also assists in addressing some of the critical organisational and skills issues that can limit the adoption of such technically and socially complex systems (Agbona et al., this volume). This is particularly effective when sourcing at least some experts from local communities. Here we see glimpses of future data-intensive plant science and related digital infrastructures being put directly in the service of social inclusion and responsiveness (similar to the blockchain schemes discussed by Kochupillai and Köninger, this volume). As other chapters in this volume indicate (e.g. Fullilove and Alimari), the possibilities for this being achieved in practice will depend heavily on institutional norms and structures, and on whether concrete support – through policy and resource allocation – can be thrown behind such efforts. In any case, the significance of investment in reliable, well-maintained, long-term data infrastructures as a fundamental requirement for the sustainable use of data-intensive tools for plant breeding cannot be underestimated.

The second component encompasses the ability for plant breeders, data and plant scientists, farmers, policy-makers and industry representatives in this domain to **explicitly confront and discuss diverse assumptions relating to conservation, biodiversity and development**. Practically, this requires implementing processes through which diverse stakeholders can come together and engage one another in ways that make a meaningful difference to how research and development are done, as in the collaborative and open-ended forms of organisation that characterise the CoEx project discussed by Louafi et al. (this volume). This typically includes consultation with social scientists and local representatives that can broker diverse concerns and help identify and debate the underpinning conceptual and normative commitments of plant breeding strategies (whether current or imagined) and how those can be reconciled to foster responsible research practice within specific communities and locations.

What do we mean by conceptual and normative commitments? These are the scientific, social, economic and other foundational concepts mobilised in agricultural research and development, which may not be explicitly recognised yet underpin ongoing practices, including how breeding strategies and related forms of data linkage are being developed and implemented. These foundational concepts are often tacit or taken for granted, but have a wide range of implications. While the large-scale mobilisation of data provides new opportunities, our analysis of social responsiveness in genetic gain-focused breeding has highlighted how data-intensive visions of agricultural research can also produce frictions when located in the wider landscape of agriculture (cf. Edwards et al., 2011). Looking beyond this specific example, additional issues include: the uneven landscapes of both scientific understanding and data flows themselves, which create discrepancies and inequalities in the extent to which data-intensive methods can be applied and can work productively for different groups (Kochupillai and Köninger, this volume; Zampati, this volume); the conceptual and cultural gulf between farming communities and research scientists when it comes to agricultural strategies (Louafi et al., this volume); and indeed the lack of training for scientists themselves to recognise and understand alternative narratives of agricultural development (and where data science can fit in these).

This is important for responsible research practice in plant data linkage for at least three reasons. First, because unquestioned, dogmatic adherence to specific normative commitments can lead to aspects of research practice becoming centralised and entrenched (materially as well as culturally) as the necessary or right way for things to be done, and block off alternatives (Scott, 1998). Second, because scientific research does not just exist in its own bubble; it feeds into much wider imaginaries of society, economy, development, and so on, which in turn also influence the ways we imagine and conceptualise science (Jasanoff, 2004). And third, because the extensive and highly diversified impact of plant breeding and agronomic strategies on planetary health makes it imperative to continue to look for alternatives and/or localised solutions, both for how science is done and for agricultural development, and to consider whether such alternative and/or localised approaches may improve current practice.

Following this, the third component we identify as crucial to an integrated and responsible approach to plant breeding is **interdisciplinary and transdisciplinary collaboration**, particularly involving historical, philosophical and social studies of science, to consider critically the implications of entrenching concepts into infrastructures – and possible alternatives. Within this volume, many examples have been given of ways to broker social and scientific considerations within data-intensive breeding. Most chapters have pointed towards ways to remain evidence-based and build on innovative data-intensive tools, while at the same time grounding novel forms of data linkage on an understanding of the geographically and conceptually diverse histories of agricultural policies and technologies. Among the many examples of such work available beyond this volume, one of the most relevant is Jonathan Harwood's (2012) effort to uncover a forgotten history of public plant breeding in southern Germany, predating the Green Revolution. Harwood uses the case to think about issues of who is supported by agricultural research and development and in what ways, particularly through a comparison with Green Revolution breeding and the growth and decline of participatory breeding methods in the CGIAR throughout the 1990s (an example that resonates with the case we have presented above).[7] An additional example is the recent Nuffield Council on Bioethics (2021) report on genome editing and farmed animal breeding, which draws on expertise from a range of disciplines across the biosciences, social sciences and humanities. Reflecting the concerns in this chapter for how data are assembled and indicators put to work in breeding practice, the authors analyse the scope and purposes of indices used to evaluate breeding animals. Among the recommendations made in the report are the need to expand the scope of the indices to include traits of public or social as well as economic value, for example those related to health traits or traits that can impact climate emissions (2021: 155–160, 192–3). The kinds of conceptual and normative considerations raised in these examples, and throughout this chapter, can crucially inform research and policy decisions around how to set up infrastructures, data governance and institutional goals for agricultural development and food security. Responses are likely to involve elements of design of socio-technical systems, thus intersecting strongly with the design of technical infrastructures whose significance we just emphasised.

In closing, it is important to stress that consideration of responsibility and social responsiveness introduced through a focus on the conceptual and normative dimensions of plant and agricultural data linkage does not produce clear, unambiguous conclusions. Insights tend to be context-specific, and thus require detailed attention to and knowledge about how research and development is set up in practice.

---

[7] Harwood has also made complementary arguments about the relevance of history of science and development to policy (Harwood & Sturdy, 2010; Harwood, 2018).

Historical, philosophical and sociological studies of science provide excellent background knowledge on these aspects; but they need to be complemented by practical and tacit knowledge held by domain experts – an interdisciplinary dialogue that this volume has attempted to contribute towards establishing. Moreover, tensions and disagreements are unlikely to be resolved easily, with disagreements over the relative value of centralised, formal breeding methods versus decentralised, participatory methods running for several decades now. In data-intensive science as in other realms of research, responsibility involves opening up such matters to public debate and the option of co-producing future strategies with relevant stakeholders and publics.

References

Almekinders, C. J. M., Beumer, K., Hauser, M., Misiko, M., Gatto, M., Nkurumwa, A. O., & Erenstein, O. (2019). Understanding the relations between farmers' seed demand and research methods: The challenge to do better. *Outlook on Agriculture, 48*(1), 16–21. https://doi.org/10.1177/0030727019827028

Arnaud, E., Laporte, M. A., Kim, S., Aubert, C., Leonelli, S., Cooper, L., Jaiswal, P., Kruseman, G., Shrestha, R., Buttigieg, P. L., Mungall, C., Pietragalla, J., Agbona, A., Muliro, J., Detras, J., Hualla, V., Rathore, A., Das, R., Dieng, I., et al. (2020). The ontologies community of practice: An initiative by the CGIAR platform for big data in agriculture. *Patterns, 1*, 100105. https://doi.org/10.1016/j.patter.2020.100105

Ashby, J. A., & Polar, V. (2019). The implication of gender relations for modern approaches to crop improvement and plant breeding. In C. E. Sachs (Ed.), *Gender, agriculture and agrarian transformations: Changing relations in Africa, Latin America and Asia*. Routledge.

Ashby, J. A., Polar, V., & Thiele, G. (2018). *Critical decisions for ensuring plant or animal breeding is gender-responsive* (CGIAR Gender & Breeding Initiative, Brief 1). CGIAR.

Atlin, G. (2016, April 29). *Modernizing plant breeding programs to deliver higher rates of genetic gain in the developing world*. ICRISAT seminar. Available at: https://youtu.be/r2Zh64QD8TQ

Atlin, G. N., Cooper, M., & Bjornstad, A. (2001). A comparison of formal and participatory breeding approaches using selection theory. *Euphytica, 122*, 463–475. https://doi.org/10.1023/A:1017557307800

Atlin, G. N., Cairns, J. E., & Das, B. (2017). Rapid breeding and varietal replacement are critical to adaptation of cropping systems in the developing world to climate change. *Global Food Security, 12*, 31–37. https://doi.org/10.1016/j.gfs.2017.01.008

Ceccarelli, S. (2015). Efficiency of plant breeding. *Crop Science, 55*, 87–97. https://doi.org/10.2135/cropsci2014.02.0158

Ceccarelli, S., & Grando, S. (2007). Decentralized-participatory plant breeding: An example of demand driven research. *Euphytica, 155*(3), 349–360. https://doi.org/10.1007/s10681-006-9336-8

Cernea, M. M., & Kassam, A. H. (Eds.). (2006). *Researching the culture in agri-culture: Social research for international development*. CABI Publishing.

CGIAR Gender and Breeding Initiative. (2018, November 12–13). *Gender-responsive product profile development tool*. Workshop report. Available at: https://cgspace.cgiar.org/handle/10568/99094

Chenu, K. (2015). Characterizing the crop environment – Nature, significance and applications. In V. O. Sadras & D. F. Calerini (Eds.), *Crop physiology: Applications for genetic improvement and agronomy* (2nd ed.). Academic Press.

Cleveland, D. A., & Soleri, D. (Eds.). (2002). *Farmers, scientists and plant breeding: Integrating knowledge and practice*. CABI Publishing.

Cobb, J. N., Juma, R. E., Biswas, P. S., Arbelaez, J. D., Rutkoski, J., Atlin, G., Hagen, T., Quinn, M., & Hwa Ng, E. (2019). Enhancing the rate of genetic gain in public-sector plant breeding programs: Lessons from the breeder's equation. *Theoretical and Applied Genetics, 132*, 627–645. https://doi.org/10.1007/s00122-019-03317-0

Covarrubias-Pazaran, G. E. (2020). *Genetic gain as a high-level key performance indicator*. CGIAR excellence in breeding platform, breeding scheme optimization manual. https://excellenceinbreeding.org/toolbox/tools/eib-breeding-scheme-optimization-manuals

De Haan, S. E., Salas, C., Fonseca, M., Gastelo, N., Amaya, C., Bastos, V. H., & Bonierbale, M. (2019). *Participatory varietal selection of potato using the mother & baby trial design: A gender-responsive trainer's guide*. International Potato Center. https://cgspace.cgiar.org/handle/10568/106633

Edwards, P. N., Mayernik, M. S., Batcheller, A. L., Bowker, G. C., & Borgman, C. L. (2011). Science friction: Data metadata and collaboration. *Social Studies of Science, 41*(5), 667–690. https://doi.org/10.1177/0306312711413314

Escobar, A. (1995). *Encountering development: The making and unmaking of the Third World*. Princeton University Press.

Fadda, C., Mengistu, D. K., Kidane, Y. G., Dell'Acqua, M., Pe, M. E., & van Etten, J. (2020). Integrating conventional and participatory crop improvement for smallholder agriculture using the seeds for needs approach: A review. *Frontiers in Plant Science, 11*, 559515. https://doi.org/10.3389/fpls.2020.559515

Harwood, J. (2012). *Europe's green revolution and others since: The rise and fall of peasant-friendly plant breeding*. Routledge.

Harwood, J. (2018). Another Green Revolution? On the perils of 'Extracting lessons' from history. *Development, 61*, 43–53. https://doi.org/10.1057/s41301-018-0174-5

Harwood, J. (2020). Whatever happened to the Mexican Green Revolution? *Agroecology and Sustainable Food Systems, 44*(9), 1243–1252. https://doi.org/10.1080/21683565.2020.1752350

Harwood, J., & Sturdy, S. (2010). What can development policy learn from the history of development? *Food Security, 2*, 285–290. https://doi.org/10.1007/s12571-010-0067-2

Holt-Giménez, E. (2008). Out of AGRA: The Green Revolution returns to Africa. *Development, 51*(4), 464–471. https://doi.org/10.1057/dev.2008.49

Jasanoff, S. (Ed.). (2004). *States of knowledge: The co-production of science and social order*. Routledge.

Jasanoff, S., & Kim, S.-H. (Eds.). (2015). *Dreamscapes of modernity: Sociotechnical imaginaries and the fabrication of power*. University of Chicago Press.

Leonelli, S. (2021). Data science in times of pan(dem)ic. *Harvard Data Science Review, 3*. https://doi.org/10.1162/99608f92.fbb1bdd6

Leonelli, S. (2022). Process-Sensitive Naming: Trait Descriptors and the Shifting Semantics of Plant (Data) Science. *Philosophy, Theory and Practice in Biology*. https://doi.org/10.3998/ptpbio.16039257.000000

McEwan, M. A., Almekinders, C. J. M., Andrada-Piedra, J. J. L., Delaquis, E., Garrett, K. A., Kumar, L., Mayanja, S., Omondi, B. A., Rajendran, S., & Thiele, G. (2021). "Breaking through the 40% adoption ceiling: Mind the seed system gaps." A perspective on seed systems research for development in One CGIAR. *Outlook on Agriculture, 50*(1), 5–12. https://doi.org/10.1177/0030727021989346

Nuffield Council on Bioethics. (2021). *Genome editing and farmed animal breeding: Social and ethical issues*. Report. Nuffield Council on Bioethics https://www.nuffieldbioethics.org/publications/genome-editing-and-farmed-animals

Orr, A., Cox, C. M., Ru, Y., & Ashby, J. (2018). *Gender and social targeting in plant breeding*. CGIAR Gender & Breeding Initiative, working paper 1. http://www.rtb.cgiar.org/gender-breeding- initiative/resources/

Persley, G. J., & Anthony, V. M. (Eds.). (2017). *The business of plant breeding: Market-led approaches to new variety design in Africa.* CABI.

Ragot, M., Bonierbale, M., & Weltzein, E. (2018). *From market demand to breeding decisions: A framework.* CGIAR Gender & Breeding Initiative, working paper 2. http://www.rtb.cgiar.org/gender-breeding-initiative/resources/

Ramirez-Villegas, J., Milan, A. M., Alexandrov, N., Asseng, S., Challinor, A. J., Crossa, J., van Eeuwijk, F., Ghanem, M. E., Grenier, C., Heinemann, A. B., Wang, J., Juliana, P., Kehel, Z., Kholova, J., Koo, J., Pequeno, D., Quiroz, R., Rebolledo, M. C., Sukumaran, S., et al. (2020). CGIAR modelling approaches for resource-constrained scenarios: I. Accelerating crop breeding for a changing climate. *Crop Science, 60*(2), 547–567. https://doi.org/10.1002/csc2.20048

Rosset, P. M., & Altieri, M. A. (2017). *Agroecology: Science and politics.* Practical Action Publishing.

Sachs, C. E. (Ed.). (2019). *Gender, agriculture and agrarian transformations: Changing relations in Africa, Latin America and Asia.* Routledge

Scott, J. C. (1998). *Seeing like a state: How certain schemes to improve the human condition have failed.* Yale University Press

Spielman, D. J., & Smale, M. (2017). *Policy options to accelerate variety change among small-holder farmers in South Asia and Africa South of the Sahara.* IFPRI discussion paper 01666. https://www.ifpri.org/publication/policy-options-accelerate-variety-change-among- small holder-farmers-south-asia-and-africa/

Van der Burg, M. (2019). "Change in the making": 1970s and 1980s building stones to gender integration in CGIAR agricultural research. In C. E. Sachs (Ed.), *Gender, agriculture and agrarian transformations: Changing relations in Africa, Latin America and Asia.* Routledge

Van der Burg, M. (2021). Gender integration in agricultural research for development. In C. E. Sachs, L. Jensen, P. Castellanos, & K. Sexsmith (Eds.), *Routledge handbook of gender and agriculture.* Routledge

van Etten, J., Abidin, E., Arnaud, D., Brown, E., Carey, E., Laporte, M.-L., López-Noriega, I., Madriz, B., Manners, R., Ortiz-Crespo, B., Quirós, C., de Sousa, K., Teeken, B., Tufan, H.A., Ulzen, J., & Valle-Soto, J. (2020). *The tricot citizen science approach applied to on-farm variety evaluation: Methodological progress and perspectives.* CGIAR Research Program on Roots, Tubers and Bananas (RTB). RTB Working Paper. No. 2021-2.

Westengen, O. T., & Winge, T. (Eds.). (2020). *Farmers and plant breeding: Current approaches and perspectives.* Routledge

Williamson, H. F., & Leonelli, S. (2022). Accelerating agriculture: Data-intensive plant breeding and the use of genetic gain as an indicator for agricultural research and development. *Studies in History and Philosophy of Science, 95*, 167–176. https://doi.org/10.1016/j.shpsa.2022.08.006

Xu, Y., Liu, Z., Fu, J., Wang, H., Wang, J., Huang, C., Prasanna, B. M., Olsen, M. S., Wang, G., & Zhang, A. (2019). Enhancing genetic gain through genomic selection: From livestock to plants. *Plant Communications, 1*(1), 100005. https://doi.org/10.1016/j.xplc.2019.100005

Printed in the United States
by Baker & Taylor Publisher Services